Plastics, Environment, Culture, and the Politics of Waste

PLASTICS, ENVIRONMENT, CULTURE, AND THE POLITICS OF WASTE

Edited by Tatiana Konrad

EDINBURGH
University Press

Edinburgh University Press is one of the leading university presses in the UK. We publish academic books and journals in our selected subject areas across the humanities and social sciences, combining cutting-edge scholarship with high editorial and production values to produce academic works of lasting importance. For more information visit our website: edinbu rghuniversitypress.com

Edinburgh University Press Ltd
The Tun – Holyrood Road
12(2f) Jackson's Entry
Edinburgh EH8 8PJ

Typeset in 11/13pt Sabon LT Pro
by Cheshire Typesetting Ltd, Cuddington, Cheshire, and
printed and bound in Great Britain

A CIP record for this book is available from the British Library

ISBN 978-1-3995-1173-5 (hardback)
ISBN 978-1-3995-1175-9 (webready PDF)
ISBN 978-1-3995-1176-6 (epub)

Contents

List of Figures

Notes on Contributors

Sasha Adkins, PhD, MPH, teaches environmental health at the School of Environmental Sustainability at Loyola University Chicago and is the author of *From Disposable Culture to Disposable People: The Unintended Consequences of Plastics*.

Nathan Beck recently completed his MA in Comparative Cultural Analysis at the University of Amsterdam. His research interests include ecocriticism, posthumanism, memory, and speculative fiction.

Amanda Boetzkes is Professor of Contemporary Art History and Theory at the University of Guelph, Canada. Her research examines the politics, aesthetics, and ecologies of contemporary art through the lens of human waste, energy consumption and expenditure, and most recently, climate crisis and glacier melt in the circumpolar North. She is the author of *Plastic Capitalism: Contemporary Art and the Drive to Waste* (2019) and *The Ethics of Earth Art* (2010). She is co-editor of *Heidegger and the Work of Art History* (2014) and a forthcoming volume on *Art's Realism in the Post-Truth Era* (2023).

Angela Cope, PhD, is a Science and Technology Studies and History of Technology early scholar. Her research has to do with the history of plastics and disposability, particularly with respect to the postwar North American devaluation of plastics from its utopian to detrital forms, with a focus on small and expendable dime-store toys. Her other interests include modular architecture, midcentury modernism as a futurist trope, and minimalism as a capitalist pursuit. She is currently affiliated with York University in Toronto, Ontario Canada, where she lives with her family, her dog, and too much stuff.

Brittany Y. Davis, PhD, is a Human-Environment Geographer and Independent Researcher.

Louise Dennis, PhD, is the Curator of MoDiP, the Museum of Design in Plastics, based at the Arts University Bournemouth, UK. MoDiP is the only UK accredited museum with a focus on plastics and has been awarded Designated Outstanding Collection status by Arts Council England. Louise's research explores the value of the museum and its specialist focus, with a particular interest in the various relationships surrounding materiality and museum objects. Such relationships exist within contemporary, historical, cultural, and curatorial contexts. Through her work with the museum, she aims to show the value of plastics materials when used appropriately, and hopes that, by learning from the past, manufacturers, designers, and consumers can make better informed choices.

Jeff Diamanti is Assistant Professor of Environmental Humanities at the University of Amsterdam and is cross-listed between Literary & Cultural Analysis and Philosophy. His first monograph, *Climate and Capital in the Age of Petroleum*, is out with Bloomsbury (2021).

Dana Feldman is a multidisciplinary artist based in Toronto, Canada. Her work explores themes of gender and sexuality, interpreting symbolism, and deconstructing structures of social culture. Feldman often works in oil on canvas painting and is currently captivated by ceramics, learning to wheel-throw clay. She recently received her MA at the University of Guelph, researching food in contemporary art practices to re-evaluate gender performativity and the imposition of femininity in North America. Feldman is also the co-founder of Community Garden Art Collective (CGAC): a collective and platform for digital art collaborations and exhibitions—spreading seeds soon to grow.

Andrija Filipović is Associate Professor of Philosophy and Art & Media Theory at the Faculty of Media and Communications in Belgrade. He is the author of *Ars ahumana: Anthropocene Ontographies in the 21st Century Art and Culture* (2022), *Conditio ahumana: Immanence and the Ahuman in the Anthropocene* (2019), and monographs on Brian Massumi (2016) and Gilles Deleuze (2015). His articles have appeared in *Sexualities, The Comparatist, Contemporary Social Science, NORMA*, and a number of edited volumes. His research interests include environmental humanities, queer theory, and contemporary continental philosophy. He is Executive Editor of *AM: Journal of Art and Media Studies*.

Donna A. Gessell is Professor Emerita of English at the University of North Georgia, where she recently taught courses in literature, linguistics, and composition. She also serves as co-editor for the peer-reviewed journal *Graham Greene Studies*. Her recent publications include "*Red Sparrow*: Cold War Redux and the Treatment of Corruption," in *Cold War II: Hollywood's Renewed Obsession with Russia*; "The [Latinx] Darling: Lorraine López Reads the Canon," in *South Atlantic Review*; and, co-authored with Creina Mansfield, "Making Sense of Graham Greene's Panama: A Fuliginous Process," in "A Sort of a Newsletter," *Graham Greene Birthplace Trust*.

Renée Hoogland is a researcher and writer based in the Netherlands. Her research interests lie at the interstices of the environmental humanities, citizen sensing, literary theory, and cultural anthropology. Based within the humanities, Renée has written on the cultural category of ecological intimacy, infrastructural disruption, and garden mediations. Her current project tracks the political ecology of the North Sea, bringing into focus new media ecologies and aesthetic forms of material participation that detail the intimacy and dissonance between ecology and economy in an increasingly polluted and warming ocean.

Cymene Howe is Professor of Anthropology at Rice University. She is the author of *Intimate Activism* (Duke 2013) and *Ecologics: Wind and Power in the Anthropocene* (Duke 2019) and co-editor of *The Johns Hopkins Guide to Critical and Cultural Theory* and *Anthropocene Unseen: A Lexicon* (Punctum 2020). Her research focuses on climatological precarity and her work has been funded by the US National Science Foundation, the Fulbright Commission, and the Andrew W. Mellon Foundation. She has also been awarded the Berlin Prize for transatlantic dialogue in the arts, humanities, and public policy from the American Academy in Berlin.

Lynn Keller is Professor Emerita of English and former director of the Center for Culture, History, and Environment at the University of Wisconsin–Madison. She is the author of *Re-making It New: Contemporary American Poetry and the Modernist Tradition*; *Forms of Expansion: Recent Long Poems by Women*; *Thinking Poetry: Readings in Contemporary Women's Exploratory Poetics*; and *Recomposing Ecopoetics: North American Poetry of the Self-Conscious Anthropocene*, along with numerous articles on contemporary poetry. She co-edits the University of Iowa Press scholarly series on Contemporary North American Poetry and in retirement continues to write about ecopoetics.

Tatiana Konrad is a postdoctoral researcher in the Department of English and American Studies, University of Vienna, Austria, the principal investigator of "Air and Environmental Health in the (Post-) COVID-19 World," and the editor of the "Environment, Health, and Well-being" book series at Michigan State University Press. She was a Visiting Fellow at the University of Chicago (2022), a Visiting Researcher at the Forest History Society (2019), an Ebeling Fellow at the American Antiquarian Society (2018), and a Visiting Scholar at the University of South Alabama (2016). She is the author of *Docu-Fictions of War: US Interventionism in Film and Literature* (University of Nebraska Press, 2019), the editor of *Cold War II: Hollywood's Renewed Obsession with Russia* (University Press of Mississippi, 2020) and *Transportation and the Culture of Climate Change: Accelerating Ride to Global Crisis* (West Virginia University Press, 2020), and a coeditor of *Cultures of War in Graphic Novels: Violence, Trauma, and Memory* (Rutgers University Press, 2018).

David LaRocca, PhD, is the author or contributing editor of more than a dozen books, among them: *Metacinema, Movies with Stanley Cavell in Mind, The Thought of Stanley Cavell and Cinema, The Philosophy of Charlie Kaufman, The Philosophy of War Films,* and *The Philosophy of Documentary Film.* His articles have appeared in journals such as *Afterimage, Cinema, Conversations, Epoché, Estetica, Liminalities, Post Script, Transactions, Film and Philosophy, The Senses and Society, The Midwest Quarterly, Journalism, Media and Cultural Studies, The Journal of Aesthetic Education,* and *The Journal of Aesthetics and Art Criticism.* He studied rhetoric at Berkeley, was Harvard's Sinclair Kennedy Traveling Fellow in the United Kingdom, and participated in an NEH Institute, a workshop with Abbas Kiarostami, Werner Herzog's Rogue Film School, and the School of Criticism and Theory at Cornell. He has taught philosophy and cinema and held visiting research or teaching positions at Binghamton, Cornell, Cortland, Harvard, Ithaca College, the School of Visual Arts, and Vanderbilt. www.DavidLaRocca.org, DavidLaRocca@Post.Harvard.Edu

Victoria Oana Lupascu is an Assistant Professor of Comparative Literature and Asian Studies at University of Montréal. Her areas of interest include medical/health humanities, visual art, twentieth- and twenty-first-century Chinese, Brazilian, and Romanian literature, and Global South studies. Her work explores different types of medical and visual narratives to unveil histories of cultural, economic, and social disposability. She has published articles in peer-reviewed journals, such as *Humanities* and *CLEAR,* and chapters in edited collections.

Chantelle Mitchell is a project assistant at the University of Vienna, originally from so-called Australia. She has written for *Stilts Journal*, Heart of Hearts Press, *Plumwood Mountain*, *The Lifted Brow*, and *Marrickville Pause*, and presented performance lectures for ACCA, the Ian Potter Museum, Bus Projects, and Free Association. Chantelle maintains a research-based practice with Jaxon Waterhouse, which has seen them present their work through Macquarie University, the Australian Centre for Contemporary Art, the international Temporal Belongings, STREAMS, and Atmospheric Humanities conferences. Their written work has appeared in *Green Letters*, *Performance Philosophy*, *unMagazine*, *e-flux*, *art+Australia*, and *Unlikely Journal*, and they have presented numerous exhibitions in artist-run spaces across Australia.

Patrick D. Murphy is Professor Emeritus and former Chair of the Department of English at the University of Central Florida. Prior to that he taught at Indiana University of Pennsylvania. Founding editor of *ISLE: Interdisciplinary Studies in Literature and Environment*, he is the author of *Transversal Ecocritical Praxis, Persuasive Aesthetic Ecocritical Praxis, Literature, Nature and Other*, and other books. He has also edited or co-edited a dozen books, including *Ecofeminist Literary Criticism* and *The Literature of Nature: An International Sourcebook*. Murphy began his ecocritical studies in 1983 with a master's thesis on spirituality and place in the poetry of Gary Snyder and Wendell Berry. He now lives in Galveston near the rising water's edge.

Danielle O'Steen, PhD, is an art historian and curator based in Philadelphia. She was the inaugural curator at the Kreeger Museum in Washington, DC. O'Steen has held fellowship positions at the Smithsonian American Art Museum and the Hirshhorn Museum and Sculpture Garden, and worked in the curatorial departments at the Phillips Collection, the National Portrait Gallery, and the Baltimore Museum of Art. She holds a PhD in Art History from the University of Maryland. Her dissertation, "Plastic Fantastic: American Sculpture in the Age of Synthetics," considers the role of plastics as a sculptural medium in the United States.

Lily Baum Pollans is Assistant Professor of Urban Policy and Planning at Hunter College in New York City, and the author of *Resisting Garbage: The Politics of Waste Management in American Cities*, published by the University of Texas Press. She studies how cities resist the extraction, consumption, and pollution demanded by our current economic system. Her work, which focuses on solid waste management, traces the complications and potential that such urban-scale resistance generates, and finds reason for optimism in some subtle, yet radical, choices that city governments are making around the world.

Emily Potter is Associate Professor in Writing and Literature, Deakin University. Her research concerns literary modes of place-making, environmental imaginaries, and Australian literary history. She is co-author of *Plastic Water: The Social and Material Life of Bottled Water* (MIT Press). Her most recent book is *Writing Belonging at the Millennium: Notes from the Field on Settler Colonial Place* (University of Chicago Press).

Kirsten Seale, PhD, is a writer and researcher living on Gadigal Land. She is the author of *Markets, Places, Cities* (Routledge) and co-editor of *Informal Urban Street Markets* (Routledge). She has published extensively on the intersections between cities, cultural production, and waste.

Mark Simpson is a settler scholar and professor in the Department of English and Film Studies at the University of Alberta (Treaty Six/Métis Territory), where he investigates US culture, energy humanities, and mobility studies. Recent examples of his scholarship have appeared in journals such as *South Atlantic Quarterly*, *Radical Philosophy*, *Postmodern Culture*, and *English Studies in Canada*, and in volumes from presses such as Minnesota, Fordham, Toronto, McGill-Queen's, and Oxford. He is Principal Investigator for "Transition in Energy, Culture and Society," a multi-year research project with Future Energy Systems at the U of A.

Jaxon Waterhouse is a writer, publisher, and artist living on unceded Ngarluma Country in regional Western Australia. He is the editor of Heart of Hearts Press, which has seen numerous publications and artist books between 2020 and 2021. In 2021, he presented the immersive digital artwork *Quest for the Night Parrot*, alongside numerous exhibitions across Australia. Jaxon maintains a research-based practice with Chantelle Mitchell which has seen them present their work at a number of national and international conferences. Their written work has appeared in *unMagazine, e-flux, art+Australia* and *Unlikely Journal*, and they have presented numerous exhibitions across Australia.

Foreword

Plastic is Plastic

Cymene Howe

Plastic is a beautiful thing. We know this because we have surrounded ourselves with it. Look around you. It appears in every corner of our world. It is handles and wheels, boxes and bread bags; trinket, talisman, tape. It is the warp and weft of petromodernity. Plastic is magical, absorbing shapes in miraculous ways. Morphing into gigantic cylindrical tanks and delicate tools, it also bulges to become an inflatable pool on the hottest day—one that is now becoming hotter still. Plastic holds air like a lung. Except better. Longer. A balloon for your party. A colored orb that flies. Plastic secures. And its transmogrificational capabilities seem to know no limit. It is an infinitely mutable, changeable thing, sucking up color and inspiration. The human mind thrives in plasticity. It has the ability to adapt and mutate and become differently, to trans-figure. And what entity is more transfigural than plastic? A wonder born from the earth. A wonder born from chemicals. A wonder of fabulation and form. An invitation to invention. A handsome material to craft a more beautiful world. Full of fresh possibilities. A palette of color. The persuasive shapeshifter. We are the plastic fantastic lovers.

Plastic is a terrifying thing. We know this because we have surrounded ourselves with it. Look around you. It now resides in the deepest recesses of the ocean. Little bits of it littering the lowest canyon on earth.[1] It is touching us always. Often. Woven into our lives at every stitch. Piercing the organic world with dexterity, duration, and ubiquity. At every turn it appears. You might be able to avoid it, but it would be difficult. Try to go a day without contacting plastic. Without having it somewhere in your service, in your head. Hello, toothbrush. Leavened from oil through the cruel magic of petrochemical engineering, it appeared on the horizon so innocently. As Bakelite. The first synthetic plastic. Born twenty years before the great crash.[2] Such a pretty name, like a fluffy pink cupcake you might give to a child. *Polyoxybenzylmethylenglycolanhydride*: this name is not so pretty. The one that is made into pipes and insulators and guns. How much plastic does it take to make a war?

Plastic is an impervious thing. It was made to capture and to hold, to encase. Rigid or soft, it depends; but its potential to seize and to clutch has captivated us since it came onto the scene. It was a thing that looked like glass but did not break. It would be a thing that could look like any other thing, molded to the specificities of need and desire. Sheer plastic wraps neatly around little white trays of raw meat. It is impervious to the airborne germ. It protects the freshness of food. Plastic lets us eat old flesh. A syringe must be watertight, airtight. *Medical grade* is a seal of approval. Plastic seals. The blood bag must hold the blood tightly. But with give. The body bag should hold the body well. And with respect. The syringe should hold the treatment, with allopathic intention. And this is where plastic excels. Hygiene. Zip. Locked.

Plastic is a leaky thing. Made from goo. It becomes goo again. It breaks down. Fails under the pressures of liquid and the rays of solar saturation. It evaporates, but so slowly. It is particles in the making and the unmaking. It permeates. Coming into bodies discreetly through invisible channels. Or indiscreetly inserted into our cyborgian selves: prosthetics, replacement parts, microplastic ingestions.[3] Max Liboiron and their team tell us that plastic is pollution and that pollution is colonialism: invading vectors of petrochemical imperialism coursing into bodies and changing them.[4] Sickening them. Fish bodies. Human bodies. Oceanic bodies. Astral bodies. Plastic is trans-corporeal.[5] It crosses the line.

Plastic is a regenerative thing. Plastic all you like. It recycles—sort of. Its industry lobbyists have known for a very long time that plastic does not, in most realities, become anew. Plastic is the new cigarette, the new climate change, the new fossil fuel. They knew its harms. Only 10 percent of all plastic made in the last three decades has been recycled. It is too expensive to pick up. It is too dirty when it is. It costs to melt. It degrades. Its resurrection can only happen once, maybe twice, in a perfect world. And making new plastic is such a cheap thing to do. Some are becoming suspicious of plastic's ability to rebirth itself.[6] The triangle of arrows embossed into plastic containers is ringing false. "We must advertise our way out of this," cries the plastics industry. From the minds of Dow and Chevron, the message is clear: plastic is okay. Really. It is forever, but in a good way. Plastic is so light and so durable. It transports so easily. Nearly weightless. That means lower emissions. From the minds of the plastics industry: plastic makes better the air we breathe.

Plastic is plasticity. Do not think of it as one thing with one effect, one trajectory. "Plastic in the singular misses things."[7] It fails to see the struggle in plastic activism, the positioning of plastic science, the nuances of plastic policy and all the other plastic relations. It is not static. It moves like a "force", like the oil out of which it comes.[8] Plastic appears to become trash. So very quickly. Some call it "single-use": the bottle, the bag, the cellophane wrapper. The *use* indicates its utility. Its function. Use value.[9] But you could easily also call it "pre-trash," indicating its ultimate place in the world. (One man's trash is another man's treasure! That is sometimes true.) Waste is relative, we might like to think. Garbage is in the eye of the beholder. Until it has debilitating effects. Plastic can be made into great art. And it has been.[10] Plastic is both visibility and (conspicuous) invisibility.[11] Its sheer sheerness is sexy. Transparent. Filmy. See-through. And also kaleidoscopic. Like oil on water. It glistens. A model substance "to construct a culture of limitless desire."[12] Plastic is becoming. It flatters you. Plastic becomes you.

Plastic is multiple. Multiplied. It has so many names. So many identities to identify. We would all be hard pressed to name them all. For one, there is "Blue Board". DuPont calls it Styrofoam. (That is trademarked, by the way.) With a capital S. Timothy Morton calls Styrofoam a *hyperobject*. A thing that exists in our world but also far, far, far beyond "us." It is white and soft to the touch. Like air. Floaty. Like honeycomb, except more eternal. Lost at sea, you might rather have a bit of Styrofoam than a bite of honeycomb. Fourteen million tons of it are born every year.[13] And it likes to go, go, go. That is why it is so "hyper." We can never hope to see it all. But when we have a piece, hold it in our hand, we are touching the hyperobject. Delicate as a bird, and yet seemingly indestructible, it crumbles and voyages and continues, exceeding the spatiologics of the human and lasting long after we are gone. Timeless. An ontological symphony of an object oriented toward us.[14] A toxic companion for the centuries. For millennia. Who says the world is for the living? The world might just as well be for the Blue Board. The blue planet.

Plastic is geomorphological. It changes rocks. It carries its own geological force. We might call our era the Plasticene. Plastic debris swirling in and across the World Ocean is a "slurry":[15] a harried mixture of solid and liquid forms. Almost all the world's seabirds have eaten it. Ninety percent have inadvertently dined on plastic fragments.[16] So do we. Plastic slurries make up a "plastisphere." This is a surrealist habitat for microbes and their companions.[17] It becomes gyres. Like a dance of oil in water in perpetual motion. Vortexes. Spirals of color and form. Stations in the crossing. 300 million tons of plastic—and counting—are created every year.[18] The volume of it everywhere is enormous. With all the plastics ever made, says the head scientist of the Anthropocene Working Group, "we have enough to wrap the entire world in plastic film."[19] Think of it! All these plastic deposits could be, he says, a great way to mark the onset of the Anthropocene.[20] The time of human imprint. The time of plastic penetration.

Plastic is a pretender. It becomes what we make of it. And in that process, it also makes us its own. To believe in plastic is to believe in the infinite and the imagined unimaginable. Where would we be without it? In a different world. A better world? Another world. Before the unearthing and the impulse to be made malleable. Mercurial. Without commitments. A polymorphous polymer. Like little drops of venomous petrol rain. Immersive. Everywhere.

Notes

1. Rebecca Morelle, "Mariana Trench: Deepest Ever Sub-Dive finds Plastic Bag," *BBC*, May 13, 2019, https://www.bbc.com/news/science-environment-48230157 (accessed October 21, 2021).
2. Of the stock market in 1929 and the ensuing Great Depression.
3. Donna Haraway, "A Cyborg Manifesto: Science, Technology and Socialist Feminism in the Late Twentieth Century," in *Simians, Cyborgs and Women: The Reinvention of Nature* (New York: Routledge, 1985).
4. Max Liboiron, *Pollution is Colonialism* (Durham, NC: Duke University Press, 2021).
5. Stacy Alaimo, *Bodily Natures: Science, Environment and the Material Self* (Bloomington: Indiana University Press, 2010).
6. Konrad, this volume, 2.
7. Liboiron, *Pollution is Colonialism*, 27.
8. See Hawkins in Konrad, this volume, 26.
9. Marx's claim in *The Economic and Philosophic Manuscripts of 1844* was that the "use value" of a product made with human labor could be objectively determined. Its intrinsic character could be understood through its satisfaction of human need.
10. See, for example, O'Steen; Lupascu; Beck and Diamanti; LaRocca, this volume.
11. See Meikle and Hawkins in Konrad, this volume, 6.
12. See MacLeod in Konrad, this volume, 3.
13. Morgan Meis, "Timothy Morton's Hyper Pandemic," *New Yorker*, June 2020, https://www.newyorker.com/culture/persons-of-interest/timothy-mortons-hyper-pandemic (accessed October 21, 2021).
14. On object-oriented ontology, see Timothy Morton, *Realist Magic: Objects, Ontology, Causality* (New York: Open Humanities Press, 2013).
15. See Jennifer Gabrys, this volume, 5.
16. Chris Wilcox, Erik van Sebille, and Britta Denise Hardesty, "Threat of Plastic Pollution to Seabirds is Global, Pervasive, and Increasing," *PNAS* 112, no. 38 (2015): 11899–904.

17. Erik Zettler, Tracy Mincer, and Linda Amaral-Zettler, "Life in the 'Plastisphere': Microbial Communities on Plastic Marine Debris," *Environmental Science and Technology* 47 (2013): 7137–146.
18. Christina Reed, "Plastic Age; How It's Reshaping Rocks, Oceans and Life," *New Scientist*, January 28, 2015, https://www.newscientist.com /article/mg22530060-200-plastic-age-how-its-reshaping-rocks-oceans -and-life/#ixzz79mEnKY8J (accessed October 21, 2021).
19. Ibid.
20. Jan Zalasiewicz, Mark Williams, Colin L. Waters, Anthony D. Barnosky, and Peter Haff, "The Technofossil Record of Humans," *Anthropocene Review* 1, no. 1 (2014): 34–43.

Introduction
The Petroproduct: On Plastics, Capitalism, and Oil

Tatiana Konrad

Plastics: they surround us, they shape us, they are *us*. At first sight, such a conclusion might seem far-fetched and grotesque, yet plastics have long since shed their identity as pure chemical chains and transformed into much more substantial, ubiquitous, political and cultural objects.

Cheap but durable, plastics are some of the most widely produced materials in today's world. The invention and improvement of plastic to the form that is known today—light and stable, among other characteristics—dramatically transformed human lives in the twentieth century. Plastics as such do nothing without human agency. People make them, use them, dispose of them, proliferate them, and ponder their significance. In Science and Technology Studies (STS), plastics are materials that one can work with, whose accumulation on the planet one can assess and control, and whose polluting effects one can measure.[1] One of the most prolific researchers in discard studies, Max Liboiron, emphasizes the scale of the problem that humanity has to address through the sheer amount of plastic on the planet; she claims, "Plastics and their chemicals are challenging regulatory models of pollution, research methods and modes of action because of their ubiquity, longevity, and scale of production."[2] Plastic's widespread integration into human lives has transformed its meaning; through its omnipresence, plastic has become one of the most distinct, defining elements of today's petrocultures.[3] This book emphasizes the utility of plastic as the frame of reference for cultural analysis. Going back to the time when plastics were first created and introduced to Western lives as an early product of petrocapitalism, and finishing with the current moment, when humanity faces plastic-related environmental and health crises, the contributors to this edited collection ponder the cultural meanings of plastics that have both sustained and revealed the proliferation of plastics in petromodern times.

Plastics were an integral component in the formation of the pet-rocapitalist world, and they remain so. Western societies, where plastics were first introduced, have largely privileged the illusion of convenience that plastic creates and spread this idea throughout the globe, further colonizing non-Western nations. In this process, the impending problem of plastic accumulation had not been considered. Plastics significantly improved *cultural* environments—at least that is how it had been advertised to global populations—whereas, through the phenomenon known as *waste*, plastics meanwhile contributed to the transformation and destruction of the *natural* environment. Specifically, the year 2000 is identified by scholars as the moment when the world became aware of a "plastic crisis."[4] Seeing plastic immedi-ately as waste, however, means neglecting the complex processes that precede the moment of discard: namely, production and consumption. Plastics as such are not waste, but the "throwaway society" that was built through the resignification of plastics in the twentieth century defines plastic objects as waste according to certain characteristics.[5] For example, a plastic cup can be reclassified as garbage if it has been touched by someone else or used once, or simply because it is cheap and is not needed at the moment. Waste is thus a "relative" charac-teristic applied in culturally predetermined contexts.[6] The precarious definition of plastic objects—from necessities to trash—and the pre-carity of the moment at which such a definition was produced high-light the following: plastics are not only culturally predetermined but themselves come to define culture. Using plastic as a prism through which to understand cultural values and preferences can help define the petrocapitalist world built through fossil fuels and their byprod-uct plastics, as well as critically reflect on what plastics reveal about human experience and the complexity of the task of moving toward a greener and healthier environment.

Plastics, Environment, Culture, and the Politics of Waste examines plastic as a distinct cultural, political, and environmental phenom-enon. It outlines the intricate relationship with plastic that humanity has been building over the course of the twentieth and twenty-first centuries, drawing on examples from history, the arts, and litera-ture, as well as examining the place of plastics in the current health, environmental, and energy crises. The aim of this book is to reveal the complex nature of plastics, from their rapid incorporation into our advancing ways of life, to the reenvisioning of plastics' role in human life and how, through abundant production, consumption, and disposal of plastics, humanity has initiated a toxic invasion of natural environments and human and nonhuman bodies. Bringing together various perspectives from the humanities, this edited collection contributes to the ongoing research on plastics and petrocultures and emphasizes the crucial significance of addressing the plastic crisis through culture.

Shaping the culture of (plastic) consumption

Can we imagine our world without plastics? No—simply because our world is a *plastic* world. We are at a point of existence when everything that surrounds us is plastic (or composed of it to a certain degree), from pens to houses. But we are also at a point when the health of the environment matters especially: the deteriorating well-being of the planet is steadily recognized as a vital issue, with *health* as such being dramatically reimagined through the unexpected global coronavirus pandemic. That health and plastics are intricately connected becomes apparent through the consequences of plastic's invasion of our lives. Indeed, there was a time when nothing was plastic.

When, at the beginning of the nineteenth century, scientists found ways to manipulate rubber, transforming it into different products, they pioneered the technology of making useful products out of polymers. Yet the quality of synthetic polymers created at that time was simply too bad, unsuitable to use in the manner of today's plastic materials.[7] The history of plastics as we know it began in 1907, when "the first fully synthetic plastic—Bakelite" was invented.[8] With the help of this material, humanity would change our bodies and our culture in profound ways, bringing along progress and destruction at the same time. In the United States shortly before the end of World War II, the world-famous chemical company DuPont prioritized research to find a material that could satisfy the growing needs of consumers. That material was plastic—"an ideal substance from which to construct a culture of limitless desire: it is inexpensive, lightweight, and capable of being moulded into any conceivable shape."[9] After the war, plastics became an integral part of human life, radically transforming it.[10]

This book recognizes the materiality of plastic.[11] Being conscious of the material agency of plastic, it emphasizes that plastic pollution is a result and reflection of human activities. By itself, plastic has not brought about progress, destruction, pollution, and multiple other crises. Therefore, this book proposes the crucial necessity of shifting focus away from plastic as the frame of reference and bringing the gaze back to human social relations, which are responsible for transforming and destroying the environment through multiple means, including through plastic. Unpacking cultural, environmental, social, and political meanings of plastic, the authors in this book foreground the role of humans in formulating those meanings and imposing them on plastic while building a conspicuously plastic world.

We have entered the so-called "Plastic Age"[12]—to use a term that fully grasps the transformative power of plastics on human and non-human lives—and we recognize plastic as a supermaterial. Perhaps the most attractive and addictive characteristic of plastic is that it is cheap and abundant, and, as Susan Freinkel ironically notes, "when in human history has that ever been a bad thing?"[13] But along with its affordability, plastic also promised an entirely new lifestyle: "Easier to shape and color, plastic has given everyday life a sense of greater possibility and plasticity."[14] Moreover, plastic was one way to promote hygiene and prevent spread of diseases, thus "separating people from the infectious biotic world."[15] It was the material of the future, the material that denoted and connoted progress. It was, nonetheless, exactly because it was cheap and easy to work with that plastic soon was reenvisioned as a material that does not last. This does not mean that plastic is bad in quality (although, certainly, depending on the purpose of the object, plastic might not be the ideal material), but rather that plastic products do not have to be valued. Already by the mid-twentieth century, plastic was reinterpreted as "impermanent and eminently disposable."[16] With such a recoding of plastic, quite inevitably, consumption of this material started to grow exponentially, and from zero it easily became 600 billion pounds annually.[17] Consumption of plastic became an inevitable, inescapable part of capitalism. With the rise of production and consumption of plastic products, the amount of waste has inevitably risen, too. In the Plastic Age, waste as such has undergone a dramatic transformation, with plastic garbage and plastic pollution accompanying the phenomenon of plastic progress. This, in turn, has influenced the meaning of plastic. Freinkel notes that plastics now conspicuously "carry . . . a negative set of associations" and "stir . . . visceral disgust."[18] The ubiquity of plastics and the plastic-like nature of the world that we have consciously created—"Plasticville"[19]—have largely prompted us to rethink the meaning of a world made out of and with the help of plastics, sketching out a radically different image of plastic. From affordable and plentiful, life with plastics has been revealed as a dead end. As Amanda Boetzkes and Andrew Pendakis put it, "Plastic weaves itself into every facet of our contemporary reality, it does not simply surround us, it is an epistemology and the reflection of a galling political impasse."[20] Plastic—from its easy availability to the growing plastic waste—is today's Sword of Damocles, destroying the environment and ready to cause even more harm if we do not rethink our way of living with it.

Pollution from plastic is one of the major threats to the environment. It causes "poisoning, disease, and death."[21] Plastic garbage, for example, has largely contributed to "the transformation of the oceans."[22] Jennifer Gabrys elaborates on the problem as follows: "The seas and oceans have become a slurry of plastic. There are now estimated to be up to 100 million tons of debris in the five gyres where plastic debris collects in still ocean currents from the Pacific to the North Atlantic"—this includes various plastic objects *and* microplastic.[23] Plastic pollution also negatively affects wildlife, transforming ecosystems and natural habitats. Such negative consequences—while they might not have been predicted when the material was first introduced—are largely the result of one of the cultural meanings that plastic has gained, namely that it is *disposable*. The throwaway characteristic of plastic emphasizes how cheap it is, how low its quality as compared to other materials, and how easily it can be acquired if needed. Thinking that plastic objects are replaceable—and doing precisely that—is a straightforward outcome of capitalism. Yet, when one plastic object is replaced with another, what happens to the former? While production and consumption directly address the availability of the replacement, these processes fully neglect the object that has been replaced. This, in turn, leads to the accumulation of plastics. Importantly, while there are piles of physical plastic waste, Gabrys reminds us that "plastics are accumulating in many different ways, as they break down, enter food chains as plasticizers and generate alterations in the eating patterns of diverse organisms."[24] Plastic build-up is fundamentally an environmental process. Conceptualizing plastic products as one-time-use yet ever-available objects is the major failure of capitalism that has contributed to environmental degradation. Thinking of plastics as throwaway materials, as Gay Hawkins observes, sustains "the accumulation of plastics in environments and bodies, human and nonhuman."[25]

Oddly enough, plastic has come to mean both "durability" and "disposability," with the latter being the presiding characteristic.[26] Plastic's classification as disposable is a vivid illustration of "how materials come to participate in enacting a temporality of the present."[27] This is a very peculiar and crucial preference that we have made with regard to plastic: no longer is it recognized for its ability to withstand decay and preserve quality (an important and necessary characteristic in a post-industrial world). Instead, it has come to mean something entirely different and, essentially, incompatible with its longevity: objects that can be wasted despite their good condition. The inevitable disposability of plastic objects has become a cultural, political, economic, and environmental process that redefined plastic as *needful* and *needless* at the same time. Consider, for example, plastic packaging for food. Plastic that covers food is viewed as *needful* for a variety of reasons, including because it makes that food "hygienic," "branded," and "convenient."[28] This very plastic, however, becomes *needless*, sometimes even before the first bite. Thinking of plastic as disposable not only charges the material itself with a very strong meaning, but it also causes a dramatic shift in culture, foregrounding "the cultural and ecological implications of more and more things produced for a single use."[29]

Treating plastic products as something we simultaneously need and do not need has led to two major interpretations of the material. Jeffrey L. Meikle contends that plastic "has been naturalized."[30] Hawkins, in turn, foregrounds plastic as a "mundane material," and argues that "in rendering plastic mundane its use was not simply routinised but normalised. That is, plastic was equipped with certain normative capacities."[31] Both scholars accentuate the ubiquity of plastic as a norm, a defining characteristic of the world that we have built. But through both naturalness and mundanity, Meikle and Hawkins effectively reinforce the *visibility* and conspicuous *invisibility* of plastic. It has become so natural for us that we no longer notice it as something extraordinary: we buy it, work with it, and ultimately get rid of it, without even being conscious of what this series of manipulations means culturally, politically, economically, and environmentally. Plastic becomes visible to us only in a form of waste; yet even in this case, invisibility might prevail, depending on where this waste is located, whom or what it threatens, its size, and so on. Plastic can thus be aptly described as a "hyperobject"—Timothy Morton's term for "things that are massively distributed in time and space relative to humans."[32] With regard to plastics, Morton focuses exclusively on "plastic bags"[33]—I suggest expanding this to plastics in general. This "hyper"-nature of plastics is defined by their cultural, political, economic, and environmental roles, and it can perhaps most successfully be grasped through the issues of visibility and invisibility.

To date, concerns about the environmental consequences of plastic pollution have almost exclusively focused on single-use product packaging. While this accomplishes important political work, unveiling the anti-environmental nature of perpetual, thoughtless buying and discarding, it misrepresents the ubiquity of plastic in the world, drastically reducing it to easily detectable, disposable objects. Environmental discourse and policy have hardly addressed the durable and largely unseen plastics embedded in numerous objects, including buildings, vehicles, electronics, appliances, and clothes. Plastics are not only cheap but also lightweight, strong, and malleable, and they have useful insulating and filtering properties. Such qualities make plastic uniquely and remarkably versatile and, indeed, handy, particularly in industries and technology. Such forms of durable but unseen plastic are much more difficult to recycle, and without them contemporary life would look dramatically different. How to interrupt the reliance on such plastics is one important issue that humanity should ponder as we navigate our way to a healthier environment.

As plastic becomes visible, for example as waste, it gains a new sociopolitical, cultural, and environmental significance. This does not, of course, mean that plastic is insignificant before becoming waste (indeed, its significance is predetermined before the object is even produced); it is just that this significance is now conspicuous and tangible, and thus harder to ignore. Hawkins makes a crucial observation, claiming that plastic turns into waste quickly, yet "[w]aste is not displaced to another time or space after consumption, nor is it an effect; it is a material presence animating and ordering interactions in particular ways."[34] Plastic waste is not just the result of our interaction with plastics but it is a pre-programmed, anti-environmental phenomenon in capitalism. Plastic waste tells us not only about capitalist production and consumption and the materiality of our culture, but also about the environment that we neglect, exploit, transform, and destroy. In her recent book *Plastic Capitalism: Contemporary Art and the Drive to Waste*, Boetzkes argues that "garbage stands as both signifier of an ecological condition and the materialization of that condition."[35] Plastic waste is the reason for environmental degradation, and it is environmental degradation in itself. Plastic waste is the result of the plastic production-consumption process, but it also exists outside this process. As such, "it acquires a crucial role in reorganizing social values."[36] In other words, as plastic becomes visible through waste—a phenomenon that we seem to take as a surprise, preposterously expecting tons of disposed plastic to simply disappear—the issue of sustainability becomes particularly pressing. Indeed, plastic waste does not disappear anywhere, and no matter how much we want to believe that our current petro-dependent lifestyle cannot change (let alone disappear), environmental degradation, including through plastic pollution, is the urgent issue to address, and sustainability is frequently presented as the solution.

Boetzkes notes: "Sustainability has become a topic, and even an epistemological standpoint, that pervades every possible domain of culture, from government parties to scientific research, the humanities, architecture, and design."[37] Saving the world today to secure the future of next generations, human and nonhuman alike, one might think would be possible through a sustainable attitude toward the environment. Yet Boetzkes challenges sustainability as *the* solution. She emphasizes: "Greening the world is all too often equated with suppressing the excesses of late capitalist culture while at the same time maintaining economic growth."[38] Sustainability, as Boetzkes views it, however, does not "offer an ethical alternative" and is, in principle, used "to maintain an economy founded on the depletion of resources."[39] Fundamentally, Boetzkes is absolutely correct about the dual nature of sustainability, at least in the way it has been advertised by politicians and other individuals who call for a greener economy. But it is also clear that while we have to solve the energy question now, and effectively address the problem of our dependence on fossil fuels—of which plastic is a major part, as I will further elucidate in the next section—we cannot expect plastic and plastic waste to disappear in the nick of time, or ever. In light of this, recycling *might* be seen as a solution—but even that is only a mirage. Hawkins makes a pivotal statement: "the idea that recycling represents the straightforward requalification and commodification of plastic waste has to be challenged."[40] Recycling does not destroy plastic waste—it turns that waste into new products. Through that process, we learn another truth about plastic: "It emerges as remarkably *non*-disposable."[41] And while recycled plastic has proved to have "agency" that, in turn "has to be enacted," directly influencing political, economic, cultural, and environmental decisions, and creating "new subjectivities—specifically, the recycler,"[42] recycling has also revealed plastic's omnipresence and irradicability.

We have created a world that is convenient for us, largely with the help of plastics. As capitalism has been developing, plastics have come to be seen as a way to maintain, promote, and celebrate the new system. Even the credit card that we use to purchase various goods and services is plastic.[43] We have created the material that puzzles us with how powerful it is, particularly through "the issue of multiplicity."[44] Gay Hawkins, Emily Potter, and Kane Race illustrate the multiple meanings that, for example, bottled water acquires: "Bottled water obviously has different meanings in different settings; its social life is complex and extensive. The same polyethylene terephthalate (PET) bottle of water can exist as a product, as a personal health resource, as an object of boycotts, as part of accumulating waste matter, and much more."[45] Just like water, however, plastic gains numerous, entirely new meanings in addition to the original definition of plastic as a chemical substance. Production and consumption have not only rendered plastic normal—necessary and unnecessary at the same time—but also generated a new world. It is the world where bottled water is perceived as "*ordinary*" while simultaneously "threaten[ing] the universal human right to water."[46] It is the world where "an ethical communicability with the planet" is minimal or nonexistent.[47] It is the *petro*world that is persistently governed through oil products, including plastics.

Understanding plastics through oil

Most plastics are made out of fossil fuels, mainly from oil.[48] Boetzkes singles out two major ways to perceive the connection between plastics and oil: "not only are they [plastics] sourced from petrochemicals that claim 10 per cent of global fossil fuel consumption, but also they are considered a potential resource for oil as well. Since the early nineties, chemists have experimented on a process called 'coliquefaction' which combines and heats waste plastic with coal in order to recuperate oil."[49] Compared to other byproducts of oil, plastics are "the most wasteful and toxic"; they are also "the site of the most utopian technological innovation."[50] Bioplastics, i.e., "*biodegradable plastics whose components are derived entirely or almost entirely from renewable raw materials,*" exist, too, but are considered a product of the future.[51] Because plastics continue to be produced from fossil fuels, they inevitably become part of the energy issue, making our dependence on cheap energy even more abject and desperate, and foregrounding the political, cultural, and environmental impasse toward which we are steadily moving (and which, in a way, we have already reached). Understanding plastic through oil, as a *petro*product, expands its meaning as a material that has become a natural, dangerously invisible component of our environment—it is around us, it is in us, it is *us*. It also reinforces the significance of the energy debate and the necessity of finding alternative ways of existing and progressing as a society, avoiding the dead end that we are approaching in an ever-accelerating fashion.

I am certainly not the first to foreground the profound role that fossil fuels have played in the formation of industrialized societies: from the large reliance on coal to the 1859 oil discovery in Titusville, Pennsylvania, and beyond, humanity achieved *progress* through fossil fuels.[52] Neither am I the first to emphasize the crucial role that oil has been playing since the twentieth century, creating the world as we know it. But in this book, the contributors and I want to throw into relief the ubiquity of oil (and fossil fuels in general) in our lives, specifically through *plastics*, thus continuing the vigorous debates of energy humanities scholars. The Petrocultures Research Group at the University of Alberta, codirected by Imre Szeman and Sheena Wilson, for example, has done prolific work exposing myriad ways in which oil has shaped, and continues to shape, the world as we know it. They coined the term "petroculture" to expose the true nature of the world around us, and, essentially, who *we* are. They explain: "We use this term ['petroculture'] to emphasize the ways in which post-industrial society today is an oil society through and through. It is shaped by oil in physical and material ways, from the automobiles and highways we use to the plastics that permeate our food supply and built environments. Even more significantly, fossil fuels have also shaped our values, practices, habits, beliefs, and feelings."[53] In addition to that, they underscore the dramatic anti-environmental and other outcomes of the fossil-fueled progress, which include climate change and growing amounts of plastic trash.[54]

The post–World War II progress came because there was "a prodigious and reliable supply of crude oil."[55] The ubiquitous presence of oil in our lives is frightening: this natural resource has not only opened ways for humanity to reinvent itself, but also ultimately invaded our planet to such an extent that imagining a world without oil—just like imagining a world without plastic—seems simply impossible. Although, of course, such work must be done and the transition to renewable energy is the only right decision humanity can make today. Our complex relationship with oil has been developing throughout the twentieth century, as Sheena Wilson, Imre Szeman, and Adam Carlson claim in their introduction to *Petrocultures: Oil, Politics, Culture*: "[o]il and its outcomes—speed, plastics, and the luxuries of capitalism, to name a few—have lubricated our relationship to one another and the environment."[56] During the twentieth century oil was reimagined as *essential* to our very existence, "imbricat[ing] into every aspect of our daily lived realities."[57] It is because of this mystic ubiquity of oil in our lives, which has lasted for such a long time, that I, in line with other energy humanities scholars, insist on reenvisioning oil not just as a natural resource, chemical substance, or viscous liquid, but as *our world* and *ourselves*. In their introduction to *Energy Humanities: An Anthology*, Imre Szeman and Dominic Boyer straightforwardly conclude: "The list of products made from petroleum includes . . . everything."[58] Plastic is on that list, too.

Oil enters our lives in myriad ways, including through plastics. And here I would like to go back to Morton's concept of the "hyperobject" to emphasize the parallel nature of oil and plastics. Boetzkes contends that "[t]he plastic object in its state of entanglement does not merely exist on an individual scale," and argues that Morton's term "hyperobject" aptly and more precisely describes the nature and condition of plastic when oil is a "hyperobject" whereas a plastic object is an "object."[59] Boetzkes's claims are correct when looking at plastic as a material made from oil: first comes oil, and then petroproducts;

without oil, plastic would not exist. In the words of Boetzkes, "The plastic object both obscures and exposes global oil, and thereby sits at the crux of the dilemmas of visibility in which petrocultures are bound."[60] I, however, would like to look at plastic as a material whose cultural, political, and environmental influences are so fundamental that we can call plastic a "hyperobject," i.e., not merely a byproduct of oil but as equally influential and problematic a phenomenon as oil itself. I concur with Hawkins, who claims that plastic has "become a force in the world, fueling new economic, ecological, and political realities."[61] Plastic is a *force*, just as oil is.

That plastic has a much more complex meaning than simply "an oil-based material" can perhaps most effectively be grasped through the issue of plasticity. I see plasticity both as the state of the world that we have built, and a phenomenological ambience that characterizes the state of things: plasticity is about being made of plastic, and about not being able to escape the plastic condition. Boetzkes argues that plasticity is what "ensures the robustness of the oil industry."[62] She also emphasizes that in order to perceive oil as such we should concentrate on very concrete phenomena of petroculture: "The equipmental being of petroculture occurs as a permeation and proliferation of objects. That is to say, it is by turning to its [oil's] sub-industry, plastics, that one can see how petrocultures are interwoven with a plastic condition."[63] For Boetzkes, therefore, petroculture is largely "plastic mesh."[64] The peculiar form of petroculture's plasticity is directly enabled through the interaction between oil and plastic that "combine in a common aesthetic and economic regime": "Together, they produce an episteme, invading substance, ways of being, the terms of exchange, and systems of signification."[65] Plastic is a unique byproduct of oil that has completely lost its subordinate position to this natural resource and become a signifier and author of our reality, enabling our complex, plasticized interaction with oil.

The destructive influence of plastic pollution on the environment is colossal. In addition, it affects our own bodies in a dangerously invasive way. In a world that is built on and sustained by fossil fuels, we are not in command, but rather enslaved by our dependence on plastics, on cheap, disposable goods, on energy. Veiled by criteria such as convenience, comfort, access, and excess, our reality is one where humanity destroys everything around us, including ourselves, causing tremendous bodily degradation and extinction. Commenting on how plastics affect us, Szeman and Boyer explain: "we unwillingly ingest plastics into our bodies throughout our lives, passing it on to next generations right at birth through the milk infants ingest from the bodies of their mothers. Today, we don't just depend on oil for energy;

by converting it into plastics, and by ingesting those plastics even without noticing it, we spend our lives *becoming* petroleum, whether we want to or not."[66] Frederick Buell also asserts: "We now literally eat oil."[67] Just as plastic pollution transforms ourselves, it transforms the environment. Through plastic waste we both notice how profoundly dependent we are on oil and can imagine the ecological present and future—"petrochemical cultures buried in their own detritus."[68] The ubiquity of plastic in our lives—from the objects that we use to discarded trash—both visualizes and obscures energy. Through plastic, humanity can also perceive what Stephanie LeMenager terms "petromelancholia," i.e., our realization (and fear) that cheap energy might no longer be accessible, and thus the world as we know it could shatter.[69] At the same time, plastic's durability secures the long-lasting presence of oil in the anthropocentric world, for even turning into useless waste, it stays with us as a grave reality of our petroculture. Plastic—and this exactly illustrates its nature as a "hyperobject"— becomes a part of the environment to such an extent that it begins to "signal the intertwinement of the global oil economy with our current ecological condition, and the extension of this entanglement into the geological future."[70] The legacy of plastic is thus even more durable than the material itself, and tells us about distinct political, economic, environmental, and cultural choices that we have already made and continue to make.

Choice as such has been a defining characteristic in the formation of petroculture. Our reliance on fossil fuels has manifested and damaged the planet in many ways. Discussing, for example, the anti-environmental effects of fossil-fueled transportation, Brian C. Black and I argue: "We *chose* to destroy the planet; we *chose* to carbonize."[71] We know now, as Ross Barrett and Daniel Worden succinctly put it in their introduction to *Oil Culture*, that oil "does things."[72] These things are multiple: their ramifications are hard to face because the temporal reality that they create appears in a variety of ways as comfortable and thus *good*. Yet the alarming call from environmental humanities and energy humanities scholars is to realize, indeed, that the *temporality* of immediate comforts precludes the possibility of the *future* that our dependence on oil fraudulently promises. Thus, Wilson, Szeman, and Carlson lament: "At the heart of this newfound awareness of oil's importance to our sensibilities and social expectations . . . is our recognition that over the course of our current century we will need to extract ourselves from our dependence on oil and make the transition to new energy sources and new ways of living."[73] They continue: "Oil transformed life over the century in which we came to depend on it; the looming threat of its absence from our lives means that it will transform us again, from people who are at home and comfortable in the petrocultures we have devised for ourselves to people who will have to shape ourselves to fit contexts and landscapes we can barely imagine, even if we need to do so—and quickly."[74] In line with these scholars, Black and I emphasize the importance of "long-term eco-thinking that demands that profound cultural, social, and political changes must occur immediately."[75] Searching for ways to eradicate plastic pollution demands envisioning an alternative world. Humanity cannot yet imagine a plastic-free world. Neither does this book sketch out a possible alternative. Instead, it provides a historical, environmental, cultural, and political analysis of plastic and comprehensively studies the conundrum of the social relations and materiality of plastic. In doing so, it contributes to the attempts of environmental humanities scholars to understand, elucidate, reenvision, and ultimately change the world as we know it to save the planet for future generations of humans and nonhumans. *Plastics, Environment, Culture, and the Politics of Waste* is a further attempt to reimagine our culture as a petroculture and to emphasize the role of plastics in the formation thereof. Through a thorough engagement with plastic, this book foregrounds the mutating role of oil, for plastic exists *as* oil, but *beyond* it, too.

Organization of the book

This book is divided into five parts, uncovering the multiple meanings of plastic and a variety of ways in which it has been used to invade, construct, and simultaneously destroy the world around us.

Part 1, "Plastic Lives," explores the ways in which we have historically come to interact directly with plastic on a daily basis. The section opens with Louise Dennis's "Plastics: What Are They Good For?" In order to understand the relationships individuals have with plastics today, it is useful to reckon how people related to those materials in the past. Thus, the author traces the history of consumption to the 1950s, when the post–World War II society was encouraged to consume products to facilitate economic growth and to improve their lives more generally. Disposability, the notion of using something once and then throwing it away, grew to become a sign of wealth and cleanliness as consumers were encouraged to use disposable products. This has led to certain new meanings of plastic: the popular understanding that plastics are low value and therefore disposable, Dennis argues, has been built up over a history of misuse of long-lived materials as short-lived products. The author examines the value placed on plastics through an exploration of their uses and misuses, their consumption and, significantly, their conspicuous non-consumption, and how consumers relate to products made from them at the end of their useful life.

In "How Hula Hoops Changed Hygiene: From Damp-Cloth Utopianism to Chemical Cleaning," Angela Cope looks at the ways that high-density polyethylene changed hygienic practices in the post–World War II years. Domestic cleanliness was extremely important in the early postwar years (1946–55)—the time when plastics were marketed as "sanitary." In 1956, as the polio vaccine was disseminated, sanitation narratives began to disappear, for example, from the toy industry. At the same time, the Ziegler process, which produces high-density polyethylene (HDPE), was invented. Cope examines HDPE's first introduction to the plastics industry and the way that most North American manufacturers came to know the material: through the childhood craze of the hula hoop. Whereas the scaling-up of polystyrene (PS), polyvinyl chloride (PVC), and low-density polyethylene (LDPE) were wartime efforts, manufacturing of HDPE scaled through a toy craze in 1958. Soon afterwards, HDPE became *the* material used for packaging household goods like shampoo, laundry detergent, and cleaning chemicals. By then, Cope asserts, plastics had completed their journey from the heights of hope for an unbreakable, permanent modernity to the ultimately transient material we know today: the stuff of one-season toys, food wrappers, and landfill.

Lily Baum Pollans's concluding essay in this section, "The Anti-Plastic City: How Local Governments Became the Frontline for Preventing Plastic Pollution," focuses on attempts to regulate material consumption of plastics in the United States. For example, in 2016, New York City passed a landmark ordinance to levy a five-cent fee on single-use plastic shopping bags. The ban had widespread support in the city, but the state legislature, under pressure from industry lobbyists, quickly organized a preemption bill. New York City was unable to implement the fee. Pollans notes that, historically, attempts to regulate material consumption in the US have only been successful in moments of extreme crisis. Currently, however, fourteen states have preemption laws, similar to the one New York approved in 2017, that do not allow local governments to regulate plastic consumption. Despite widespread preemption, bans and fees on single-use plastics have become symbols of responsible municipal environmental governance. In this essay, Pollans explores why cities are regulating plastic consumption and what it really means for both the environment and the waste regime. Pollans uses the case of New York City's failed attempt to regulate single-use plastics to explore municipal aspirations and the limits of municipal-scale environmental governance.

Part 2, "Plastic Proliferation," addresses the problem of the excessive increase of plastic in the environment. It begins with Amanda Boetzkes and Dana Feldman's "From Plasticity to the Aesthesis of Queer Toxicity." This essay considers how plastic produces reaction formations: schemas, behaviors, and ultimately a topology of affect. This topology precisely characterizes the dilemma of plastic, that it can be both the condition of thinking as well as the character of thought; plasticity produces plastic thoughts and plastic orientations. In materialist terms, plastic matter replicates the capitalist logic that brought it into being, and this very logic continues to reproduce itself as a planetary topology that retroactively affirms the supremacy of capitalism. Boetzkes and Feldman ask: What, then, counts as a plastic resistance to plastic? They argue that in retrieving plastic's antithesis, its material recalcitrance and dysfunctionality, we might arrive at the limit of the plastic paradigm and thereby imagine its unlimitation. Boetzkes and Feldman consider how resistance to plastic is emerging from a reflection on forms of planetary resistance to its topological spread. Additionally, they suggest that plastic's toxicity is a signal to its queer ontology. The authors, therefore, differentiate ontological toxicity from other forms of subjective distortion, diffraction, and play with normativity.

Chantelle Mitchell and Jaxon Waterhouse approach the petrocultural realities of plastic waste in their essay, "Microplastics in Arctic Sea Ice: A Petromodern Archive Fever." Following the spread of plastic waste into every corner of the globe, they view its "storage" in sea ice and animal bodies as an alternate archive, an obfuscation of ecological agency and the diminution of our own bodily autonomy. Approaching this through a Derridean "hauntological" frame, the authors posit that the presence and persistence of microplastics in sea ice are a manifestation of a global agential community of commodities archiving itself. This palimpsest of ecological histories (in terms of how traces of petrocultures write over one another) subsists as traces and samples of new industrial, cultural, and ecological realities. Considering the personal and affective impacts of this petromodern archival landscape, Mitchell and Waterhouse frame the ensuing melancholia as an Anthropocene equivalent of the Derridean archive fever: a sickness unto the death of ecosystems.

The section proceeds with Andrija Filipović's "Jugoplastika: Plastics and Postsocialist Realism." This essay examines plastic—its production, dissemination, and removal—as a complex ontopolitical problem in the times of postsocial realism, a condition of possibility of imagining and living (and dying) in the spacetimes of perpetual transition toward contemporary European society. Focusing on Jugoplastika, a factory for the production of petrochemical products, Filipović shows the ways in which plastic becomes a pluritemporal, multimaterial, ontopolitical hyperobject, from the finished product, such as toys for children, to the waste deposited at the largest landfill in Serbia, near Belgrade. Occasional spontaneous ignitions at the landfill transform waste plastic into smoke, changing the urban "olfactoscape." Filipović claims that events in Belgrade's olfactoscape reveal a crumbling infrastructure built during the modernization of socialist Yugoslavia. The discourse of a "ruined" landscape and the promise of a "clean" future are inextricably connected to the flows of capital and the financialization of the environment, which is seen as necessary in the transition from a ruined postsocialist state to contemporary European society.

The final essay in this section is "Failed Infrastructures, My Little Ponies, and Wadden Plastics: Eco-Intimacies of the *MSC Zoe* Container Disaster." Here, Renée Hoogland zeroes in on the *MSC Zoe* disaster, a large cargo spill in which 342 shipping containers ended up in the North Sea in early 2019. Soon after the accident, residents of the nearby Dutch Wadden Islands witnessed enormous amounts of plastic waste that had escaped the spilled containers and washed up on shore. By engaging with this container spill, Hoogland explores the emerging intimacies in a landscape where humans exist in co-constitutive relationships with their ecologically damaged surroundings. The author begins by figuratively tracing the shipping container before it entered the surroundings of the Wadden Islands as a washed-up "lost" object. After exploring how the container disaster marked a dialectic of visibility and invisibility, Hoogland analyzes the social, artistic, and technological bids to make sense of the plastic content of these spilled steel boxes. In doing so, she not only reflects upon the ways that the materiality of plastic enters our lives in moments of infrastructural disruption but also explores how, through plastic, we interact with and invade the environment and other species.

Part 3, "Plastics in Art," investigates artistic responses to the emergence and ultimate ubiquity of plastic in our world. The section begins with Danielle O'Steen's "The Pioneers of Plasticraft: When Artists Found Plastics in the United States." O'Steen asserts that public presentations and exhibitions were integral tools for the growth of modern plastics in the United States, starting in the 1930s. She considers key collaborations between the arts and the plastics industry, particularly seen in the 1939–40 World's Fair in Flushing, New York. Modern plastics had their most public debut at the fair, creating many opportunities for manufacturers to explore creative ways to sell their products to the public. O'Steen shows how the emerging plastics industry used artwork and exhibitions as part of their corporate agendas. She argues that this 1930s moment laid the groundwork for much collaboration between the arts and plastics for the following decades—particularly in the 1960s and 1970s—when artists were turning to plastics in even larger numbers and with great enthusiasm.

Victoria Oana Lupascu's "Plastic Intimacy: Chinese Art Making as Recycling Practice" analyzes the multifarious ways in which plastic and its recycling have influenced the development of Chinese art, permanently transforming artistic practices from the beginning of the twenty-first century onwards. The connection between the West, largely defined, and China as seen through these imports, according to Lupascu, has direct consequences on human bodies and the environment, especially around recycling points. The author's focus is on Wang Jiuliang's documentary *Plastic China* (2016), which explores the meaning of plastic and constructs an almost inextricable intimacy between plastic and human bodies, and on artist Wan Yunfeng's unwearable clothes—designed out of scrap plastic—which publicly visualize the intimacy between human skin and plastic, between waste and lives. The essay draws on Wang Jiuliang and Wan Yunfeng's works to theorize and clarify ways of conceptualizing plastic as a pivotal material that redefines Chinese art practices, China's relationship with the world, its effects on human and nonhuman bodies, and the now inextricable intimacy between plastic and human skin.

The section proceeds with Nathan Beck and Jeff Diamanti's "Plastic Poetics: Challenging the Epistemologies of Plastic Waste in the Artwork of Maria Roelofsen." Using plastic and driftwood materials salvaged from the shore, artist Maria Roelofsen creates animal-like sculptures that perform a vital function in surfacing the circuitous nature of global waste and pointing to the universal human complicity in ineffective plastic disposal. Beck and Diamanti argue that the ontological demarcation of plastic as waste or toxic or disposable is temporary, and rewritten in Roelofsen's art by introducing a poetics of aesthetic play that displaces the immediate discourse on the environmentally destructive tendencies of plastic, and instead provokes a sense of beauty and renewed purpose in its structural (re)composition with organic materials. Drawing on concepts of scale, surface, and camp, Beck and Diamanti situate Roelofsen's art and its poetics amidst recent thinking by Stacy Alaimo and Amanda Boetzkes in particular to consider how cutesy and creative play can widen the discursive frame and illuminate new ways to engage in critical, aesthetic, and activist practice that help us learn to live within the far-reaching penetration of plastic life.

Finally, in "The Performance of Plasticity: Method Acting, Prosthetics, and the Virtuosity of Transformation," David LaRocca investigates plasticity in film, specifically the plasticity of performers. As film critics make their appraisals known in each new season, it is so often those actors who "transform," who "become" whom they play, who "inhabit" their roles, or who—as the saying goes—simply "disappear," leaving nothing but the character, who are most vaunted. In these moments, we seem to discover not a respect for the character so much as an infatuation with the virtuosity of the actor. Hence, LaRocca asserts, the public favors, celebrates, and stands in awe of the plastic artist—the more plastic the better. And yet, differences should be struck between Meryl Streep's accent, weight gain and loss by Charlize Theron and Joaquin Phoenix, and variations of costuming—from cosmetics and prosthetics to the digital "skins" afforded by motion capture technology. LaRocca thus critically explores this taxonomy with the aim of making a case for why these differences in plasticity matter for the aesthetic appreciation and ethical assessment of cinematic performances.

Part 4, "Plastics in Literature," compiles essays that analyze literary portrayals of plastics. It begins with Lynn Keller's "Polymeric Thinking: Allison Cobb's *Plastic: An Autobiography*." Keller claims that the distinctive approach to autobiography in Allison Cobb's hybrid text, *Plastic: An Autobiography* (2021), both reflects the conditions of this Anthropocene moment and offers a generative model for responding to them, challenging outmoded ideas of boundaries between species, nations, generations, individuals, and between bodies and their material environments. Cobb is the daughter of a physicist at Los Alamos National Laboratory, where the atomic bomb was developed, so her life story necessarily includes the story of the bomb's development. Because the deployment of the atomic bomb depended on the development of plastic, her autobiography tracks those responsible for plastic's invention. Entangled in the global accumulation of plastic waste and the trans-corporeal movement of toxins from plastic, Cobb's story incorporates the poisoning of albatrosses and of human communities where plastics are produced. Importantly, it is Cobb's way of insisting on elaborate chains of thought linking seemingly distant events and beings—what Keller calls *polymeric thinking*—that propels her journey of discovery. It draws Cobb into relation with activist communities of color in environmentally degraded and impoverished areas of Texas and Louisiana where petroleum is processed and plastics are made. Keller thus explores what Cobb reveals about the resources of polymeric thinking for confronting current environmental challenges.

The section proceeds with Emily Potter and Kirsten Seale's "Plastic City: Temporality, Materiality, and Waste in Vanessa Berry's *Mirror*

Sydney," where the authors consider the ubiquity of plastic in Australian urban and peri-urban life through the literary practices of Sydney author Vanessa Berry. Berry's ethnographic practices of mapping and tracing forgotten layers in the urban fabric, through walking, collecting, and narrating, highlight the paradoxical temporality of modernity in an urban powerhouse like Sydney, where a temporality is embedded in its materiality. As in all global cities, in Sydney plastic polymers structure both the rapidity of urban development and its cultures of inhabitation. Berry's practice, Potter and Seale claim, unsettles this trajectory by paying attention to the debris, the abandoned infrastructure, and the quieter polyphonic stories that fall away from the dominant mono-vocality of the global city narrative—worlds that are also composed of plastic. Yet the authors argue that the fragmentary nature of these histories, and their slow undoing in the urban landscape, tells a different account of the role of plastic in urban Australia. These complex, affective encounters are more intimate, more mobile, and ultimately something that we can imagine ourselves abandoning, in a necessarily different environmental future. In this way, these accounts, and Berry's embodied, local ethnographic practices, work against the erasure of the worlds from which hypermodernity arose, to contemplate a genuine postmodernity.

The section closes with Donna A. Gessell's "Better Learning through Plastic?: The *Moby-Duck* Saga." Gessell draws on an accident that took place during a 1992 storm in the Pacific Ocean, when falling cargo containers from a ship launched 28,800 plastic bath toys into the ocean. The losses, a fourth of which were yellow ducks, have had large ripple effects for both scientific research and the popular imagination. For months, stories in the popular press publicized spottings of the PVC bath toys on far-flung beaches, following their movements. Organized communication among those tracking the spill's progress caused oceanographers to revise their modeling of ocean currents; instead of separated and independent, the spill showed the currents as one huge interrelated system. As sightings of the "Friendly Floatees" continued to capture the public imagination, others who were monitoring plastics in the ocean, particularly the huge garbage patch east of Hawaii, used the phenomenon—which had morphed into rubber duckies in the popular imagination—to raise awareness of the immensity of the plastic problem. Surveying the array of stories from the original accident through to all of the resulting representations, this essay explores the evolution of the pedagogy at stake. Even though these constructions sought to educate people of all ages about environmental issues using yellow ducks to publicize the seriousness of plastic waste in the ocean, they have unintentionally normalized the phenomenon.

Part 5, "Plastics and the Future," addresses the changing role of plastic at this very moment, when the world is struck by multiple crises including the COVID-19 pandemic and the global energy crisis. The section opens with Patrick D. Murphy's "Disposable: The Dirty Word in Medical Plastics." The ability to recycle has become a justification for the continued use of numerous plastic products. But one area where plastics are virtually never recycled, and which comprises a growing percentage of total plastic waste, is the medical field. This results from another dirty word coursing through hospital hallways: *contaminated*. In this essay, Murphy explores the problem of medical waste, particularly that coming from hospitals in the United States. The essay includes consideration of the growing substitution of hydrocarbon-based plastics for natural rubber latex. Additionally, it analyzes the enormous increase in the use of disposable gloves in all medical areas, including dentistry, for example, in response to the AIDS epidemic and the persistence of HIV among the population. It looks at the impact of a pandemic, such as COVID-19, on disposable plastic consumption. Finally, it takes up the polluting problems of the primary method of medical waste disposal—incineration—and considers whether there are any alternatives to this particular use of hydrocarbons in American daily life.

Sasha Adkins and Brittany Y. Davis continue to explore the changing role of plastics in (post-)pandemic times in "Eco-Fascism and Alienation: Plastics in a Post-COVID World." The authors note that plastics have emerged from the first wave of COVID-19 stronger: not only as toxic molecules but toxic symbols. In the time of COVID-19, plastics became simultaneously the hero and villain of the COVID response. On the one hand, they provide a life-saving interposition between our bodies and a virus that has toppled economies and taken too many lives. Simultaneously, their increased use, as reusable bags and cups return in an effort to guard against pathogens that might be brought from domestic into public spaces, threatens to erase decades of work to move away from disposable culture. Adkins and Davis warn that questions remain about how the desire for a barrier between our bodies and that which is foreign will manifest in a reconfigured post-COVID world. How much damage (in terms of plastic usage) has been done that must be undone to realize life without plastics after this global pandemic? The authors argue that any such vision must necessarily confront people's desire for protection from the foreign, leading one to ponder the tradeoffs (such as xenophobic policies) that will be made to provide "safety" without disposable plastic barriers.

This section, and consequently the book itself, concludes with Mark Simpson's "Plastic in the Time of Impasse." Simpson claims that although *energy transition*—a shift from dirty to clean energy—is a mantra for the present, *energy impasse* is *the* defining condition of our age. So envisioned, according to Simpson, impasse has less to do with blockage, obstacles to circumvent or overcome, and more to do with "stuckness." In his essay, Simpson argues that plastic, petromodernity's ubiquitous offspring, materially as well as emblematically epitomizes such impasse-as-stuckness. In developing this argument, the author attends more particularly to *plastic time*: to the time signatures and temporal imaginaries instituted and normalized, in petroculture, through plastic. The time of plastic operates according to a contradictory rhythm that connects disposability with saturation and impermanence with perpetuity while seeding malleability or fungibility as the determining condition and abiding allure of objects and subjects alike. Simpson explores how a reckoning with these plastic temporalities might illuminate the urgent problem of energy impasse today—and how it might open up some perspective against or beyond the stuckness of impasse.

Notes

1. For more on this crucial topic, see the work of Max Liboiron.
2. Max Liboiron, "Redefining Pollution and Action: The Matter of Plastics," *Journal of Material Culture* (2015): 1.
3. Imre Szeman, Petrocultures Research Group, *After Oil* (Edmonton: Petrocultures Research Group, 2016), 9.
4. Catherine Alexander and Joshua Reno, "Global Entanglements of Recycling Policy and Practice," September 28, 2020, n.p., https://oxford re.com/anthropology/view/10.1093/acrefore/9780190854584.001.0001 /acrefore-9780190854584-e-18.
5. The term "throwaway society" was coined by Vance Packard in 1967. See Martin O'Brien, "Consumers, Waste and the 'Throwaway Society' Thesis: Some Observations on the Evidence," *International Journal of Applied Sociology* 3, no. 2 (2013): 19. The term "the throw-away society" has also been recently used by Michiel Roscam Abbing. See Michiel Roscam Abbing, *Plastic Soup: An Atlas of Ocean Pollution* (Washington, DC: Island Press, 2019), 6.
6. Susan Strasser, *Waste and Want: A Social History of Trash* (New York: Henry Holt and Company, 1999), 5.
7. Scott Lambert and Martin Wagner, "Microplastics Are Contaminants of Emerging Concern in Freshwater Environments: An Overview," in *Freshwater Microplastics: Emerging Environmental Contaminants?*, ed. Martin Wagner and Scott Lambert (Cham: SpringerOpen, 2018), 3.
8. Jennifer Gabrys, Gay Hawkins, and Mike Michael, "Introduction: From Materiality to Plasticity," in *Accumulation: The Material Politics of*

Plastic, ed. Jennifer Gabrys, Gay Hawkins, and Mike Michael (London: Routledge, 2013), 2.

9. Janine MacLeod, "Holding Water in Times of Hydrophobia," in *Petrocultures: Oil, Politics, Culture*, ed. Sheena Wilson, Adam Carlson, and Imre Szeman (Montreal: McGill-Queen's University Press, 2017), 270.

10. Jeffrey L. Meikle, *American Plastic: A Cultural History* (New Brunswick: Rutgers University Press, 1995), 1.

11. For more on materiality, see Daniel Miller, ed., *Materiality* (Durham, NC: Duke University Press, 2005) and Stacy Alaimo and Susan Hekman, ed., *Material Feminisms* (Bloomington: Indiana University Press, 2008).

12. See Gay Hawkins, "Plastic and Presentism: The Time of Disposability," *Journal of Contemporary Archaeology* 5, no. 1 (2018): 94–5.

13. Susan Freinkel, *Plastic: A Toxic Love Story* (Boston: Houghton Mifflin Harcourt, 2011), 7.

14. Meikle, *American Plastic*, 2.

15. Frederick Buell, "Energy Systems," in *Fueling Culture: 101 Words for Energy and Environment*, ed. Imre Szeman, Jennifer Wenzel, and Patricia Yaeger (New York: Fordham University Press, 2017), 143.

16. MacLeod, "Holding Water," 270.

17. Freinkel, *Plastic*, 7.

18. Ibid. 8–9.

19. Ibid. 1.

20. Amanda Boetzkes and Andrew Pendakis, "Visions of Eternity: Plastic and the Ontology of Oil," *e-flux* 47 (September 2013): n.p., http://www.e-flux.com/journal/visions-of-eternity-plastic-and-the-ontology-of-oil/.

21. Buell, "Energy Systems," 143.

22. Margaret Atwood, "It's Not Climate Change—It's Everything Change," in *Energy Humanities: An Anthology*, ed. Imre Szeman and Dominic Boyer (Baltimore: Johns Hopkins University Press, 2017), 142.

23. Jennifer Gabrys, "Plastic and the Work of the Biodegradable," in *Accumulation: The Material Politics of Plastic*, ed. Jennifer Gabrys, Gay Hawkins, and Mike Michael (London: Routledge, 2013), 208.

24. Ibid. 209.

25. Gay Hawkins, "Plastics," in *Fueling Culture: 101 Words for Energy and Environment*, ed. Imre Szeman, Jennifer Wenzel, and Patricia Yaeger (New York: Fordham University Press, 2017), 271–2.

26. MacLeod, "Holding Water," 270.

27. Hawkins, "Plastic and Presentism," 92.

28. Gay Hawkins, "The Performativity of Food Packaging: Market Devices, Waste Crisis and Recycling," *The Sociological Review* 69, no. S2 (2013): 66.

29. Hawkins, "Plastics," 271.

30. Meikle, *American Plastic*, 1.

31. Hawkins, "The Performativity of Food Packaging," 66; Gay Hawkins,

"The Skin of Commerce: Governing through Plastic Food Packaging," *Journal of Cultural Economy* 11, no. 5 (2018): 389.

32. Timothy Morton, *Hyperobjects: Philosophy and Ecology after the End of the World* (Minneapolis: University of Minnesota Press, 2013), 1.

33. Timothy Morton, "A Quake in Being," in *Energy Humanities: An Anthology*, ed. Imre Szeman and Dominic Boyer (Baltimore: Johns Hopkins University Press, 2017), 357.

34. Hawkins, "Plastics," 273.

35. Amanda Boetzkes, *Plastic Capitalism: Contemporary Art and the Drive to Waste* (Cambridge: MIT Press, 2019), 26.

36. Gay Hawkins and Stephen Muecke, "Introduction: Cultural Economies of Waste," in *Culture and Waste: The Creation and Destruction of Value*, ed. Gay Hawkins and Stephen Muecke (Lanham: Rowman & Littlefield Publishers, 2003), x.

37. Boetzkes, *Plastic Capitalism*, 7.

38. Ibid. 8.

39. Ibid. 8.

40. Gay Hawkins, "Made to Be Wasted: PET and Topologies of Disposability," in *Accumulation: The Material Politics of Plastic*, ed. Jennifer Gabrys, Gay Hawkins, and Mike Michael (London: Routledge, 2013), 64.

41. Ibid. 64; italics in original.

42. Hawkins, "The Performativity of Food Packaging," 67.

43. See Joe Deville, "Playing with Plastic: The Enduring Presence of the Credit Card," in *Accumulation: The Material Politics of Plastic*, ed. Jennifer Gabrys, Gay Hawkins, and Mike Michael (London: Routledge, 2013), 87–104.

44. Gay Hawkins, Emily Potter, and Kane Race, *Plastic Water: The Social and Material Life of Bottled Water* (Cambridge: MIT Press, 2015), 9.

45. Ibid. 9–10.

46. Ibid. 72 (italics in original), 133.

47. Amanda Boetzkes, *The Ethics of Earth Art* (Minneapolis: University of Minnesota Press, 2010), 182.

48. Sally Morgan, *Waste, Recycling and Reuse* (London: Evans, 2009), 21. See also Szeman, Petrocultures Research Group, *After Oil*, 67.

49. Amanda Boetzkes, "Plastic Vision and the Sight of Petroculture," in *Petrocultures: Oil, Politics, Culture*, ed. Sheena Wilson, Adam Carlson, and Imre Szeman (Montreal: McGill-Queen's University Press, 2017), 228.

50. Ibid. 228–9.

51. E. S. Stevens, *Green Plastics: An Introduction to the New Science of Biodegradable Plastics* (Princeton: Princeton University Press, 2002), 104; italics in original.

52. Sheena Wilson, Imre Szeman, and Adam Carlson, "On Petrocultures: Or, Why We Need to Understand Oil to Understand Everything Else," in *Petrocultures: Oil, Politics, Culture*, ed. Sheena Wilson, Adam

Carlson, and Imre Szeman (Montreal: McGill-Queen's University Press, 2017), 5.

53. Szeman, Petrocultures Research Group, *After Oil*, 9.

54. Ibid. 60.

55. Brian C. Black, "Oil for Living: Petroleum and American Conspicuous Consumption," *The Journal of American History* 99, no. 1 (2012): 41.

56. Wilson, Szeman, and Carlson, "On Petrocultures," 15.

57. Ibid. 16.

58. Imre Szeman and Dominic Boyer, "Introduction: On the Energy Humanities," in *Energy Humanities: An Anthology*, ed. Imre Szeman and Dominic Boyer (Baltimore: Johns Hopkins University Press, 2017), 12–13. See also Boetzkes, "Plastic Vision," 229; Kirsty Robertson, "Oil Futures/Petrotextiles," in *Petrocultures: Oil, Politics, Culture*, ed. Sheena Wilson, Adam Carlson, and Imre Szeman (Montreal: McGill-Queen's University Press, 2017), 242–63.

59. Boetzkes, "Plastic Vision," 230, 231.

60. Ibid. 231.

61. Hawkins, "Plastics," 271.

62. Boetzkes, "Plastic Vision," 222.

63. Ibid. 227.

64. Ibid. 227.

65. Ibid. 240.

66. Imre Szeman and Dominic Boyer, "The Aesthetics of Petrocultures," in *Energy Humanities: An Anthology*, ed. Imre Szeman and Dominic Boyer (Baltimore: Johns Hopkins University Press, 2017), 429–30; italics in original.

67. Buell, "Energy Systems," 143.

68. Hawkins, "Plastics," 271.

69. Stephanie LeMenager, *Living Oil: Petroleum Culture in the American Century* (Oxford: Oxford University Press, 2014), 102.

70. Amanda Boetzkes, "Plastic, Oil Culture, and the Ethics of Waste," *RCC Perspectives*, no. 1 (2016): 51, https://www.jstor.org/stable/10.2307/26 241344.

71. Tatiana Prorokova-Konrad and Brian C. Black, "Carbonization as a Choice: Environmental Ethics, Mobility, and Energy Options," in *Transportation and the Culture of Climate Change: Accelerating Ride to Global Crisis* (Morgantown: West Virginia University Press, 2020), 6; italics in original.

72. Ross Barrett and Daniel Worden, "Introduction," in *Oil Culture*, ed., Ross Barrett and Daniel Worden (Minneapolis: University of Minnesota Press, 2014), xvii.

73. Wilson, Szeman, and Carlson, "On Petrocultures," 3.

74. Ibid. 3.

75. Prorokova-Konrad and Black, "Carbonization as a Choice," 3.

Part I

Plastic Lives

1

Plastics: What Are They Good For?

Louise Dennis

The title of this chapter poses a question: Plastics—what are they good for? And if we think about the pollution that we see in our streets, waterways, and oceans we might answer "not much." However, by taking a journey through the uses and misuses of plastic materials, our consumption and significantly conspicuous non-consumption of them, as well as how we deal with them at the end of their useful life, this chapter will explore the value placed upon plastic objects and give an insight into why we are in the position we are. The appreciation or reception of materials can create a positive or a negative reaction in the user, and an individual's understanding of materials comes from their own experiential knowledge, the influence of others, and cultural perception. Empirically, individuals tend to be either nonchalant or have very strong views about this group of materials; Jeffery L. Meikle acknowledges that we have a culturally ambivalent relationship with plastics, as the materials are appreciated "as high-tech miracle and as cheap substitute simultaneously."[1] For Susan Freinkel, the proliferation and the length of time that plastics have been part of our lives indicates a "deep and enduring relationship."[2] However, relationships can be both good and bad, and plastics are probably one of the only material groups to generate such extreme diversity and debate.

Cultural relationships with plastics

In a statement ahead of her time, Marion Gough, a journalist for *House Beautiful* magazine, was keen that the public of the late 1940s should take a rounded view of plastics: not to see them as miracles, but to give them more consideration. She suggested that plastics "need fewer people joyously believing that there's no ill in the world that plastics can't cure." She also reminded the reader that there was a need for more "well-informed critics who refuse to condemn all plastics because some have failed" and that they should take the time

to learn about plastics, and what the material could do for them.[3]
Gough's views on plastics at a critical time in their use and development were forward-thinking and balanced.

At the outbreak of World War II, plastics were seen as essential to
the war effort and were used in groundbreaking technologies such as
radar systems in aircrafts and the bubble cockpit canopies of fighter
planes. All sides focused their development of plastics into their use in
conflict; with the shortages in natural materials, new synthetics were
developed to fill the gaps. The production of new materials during the
war created a surplus that did not necessarily have a natural place to
settle in peacetime. These materials gave designers the opportunity
to experiment with new forms of old products, such as Charles and
Ray Eames using glass-reinforced polyester in the production of their
chairs for Herman Miller.[4]

As Roland Barthes stated, plastics can be turned into "buckets as
well as jewels"[5] but they are *neither* buckets *nor* jewels and as such
have suffered culturally through a lack of identity. This has given
rise to the notion that plastics are inauthentic; that they cannot be
true to themselves as materials, as they do not *have* a true identity.
Initially, plastics were used as substitutes for other, usually natural,
materials—what the industrial designer Ezio Manzini would call
"nobler materials,"[6] those that could be said to have great value and
integrity, such as stone and wood. In a 1942 talk on the BBC's Home
Service, C. Frank Merriman stated that plastics should not be seen as
a "cheap substitute" for other materials, as they could often provide
a "better alternative for the job in hand."[7] In the same year, James
Hogan wrote that the biggest mistake of the plastics industry was to
disguise the material as something else.[8] Plastics had the capability to
become and do different things that natural materials were not able
to do. A. H. (Woody) Woodfull, an eminent designer who worked
within the plastics industry from the 1930s to the 1970s, urged others
not to use plastics as imitative; he stated: "If I want wood, I buy wood;
if I want plastics I buy plastics, and I have no desire to hide the fact
that what I have bought is a moulding. This then is the position with
regard to materials—give them honest usage."[9]

The cultural meanings of plastics over the twentieth century have
been "varied and complex," with the materials acting as substitutes
for luxury alternatives (such as ivory), while at the same time relating
to new technological innovations. These two areas of development
earned them "respect and admiration."[10] It was only with the emergence of mass culture during the first half of the twentieth century that
plastics came to be seen as "cheap and nasty" in the eyes of society,
associated with mass manufacture and the move away from the luxury

market. Connotations of cheapness, low quality, and inauthenticity contrasted with traditional craft materials. These ideas of plastics did not sit well with the modern movement and the dictum of "truth to materials."[11] Manzini argues that for plastics to enter high culture they had to change, take on their own form, and find their own image. He states that during the modernist movement plastics became accepted when their economic value combined with cultural value. By showing their own qualities such as clarity or colorfulness, glossy surfaces, fluid forms and being made in a single piece, they demonstrated marks of the new not imitating the old. By not pretending to be anything other than plastics, and by not denying their manufacturing processes or apologizing for what they were, plastics became accepted materials.[12]

The rise of plastics was significant during a time of depression, when "merchandisers hungrily sought color and novelty."[13] The dramatic and passionate relationship between manufacturers and plastics was publicly experimental, sometimes seeing the wrong materials used in the incorrect setting. Freinkel offers the examples of toys that broke after just a few interactions, and raincoats that fell apart when they got wet. Despite technological improvements in the 1950s, the reputational damage was firmly lodged in the minds of the public.[14]

The 1950s saw a rise in the number of objects made from plastics by a manufacturing community that was swamping the market with objects made from misapplied materials. This led to objects failing to fulfill the tasks they were designed to do. It was this swamping of "shoddily made and poorly designed" goods that eroded the status of plastics that many had worked so hard to achieve. Plastics were therefore commonly seen as "tacky, inferior and expendable."[15] However, the 1960s saw another significant era for plastics making use of this very feature. They offered a way of conveying some of the most fundamental values embraced by Pop culture. Pop culture, as Penny Sparke describes it, had two focuses: an "aesthetic of expendability" or an "expendable aesthetic"—the former offering the idea of disposability, whilst the latter could be easily disposable. Her examples are an inflatable PVC chair versus a paper chair.[16] In contrast to the celebrated disposability of the 1960s, plastics then went through the "ecological outcry of the 1970s" and were considered to be inferior to "natural" materials.[17] However, the selection of materials is more complex than "natural equals good/plastics equals bad"; the production of unbleached, undyed cotton, for instance, uses large quantities of hazardous chemicals and high levels of water during its production,[18] and paper manufacturing requires significant land and water resources.[19]

The developments in plastics saw more increases in the late 1980s than in the previous two decades, and as such the image of the material family was constantly refocusing. By this time, the brightly colored fun of the 1960s had been replaced by a more "high-tech" image. Sylvia Katz describes the materials as offering "new textures, colours and moulding possibilities," leading to a changing perception of objects which allowed the designer a new vision. Even when we think we understand the material, Katz argues, we are offered a new surprise. She cites the historically accepted electrical insulating properties of plastics being joined by electrically conductive properties in the 1980s, which have gone on to allow the form of electrical circuits to change. Katz celebrates the evolving nature of the material family, constantly replacing and improving upon itself.[20] Sparke suggests that the key elements of Pop culture reemerged in the 1980s under the label of postmodernism, which embraced plastics for their almost infinite range of possibilities, with the ability to carry different messages. This limitless metamorphosis made the materials perfect for a "culture which thrives upon pluralism."[21]

Unseen plastics

Today, there are many unseen applications where plastics are used because of their specialist properties. These applications include leak-proof water and gas pipes and electrically insulating cable covers hidden in the floors, walls, and streets all around us. The materials used here would generate relational responses if seen, but when they are hidden and doing vital jobs it is difficult to make any kind of relationship with them. There is often a barrier, the wall or the pavement, between people and the object, the pipe or cable. There is a kind of interaction, but one that is at a distance, and not always recognized by the subject. This distance prevents a relationship from being built in the same way as if the object was in our hand. As such, these kinds of applications are not always acknowledged when the use of plastic materials are considered purely because they are out of sight and out of mind.

By the unseen, I am not just referring to products that are literally out of sight, but also to objects that are simply unnoticed. As long as they do their jobs well, they function without public consideration even in full sight. Meikle suggests that, because plastics assume many guises and satisfy so many desires, "as long as it does its job well" they become invisible;[22] or, to put it another way, our attention is not drawn to them. Plastics are ubiquitous materials and because of this they can be taken for granted and assumed to be cheap, their true value not recognized. Mimi Sheller describes a similar attitude toward aluminum when she writes that because we take it for granted, we forget it is there, and we end up throwing away aluminum products without seeing it for the "precious substance that it is."[23] The use of plastics in the automotive, electrical, water, and medical industries demonstrates this by not being the focus of today's anti-plastics campaigns. This is because in these situations, the plastic materials are doing such a good job that they do not garner any attention. Plastics come into people's vision when they do not do their job well or when they are used inappropriately, when they are particularly common, or especially visible.[24] For example, the use of a long-lived material for a short-term need causes the material to become highly visible and highly contentious, especially if it is poorly disposed of. One of the plastic family's key features, its lightness, could also be considered its main downfall.[25] Products made of plastics can often be seen floating in water, blowing in the wind, or caught in trees as litter. All materials become waste[26] and will sometimes become litter, but other materials are, perhaps, less visible than plastics as they sink to the bottom of the stream or are kicked under a bush and stay, out of sight, where they land.

The use and misuse of plastics

Packaging and single-use disposable products are two of the most prominent applications for which plastics are used. They are things that are thrown away with little concern.[27] Single-use products are those that are designed to be used once and then thrown away, products that are discarded, sometimes as litter, often as rubbish, and occasionally recycled. This notion has attracted a lot of media attention in the past few years and affects the way the public understand the material.[28] For Fenichell, one of the most significant things about plastics is the fact that the packaging they can be turned into has practically become more noteworthy than the product they encase: "the ultimate triumph . . . of style over substance, of surface over essence."[29] I understand this as the look of the packaging becoming more important than the product inside but also that the style and casing of a radio, for example, has become more important than the quality of the sound it makes, form having precedence over function. However, let us return to look at the purpose of packaging, using water bottles as an example. The first bottled water in Britain came from the natural Malvern springs in 1851. At this time, municipal water supplies were not always safe to drink owing to water-borne illnesses like cholera and typhoid, so people sought out uncontaminated sources.[30] This is a demonstration of the importance of packaging to protect its contents and provide clean, safe food and drink to the consumer. Originally, bottles would have been made of glass and had cork bungs, with PET (polyethylene terephthalate) bottles being introduced in 1973.[31] Plastic bottles are lightweight and unbreakable, unlike glass, and they use less energy to be delivered than glass bottles, with up to 40 percent less fuel used to transport drinks in plastic bottles compared to equivalent glass bottles.[32]

The initial possession of an object, or the consumption of it, makes for a powerful relationship between object and user; with an emotional attachment it becomes "mine," for example, "my bottle of water" or "my cardigan." Similarly, the dispossession of that object is a powerfully freeing act. Throwing something away has a finality that, to quote Gavin Lucas, de-constitutes and de-matters an object both structurally and symbolically.[33] By putting objects in the bin or dumping them somewhere, the relationship is actively broken as a conscious decision is made. It then becomes easy to disassociate oneself from that which is thrown away: that water bottle or cardigan is no longer "mine." This can be seen by observing people who drop litter and suggest it is not their problem as someone else will clear it up. In a 2009 report by ENCAMS, participants in a study even found benefits to littering as "it provided jobs for cleansing staff; revenue was raised from fining people; it fed birds and other wildlife."[34]

When we become aware of an issue, it can evolve into something that we notice even more readily. It works in the same way as when you get a new car and suddenly notice more of the same model than you had seen before: it is not that there are suddenly more cars of that model around, they are just at the forefront of your mind. Writing about waste, Gay Hawkins suggests that a lot can happen when we notice waste: managing its material reality is part of how we organize ourselves and our environment. How we put out the rubbish, whether we recycle, reuse, and reduce, embodies our ethics.[35] Hawkins's work looks at how we relate to waste and is significant to the story of plastic packaging and its life after use. She suggests that reducing waste is evidence of an effective economy.[36] She explores the material used to make single-use water bottles, and how it could be seen to be designed to become waste. In her conclusion to this work, she states that "the single-use PET bottle can be considered a conduit of topological relationships that mixes up plastics waste with plastics production and consumption."[37] That is to say, the waste plastic that is seen in the wrong place, doing the wrong thing, sullies popular understanding of the material itself. The focus is at the end of life of the product, before the beginning and the middle.

After the restrictions of World War II, with rationing touching many aspects of consumption, society of the 1950s was encouraged to consume products to aid economic growth, to maintain jobs, and to improve lifestyles for those further down the economic chain by creating a second-hand market.[38] The disposability of ephemeral items, and less ephemeral objects with designed-in obsolescence, was encouraged. Disposability and the notion of using something once and then throwing it away grew to become signs of wealth and cleanliness. Consumers were encouraged to use disposable products for their efficiency and to avoid contamination. The ideas of purification and convenience encouraged the development of ethical justifications for the use of disposable items.[39] The link between cleanliness and single-use packaging is strengthened by the act of throwing away the wrapper;[40] the ecological consequences were not, at that time, contemplated.[41] Hawkins writes of the notion of disposability having enabled waste to be *"ethically* insignificant."* She suggests that when commodity cultures declared that freedom was the "freedom to consume," it also meant the "freedom to waste"; she states that "before the emergence of environmentalism, the ethos of disposability framed waste as a *technical* rather than a *moral* problem."[42]

There are areas in life where the need for disposability is evident, such as in the medical sector with the requirement to prevent cross-contamination, particularly in times of pandemic and epidemic disease. The UK's National Health Service is extremely wasteful, creating a huge amount of waste each year.[43] This is partly owing to the quantity of single-use medical devices and clothing required to protect both patients and staff, which has played a vital role during the COVID-19 pandemic. Where previously reusable products were used, the fear of serious infections and diseases such as Creutzfeldt–Jakob disease has seen a systematic removal of reusable products from hospitals.[44]

However, unless disposable items including packaging are properly managed, they are likely to become noticeable waste. As a process, waste is now subjected to legislation and normative realities, by codes of conduct and wider objectives. Those objectives include the reduction of waste going to landfill, the use of recycling, and the reduced production of waste in the first place for the benefit of the global ecology. As Hawkins explains, the anti-litter campaigns of the 1960s displayed the act of littering as evidence of "environmental and moral decline."[45] Initially, plastics waste was perceived as an "aesthetic problem," with littering being an ugly thing to see on our streets and beaches; over the last couple of decades it has been recognized as a significant ecological issue.[46]

A momentous and tragic misuse of plastics was in the cladding of Grenfell Tower, London, and other similar large residential, hospital, and office buildings. On June 14, 2017, a fire started in the 24-story

Grenfell Tower block of 129 flats, which housed 350 people.[47] The fire was quick to spread through the building and killed 71 individuals.[48] The footage shown on the TV news was shocking, and the large burned-out tower standing out against the cityscape and sky was a monument to the tragedy. Over a very short period of time news reports were published suggesting how the fire was started and why it spread so quickly. The cladding was to blame, and it was "never subjected to fire safety testing, according to reports."[49] The fire is thought to have started in a fridge freezer in a flat on the fourth floor; the flames came out of a nearby window and traveled around the external walls. The fire then moved up the building, through the external cladding, and entered additional flats via open windows.[50] After this event, identical or similar cladding was found to have been used in 600 blocks or buildings in England.[51] Much of this cladding has since been removed in order for residents to "feel safe in their homes."[52]

The fire led to an independent review of building regulations and fire safety which resulted in an interim report published in December 2017,[53] followed by a final report published five months later.[54] The recommendations relating to products used in buildings are presented in four parts. The first and second parts aim to establish a more transparent testing regime with the restriction of desktop assessments in lieu of physical tests, and the clarity as to the testing of individual materials or whole systems to be more robust, as well as streamlining the "plethora of standards" to make them more understandable. Part three looks at the labeling and traceability of the materials in use rather than just the packaging, and the final part looks at creating a more effective market surveillance team.[55] With these recommendations in place, it is hoped that the combination of materials used and the design of building cladding will not result in such a tragedy again.

Conspicuous non-consumption of plastics

Conspicuous consumption can be described as an advertisement to wealth and social status,[56] and is most notably linked to the consumption of luxury goods. In this sense, it is a concept that suggests that "people with adequate means tend to consume highly visible goods to display their wealth and gain social status." For Yann Truong and Rod McColl in their article looking at the consumption of luxury goods, the pleasure derived from buying such things is not simply extrinsic, to gain reward from others and to imitate the behavior of others, but that intrinsic factors have a direct positive relationship with self-directed pleasure—that is to say, self-esteem can be lifted through quiet consumption. Their study found that there was a strong relationship between self-esteem and the consumption of luxury goods for self-directed pleasure, suggesting that an individual can maintain or boost self-esteem by purchasing luxury goods.[57] It could be said that, in the current climate, when it comes to green consumption it is the extrinsic factors that drive consumption or anti-consumption. It is my suggestion that when it comes to plastics, it is not so much *anti*-consumption, the refusal to consume completely, but the *non*-consumption of plastics. Takeaway coffees are still purchased, products are still consumed and carried home, but there is an avoidance of plastic packaging or carrier bags. When this avoidance takes place in full view, it becomes what I am calling *conspicuous* non-consumption of plastics. The reasons behind this conspicuous non-consumption of plastics could be to fit in with peers and groups, or to be seen to be doing the "right" thing. This might be in the form of carrying a reusable coffee mug or carrier bag that metaphorically, and sometimes literally, shows the purchaser's green colors. Geeta Sharma looks at consumer buying decisions and suggests that the behavior of the individual reflects their own adopted issues, and that their consumer behaviors have an impact on their satisfaction with their life. In other words, they feel that they have contributed to their cause—to protect the environment—by making the purchasing decisions they have. Sharma points out that consumers are becoming "more sensitive to their environmental attitudes, preferences and purchases" and this results in being more ecologically conscious with regard to the consumption of products and services.[58]

When looking at political motivations of the consumer, Melissa Gotlieb considers why people purchase certain items. She discusses the functional motivations, the fact that something can be reused; the experiential motivations, the fact that the product may make something more enjoyable; and, most importantly for political consumerism discourse, the symbolic motivations. She suggests that a consumer may well use a reusable bag, for example, because of a sense of civic duty to the environment, or cooperatively to feel part of a movement, or for a more contrived reason, such as to be seen to be green. These civic, cooperative, and contrived behaviors are all "generally motivated by personal values, social identification and social accreditation,"[59] or, putting it another way, how the actions make the individual feel, how they fit in with a group, and how they are seen by society. With plastics being a notable "politically incorrect symbol" of environmental disaster,[60] the intentional and overt non-consumption of plastics can be seen as an "act of consumer resistance against 'them.'"[61] The "them" described by Hélène Cherrier, Iain R. Black, and Mike Lee are other "careless consumers," the mainstream consumer, or perhaps the consumer that does not think about the environment when they purchase items. The "them" and "us" are pitched against each other, with the non-consumer feeling superior about their actions.

An example of conspicuous non-offering of plastics was observed on a visit to a café in Notting Hill, London, where our charcoal water drinks were presented with a paper straw that was keenly pointed out to us by the server, who stated with glee "no plastic straws here!" This statement was swiftly followed by the advice to remove the straw from the drink because it "falls apart after a while." The conspicuous non-plastic offering was shown not to be fit for purpose. We could have easily not had a straw in our drinks, but the addition of a paper straw overtly highlighted the non-inclusion of the plastic alternative. Avoiding plastics and replacing them with a perceived better material, simply because it is not plastic, shows that the alternative is not truly considered. For example, supermarket representatives recalled the request for plastic bottles for soft drinks to be replaced by glass and Tetra Pak, yet the Tetra Pak containers were not recycled in that particular area and the use of single-use glass increases carbon emissions significantly.[62]

In July of 2018, the coffee shop Starbucks announced that it would "eliminate single-use plastic straws from its more than 28,000 company operated and licensed stores by making a strawless lid or alternative-material straw options available, around the world."[63] This is good news if it kick-starts the reduction of unnecessary single-use materials; however, at the time of their announcement the company was offering flavoured paper straws individually wrapped in plastics for hygiene reasons.[64] The alternative would be to not use any straws, but some disability charities argue that straws are essential to their clients, who find it hard to drink from a cup or who find the use of metal straws too dangerous to their teeth. One wheelchair user wrote "straws are necessary because I do not have the hand and arm strength to lift a drink and tip it into my mouth. Plastic straws are the best when I drink hot liquids; compostable ones tend to melt or break apart."[65]

End of life practices

The problems with plastics come to light when objects that make use of the materials come to the end of their valued life.[66] The valued application of a material must inevitably be followed by its disposal at the end of its life, but what form that disposal takes is dependent on the availability of a range of options and the user's knowledge of those options. The choices for disposal can be narrowed or widened before the product has even been made. The selections made by the designer and the manufacturer in the making of a product have a significant effect on how that item ends its life. One option for disposal is through the recycling route. For this to be a success, it is much better if the materials being recycled are free from additions and impurities. The removal of contaminants is costly and time-consuming, so it is preferable if the "recycling inhibitors" are "consciously designed out" of mass-produced products. By having materials that are as close to virgin materials as possible, it becomes much easier to establish a closed-loop recycling system. For an object to become good-quality recyclate, a designer or manufacturer must consider the construction of their product. It must be easily disassembled or deconstructed and offer a clean and pure recyclate by being free of paint, coatings, labels, or adhesive.[67] Currently plastics recycling is seen as problematic for the manufacturer, recycler, and the general public. For a change to happen, the value of the materials needs to be realized; if it were easier to dismantle objects and understand what can and cannot be recycled, it would become feasible and worthwhile to make an effective job of the process.[68]

In the UK Government's "25 Year Plan" published in 2018, one of the areas of focus was the reduction of waste. They pledged to "minimise waste, reuse materials as much as we can and manage materials at the end of their life to minimise the impact on the environment," and this will occur with the elimination of "unavoidable" plastic waste by 2042. The use of the word "unavoidable" is particularly interesting, as it acknowledges that some of the uses of plastics, which eventually become waste, are "technically, environmentally, and economically" inescapable.[69]

Part of this aim is to work with the waste management and reprocessing industries to improve the percentage of plastic packaging that is gathered and recycled and to improve the standard of biodegradable bags.[70] Despite being seen as having great potential in the 1980s, biodegradability has since been assessed as less green than initially anticipated. According to Whiteley, writing in the 1990s, environmentalists had established that "so-called biodegradable plastics" did not completely disappear from the environment, leaving behind microscopic fragments that caused issues to wildlife and the landscape. The conditions needed for the process to be a success were not readily available and as such, a boycott was started in 1989 in the USA against all biodegradable plastics, as they were seen as a barrier to reuse and recycling.[71] A large proportion of waste is not well biodegraded and is not easily converted into added-value chemicals by microorganisms due to it containing "mixtures of very complex compounds and some toxic pollutants which are highly recalcitrant to degradation."[72] Research into improved bio-based plastics is currently being carried out; it is thought that BioPET will have the biggest production capacity in the coming years, with other materials such as polyhydroxyalkanoates (PHA) and polylactic acid (PLA) being on an upward trend.

However, there is also a need for regulated and standardized government policies in the development and waste management of these materials.[73] Burall discusses the idea that using biodegradable materials is promoted by governments outlawing the use of non-degradable packaging and carrier bags, but he asks if they should always be specified by designers of packaging.[74] Biodegradable plastics will release harmful gases such as methane within a landfill setting, and their compostability relies heavily on the correct level of heat and water, which can usually only be provided within an industrial setting.[75] There is a distinction between biodegradable plastics that are biodegradable in an industrial setting and those that can be composted at home.[76] This distinction is not always made clear to the consumer and, worryingly, the United Nations Environment Programme reports that there is evidence that littering behavior is influenced by products being labeled biodegradable and as such, the use of biodegradable plastics will not alleviate litter in the oceans.[77] Biodegradable plastics can also create problems if they become part of the recycling stream, as they will contaminate and undermine the structure of the resultant material;[78] starch and other natural fibers complicate the recycling process of traditional plastics.[79] Due to this potential risk to the recycling system, the European Commission requires more research and resources to be put into biodegradable products before it can promote them as an alternative.[80]

In the UK, the UCL Plastic Waste Innovation Hub[81] is carrying out a citizen science experiment into the efficacy of compostable and biodegradable plastics. This experiment aims to understand how objects labeled as "home compostable" or "home biodegradable" behave within home compost conditions, those that their manufacturers claim will break down in an environmentally friendly way. The debate around plastics that are intended to biodegrade in the natural environment is considerable and runs between scientists, plastics manufacturers, and recyclers.[82] A survey carried out by GrocerVision in 2019 discovered that consumers believe that plant-based compostables are the most environmentally friendly material for packaging.[83] There are a number of standards in place, both nationally and internationally, which describe the requirements for a bioplastic to breakdown in an industrial composter. Such a device should reach a temperature of 70° C (158° F), resulting in 90 percent of the organic matter transforming to CO_2 within a six-month timeframe.[84] The home composter will not be able to provide these conditions.

The way waste is collected is affected by its location and intended destination. Whiteley looked at recycling and suggested that the main problem with recycling plastics is the need to separate the different types. He believes that this is a big ask of the consumer. He writes that coding the materials does make it easier, but it would be far simpler to make all containers out of the same material.[85] The coding of plastics, particularly packaging, is now the norm and the need for the consumer to sort their recycling can be reduced with local authorities, such as my own in Bournemouth, simply using one bin for the collection of all materials destined to be recycled;[86] of course, it does then need to be sorted at the waste plant, which incurs a cost for the local authority. Recycling plays an important role in the reuse of materials and the reclamation of value. There are many ways to recycle plastics, and it is a relatively easy process if the product being recycled is made of a single material—however, it becomes more problematic and more expensive when the product is multilayered and has multiple components.[87] For example, with crisp packets and Tetra Pak drinks containers, the materials have to be mechanically separated or the resulting recyclate has very limited uses.

For recycled materials to be useful, they need to have a market value. Papanek highlights the lack of diversity in products made from recycled materials during the mid-1990s, suggesting that there "must be an upper limit to the number of dark grey, rough-textured countertops that can be usefully employed."[88] This was still the case even in 2009 when Jefferson Hopewell, Robert Dvorak, and Edward Kosior suggested that it was not always technically practicable to add recycled plastics to virgin materials without reducing the quality of the color, clarity, or mechanical properties of the new material.[89] However, innovation in the types of objects made out of recycled materials has improved in the last few years. The color of the new product is dependent on the color of the recyclate; if the recyclate comes from multicolored sources, the new material will be dark in color. To have a freedom of colors the recyclate needs to be paler than the end color required, which is why the classic green Sprite bottle has recently been replaced by a clear bottle "to enable bottle-to-bottle recycling."[90]

An alternative use for recovered products is to recapture the energy held within the materials, a process advocated by Professor Dame Julia Higgins, Emeritus Professor of Polymer Science and former Principal at the Faculty of Engineering, Imperial College London. She believes that with a properly managed smokestack, plastics can be burned without adding to the greenhouse effect. As oil is burned in power stations anyway, by giving it a useful life as a plastic object for a time, you are making better use of the resource.[91] There have been advancements in the conversion of plastics into energy but there are issues, especially if such materials as PVC are present, as they produce harmful gases. However, researchers have found ways to pre-treat these materials using hydrochloric acid absorbents,[92] making them safer to use.

Conclusion

As plastics can become anything, designers have been seduced into using them. This was particularly evident in the early days of plastic, when manufacturers used materials for the sake of using them rather than because they were the best material for the job in hand.[93] What plastics have done is free designers from the traditions of craft, allowing products to break away from previous limitations of design and process;[94] they can be molded into any shape, have any texture, be any color, and be made to have the strength and/or flexibility required for any specialist need. Plastic materials are used to produce safety equipment for the construction industry, reuseable food storage that is lightweight and hygienic, sports equipment that makes expensive activity affordable thereby opening it up to all, look- and feel-a-like fashion items that require no animal to be killed, and clothing that is breathable yet keeps our bodies warm and dry, to name but a few applications. However, today one might be forgiven for thinking that plastics are used in packaging—over packaging at that—and nothing else. It becomes difficult to see that plastics, as a material group, have any real use.

Some ten years ago, Anthony L. Andrady and Mike A. Neal looked at the benefits that plastics offer to society. As expected, they suggest that "well over a third of plastics consumption is in packaging applications," but perhaps more surprisingly, "another third or more [is used] in building products including common products such as plastic pipes or vinyl cladding." Plastics offer not only convenience but also other benefits including health improvements, facilitating clean drinking water and providing safe hospital equipment.[95] Products that are needed to last many years use plastic casings that will not crack, have smooth corners that will not chip, and can be reinforced to withstand applied pressure. The fact that these materials will last a long time means that the energy cost is much more efficient.[96]

Plastics are important materials that can provide essential properties in specific situations: they offer us leakproof gas pipes, insulated electrical cables, cool-touch toasters and kettles. As they are lightweight and less likely to break than other materials, they help us to prevent the overuse of additional packing materials and fuels used in transportation. We as artists, designers, makers, consumers, and human beings need to think about the overall picture of the use, misuse, and disposal of *all* materials to make informed choices. We have found ourselves in a situation where over many years disposability had become the norm, and was in fact positively encouraged; to move away from that way of life will take time. Understanding the value of an object, and the materials from which it is made, will help inform choices and processes. It is important that we, as consumers, do not think of packaging, for example, as "free": a great deal of design and manufacturing time and money has been put into making it fulfill the role it has been designed for, that is, to stop contaminants getting into our food. Plastics: what are they good for? Lots of things, as long as they are thoughtfully used for the best applications, looked after, and disposed of appropriately.

Notes

1. Jeffrey L. Meikle, *American Plastic: A Cultural History* (New Brunswick & London: Rutgers University Press, 1997), xiii.
2. Susan Freinkel, *Plastic: A Toxic Love Story* (Melbourne: Text Publishing, 2011), 8.
3. Marion Gough, "The Truth About Plastics," *House Beautiful, Plastics Special* (October 1947): 121.
4. Charlotte Fiell and Peter Fiell, *Plastic Dreams: Synthetic Visions in Design* (London: Fiell Publishing, 2009), 18, 19.
5. Roland Barthes, "Plastic," in *Mythologies* [1957], trans. Annette Lavers (Vintage Classics, 1993), 97.
6. Ezio Manzini, "And of Plastics?" *Domus*, no. 666 (1985): 54.
7. C. Frank Merriman, "Plastics," *British Plastics and Moulded Products Trader* 13, no. 155 (1942): 470.
8. James Hogan, "Pottery, Glass and Plastics," *British Plastics* 13, no. 154 (1942): 396.
9. Albert Henry Woodfull, "The Designer and Plastics (Transcript)," 1948, Museum of Design in Plastics.
10. Penny Sparke, ed., *The Plastics Age: From Modernity to Post Modernity* (London: Victoria & Albert Museum, 1990), 7.
11. Ibid. 7–8.
12. Manzini, "And of Plastics?" 54.
13. "Plastics in 1940," *Fortune Magazine* (October 1940): 89.
14. Freinkel, *Plastic: A Toxic Love Story*, 33.
15. Fiell and Fiell, *Plastic Dreams*, 20.
16. Penny Sparke, "Plastics and Pop Culture," in *The Plastics Age: From Modernity to Post-Modernity*, ed. Penny Sparke (London: Victoria & Albert Museum, 1990), 93–4.

17. Sparke, *The Plastics Age*, 11.
18. Paul Burall, *Green Design* (London: The Design Council, 1991), 39.
19. Victor Papanek, *The Green Imperative: Ecology and Ethics in Design and Architecture* (London: Thames and Hudson, 1995), 40.
20. Sylvia Katz, "Plastics in the '80s," in *The Plastics Age: From Modernity to Post-Modernity*, ed. Penny Sparke (London: Victoria & Albert Museum, 1990), 145, 149, 151.
21. Sparke, "Plastics and Pop Culture," 103.
22. Meikle, *American Plastic*, xiii.
23. Mimi Sheller, "Metallic Modernities in the Space Age: Visualizing the Caribbean, Materializing the Modern," in *Visuality/Materiality: Images, Objects and Practices*, ed. Gillian Rose and Divya Praful Tolia-Kelly (Farnham: Ashgate, 2012), 14.
24. Rosie Hornbuckle, "Utilising Waste Plastic in Design Practice," in *Plastics: Looking to the Future and Learning from the Past*, ed. Brenda Keneghan and Louise Egan (London: Archetype Publications, 2008), 164.
25. Burall, *Green Design*, 42.
26. Kenneth Geiser, *Materials Matter: Toward a Sustainable Materials Policy* (Cambridge: MIT Press, 2001), 83.
27. Gay Hawkins, *The Ethics of Waste: How We Relate to Rubbish* (Lanham, Boulder, New York, Toronto, & Oxford: Rowman & Littlefield Publishers, 2006), 29.
28. Using BoB (*BoB—Box of Broadcasts*, https://learningonscreen.ac.uk/on demand) to explore the occurrences of the term *plastic*, 2007 produced 577 results, compared to 2019, the last full year, which produced 6,678. These results will include the terms *plastic surgery*, *plastics explosives*, etc. but they still demonstrate the increase of the use of the word in the media.
29. Stephen Fenichell, *Plastic: The Making of a Synthetic Century* (New York: HarperBusiness, 1996), 6.
30. European Federation of Bottled Waters, "History of Bottled Water," http://www.efbw.eu/index.php?id=39.
31. British Plastics Federation, "A History of Plastics," http://www.bpf.co.uk/plastipedia/plastics_history/default.aspx.
32. Recycle More, "Top Facts," https://www.recycle-more.co.uk/household-zone/top-facts.
33. Gavin Lucas, "Disposability and Dispossession in the Twentieth Century," *Journal of Material Culture* 7, no. 1 (2002): 19.
34. Fiona Campbell, *People Who Litter* (ENCAMS, 2009).
35. Hawkins, *The Ethics of Waste*, 3–5.
36. Ibid. vii.
37. Gay Hawkins, "Made to Be Wasted: PET and Disposability," in *Accumulation: The Material Politics of Plastic*, ed. Jennifer Gabrys, Gay Hawkins, and Mike Michael (London & New York: Routledge, 2013), 65.

38. Thomas Hine, "Populuxe," in *The Design History Reader*, ed. Grace Lees-Maffei and Rebecca Houze (Oxford & New York: Berg, 2010), 155.
39. Hawkins, *The Ethics of Waste*, 25–6.
40. Lucas, "Disposability and Dispossession in the Twentieth Century," 12.
41. Fiell and Fiell, *Plastic Dreams*, 20.
42. Hawkins, *The Ethics of Waste*, 29.
43. "The NHS produced 408218 tonnes of waste in 2005–6," D. C. J. Hutchins and S. M. White, "Coming Round to Recycling," *British Medical Journal* 338 (March 2009): 746.
44. M. N. Chauhan, T. Majeed, N. Aisha, and R. Canelo, "Use of Plastic Products in Operation Theatres in NHS and Environmental Drive to Curb Use of Plastics," *World Journal of Surgery and Surgical Research— General Surgery* 2 (2019): 5.
45. Hawkins, *The Ethics of Waste*, 31–0.
46. David K. A. Barnes et al., "Accumulation and Fragmentation of Plastic Debris in Global Environments," *Philosophical Transactions of the Royal Society of London B: Biological Sciences* 364, no. 1526 (2009): 1985.
47. Jan-Carlos Kucharek, "After Grenfell," *The RIBA Journal* (August 2017): 35.
48. Great Britain and Department for Communities and Local Government, *Building a Safer Future: Independent Review of the Building Regulations and Fire Safety: Interim Report*, 2017, 5.
49. Lucy Pasha-Robinson, "Cladding Fitted to Grenfell Tower 'Was Never Fire Safety Tested,'" *Independent*, February 8, 2018, https://www.inde pendent.co.uk/news/uk/home-news/grenfell-tower-cladding-fire-safety-checked-acm-cladding-aluminium-composite-material-reynobond-a820 0801.html.
50. Gareth Davies, "Grenfell-Style Cladding to Be Stripped from 170 Tower Blocks as Government Foots £200m Bill," *Telegraph*, May 9, 2019, https://www.telegraph.co.uk/news/2019/05/09/grenfell-style-cladding-st ripped-170-tower-blocks-government/.
51. Nick Parker, "600 Tower 'Fire Traps': Hundreds of High Rises across the Country Have Grenfell Tower-Style Cladding," *The Sun*, June 22, 2017, https://www.thesun.co.uk/news/3864036/hundreds-of-high-rises-across-britain-have-grenfell-tower-style-cladding-leaving-thousands-of-families-living-in-fear/.
52. Davies, "Grenfell-Style Cladding."
53. Great Britain and Department for Communities and Local Government, *Building a Safer Future: Interim Report*.
54. Great Britain and Department for Communities and Local Government, *Building a Safer Future: Independent Review of the Building Regulations and Fire Safety: Final Report*, 2018.
55. Ibid. 93–5, 96, 98.
56. Laurie Simon Bagwell and B. Douglas Berheim, "Veblen Effects in

a Theory of Conspicuous Consumption," *The American Economic Review* 86, no. 3 (1996): 349–73.

57. Yann Truong and Rod McColl, "Intrinsic Motivations, Self-Esteem, and Luxury Goods Consumption," *Journal of Retailing and Consumer Services* 18, no. 6 (November 1, 2011): 556, 559.

58. Geeta Sharma, "Green Association in Consumer Buying Decisions," *International Journal of Business Management* 1, no. 1 (2014): 49–50.

59. Melissa R. Gotlieb, "Civic, Cooperative or Contrived? A Functional Approach to Political Consumerism Motivations," *International Journal of Consumer Studies* 39, no. 5 (2015): 552–3.

60. Paola Antonelli, *Mutant Materials in Contemporary Design* (New York: Museum of Modern Art, 1995), 12.

61. Hélène Cherrier, Iain R. Black, and Mike Lee, "Intentional Non-consumption for Sustainability: Consumer Resistance and/or Anti-consumption?" *European Journal of Marketing* 45, no. 11/12 (2010): 1764.

62. Libby Peake, "Plastic Promises: What the Grocery Sector Is Really Doing About Packaging" (Green Alliance, 2020), 3, https://www.green-alliance.org.uk/resources/Plastic_promises.pdf.

63. Starbucks Newsroom, "Starbucks to Eliminate Plastic Straws Globally by 2020," July 9, 2018, https://stories.starbucks.com/press/2018/starbucks-to-eliminate-plastic-straws-globally-by-2020/.

64. Aisling Finn, "Starbucks Causes a Stir Over Its New Individually Plastic-Wrapped Paper Straws," *The Sun*, August 1, 2018, https://www.thesun.co.uk/news/6914198/starbucks-causes-a-stir-over-its-new-individually-plastic-wrapped-paper-straws/.

65. Alice Wóng, "Banning Plastic Straws Is a Huge Burden on Disabled People," *Eater*, July 19, 2018, https://www.eater.com/2018/7/19/17586742/plastic-straw-ban-disabilities.

66. Hornbuckle, "Utilising Waste Plastic," 163.

67. Louis O. D'Anjou et al., "Designing with Plastics: Considering Part Recyclability and the Use of Recycled Materials," *AT&T Technical Journal* 74, no. 6 (1995): 55–6.

68. Hornbuckle, "Utilising Waste Plastic," 169.

69. H. M. Government, "A Green Future: Our 25 Year Plan to Improve the Environment," 2018, 29, https://www.gov.uk/government/publications/25-year-environment-plan.

70. Ibid. 86–9.

71. Nigel Whiteley, *Design for Society* (London: Reaktion Books, 1993), 73.

72. Auxiliadora Prieto, "To Be, or Not to Be Biodegradable . . . That Is the Question for the Bio-based Plastics," *Microbial Biotechnology* 9, no. 5 (2016): 655.

73. Ibid. 656.

74. Burall, *Green Design*, 45–7.

75. Laurel Miller and Stephen Aldridge, *Why Shrink-Wrap a Cucumber? The Complete Guide to Environmental Packaging* (London: Laurence King Publishing, 2012), 60, 63.
76. J. H. Song et al., "Biodegradable and Compostable Alternatives to Conventional Plastics," *Philosophical Transactions of the Royal Society of London B: Biological Sciences* 364, no. 1526 (2009): 2133.
77. United Nations Environment Programme, *Biodegradable Plastics & Marine Litter: Misconceptions, Concerns and Impacts on Marine Environments* (2015), 31.
78. Miller and Aldridge, *Why Shrink-Wrap a Cucumber?* 63.
79. Song et al., "Biodegradable and Compostable Alternatives," 2129–30.
80. European Commission, "Analysis of the Public Consultation on the Green Paper 'European Strategy on Plastic Waste in the Environment'—Final Report," 2013, https://op.europa.eu/en/publication-detail/-/publi cation/da9aa65e-b822-4dd8-83d8-b80310315be6.
81. "Big Compost Experiment," *Big Compost Experiment*, https://www.big compostexperiment.org.uk/.
82. United Nations Environment Programme, "Biodegradable Plastics & Marine Litter: Misconceptions, Concerns and Impacts on Marine Environments," 6.
83. Peake, "Plastic Promises: What the Grocery Sector Is Really Doing About Packaging," 6.
84. United Nations Environment Programme, *Biodegradable Plastics & Marine Litter*, 19.
85. Whiteley, *Design for Society*, 75.
86. Bournemouth Borough Council, "Household Recycling Collections—Big Bin," https://www.bournemouth.gov.uk/BinsRecycling/BinCollections/ HouseholdRecyclingCollections-BigBin/HouseholdRecyclingCollections BigBin.aspx.
87. Jefferson Hopewell, Robert Dvorak, and Edward Kosior, "Plastics Recycling: Challenges and Opportunities," *Philosophical Transactions of the Royal Society of London B: Biological Sciences* 364, no. 1526 (2009), https://royalsocietypublishing.org/doi/10.1098/rstb.2008.0311.
88. Papanek, *The Green Imperative*, 39.
89. Hopewell, Dvorak, and Kosior, "Plastics Recycling: Challenges and Opportunities," 2119.
90. Kelly Maile, "Sustainability Alert: Sprite Replaces Iconic Green Bottle with Clear Plastic," *Recycling Today*, https://www.recyclingtoday.com /article/sprite-pepsico-mcdonalds-trash-wheel-project-sustainability/.
91. Julia Higgins on *The Life Scientific*, BBC Radio 4, 2017, https://www. bbc.co.uk/programmes/b083n2jg.
92. Bidhya Kunwar et al., "Plastics to Fuel: A Review," *Renewable and Sustainable Energy Reviews* 54 (February 2016): 421–8.
93. Freinkel, *Plastic: A Toxic Love Story*, 32.
94. Hogan, "Pottery, Glass and Plastics," 396.

95. Anthony L. Andrady and Mike A. Neal, "Applications and Societal Benefits of Plastics," *Philosophical Transactions of the Royal Society of London B: Biological Sciences* 364, no. 1526 (2009): 1980.
96. Whiteley, *Design for Society*, 73.

2

How Hula Hoops Changed Hygiene: From Damp-Cloth Utopianism to Chemical Cleaning

Angela Cope

Writing this paper in the time of COVID has been strangely apropos. I always intended it to map to spring: a time for cleaning and Easter, which are my clearest memories of receiving hula hoops. I admit to hating hula hooping as a child. My brother and I usually used them for different types of games: they would be a safe spot on the floor which was otherwise lava, or a hole to jump through in the pool, or they provided a "harness" between horse and rider. These kinds of games nearly always pulled the hula hoop out of shape, bent it into an oval with dents and creases, sometimes made it entirely fold in half. The baser materiality of the hoop would never change, however: it would remain a tube of waxy plastic, brightly colored and insubstantial, easily bent and misshapen but nearly impossible to break or permanently damage despite our best efforts.

This Easter was instead the Easter that was not. Several days early I rushed to my parents' place to see them. Arriving at dusk, I walked into the house, N95 mask donned, hand sanitizer at the door, and went immediately into the shower, where I bagged my clothes and put on ones that had been freshly laundered. I then proceeded to wear a cloth mask the entire time I was there, ensuring that I did not put my parents (sixty-five and sixty-eight respectively) and my grandmother (eighty-three) at risk of COVID. Everything I touched in the house was promptly wiped with Clorox wipes, or immediately put in the dishwasher. The HDPE was omnipresent: it contained the hand sanitizer, the body wash, the dishwasher liquid, the Clorox wipes, the laundry detergent. Nearly without exception, if there is a caustic material, it is housed in HDPE. Toilet bowl cleaner. Bleach. Vim. Mr. Clean. Clorox. Lysol. Tide. Gain. Persil. Liquid soap for hand washing. Every shampoo, every conditioner, all of the body wash. The lotion applied to the cracked hands of skin washed too many times. An array of colors, shapes, and designs, HDPE is amazingly versatile in this respect. Name brands, logos, and instructions printed directly onto the bottles or tubs, in every color of the rainbow—meant to convey information about the content within—bright, cheerful, cheap, and above all else, clean.

We did not always clean this way. There were few, if any, cleaning materials which fostered the "schmear" paradigm of chemical cleaning, where the detergent is left on the surface.[1] As the convenience and popularity of this kind of cleaning has grown, so too have the indoor air quality problems associated with it.[2] A poem snippet that appeared in a 1945 advertisement in Playthings magazine indicates how very different the paradigm of cleaning has become in a mere century. The poem originally appeared in Modern American Poetry: An Introduction in 1919, and is a hyperbolic commentary on the new standards for cleanliness that were imposed during the Spanish Flu outbreak of 1918–19. It describes an Easter scene, as innocent as it is sterile:

> The Antiseptic Baby and the Prophylactic Pup
> Were playing in the garden when the Bunny Gamboled up;
> They looked upon the Creature with loathing undisguised.—
> It wasn't Disinfected and it wasn't Sterilized.[3]

Not included in the advertisement, but equally illuminating, was the next stanza, which continues to explain what happens to the unfortunate bunny:

> They said it was a Microbe and a Hotbed of Disease;
> They steamed it in a vapor of a thousand-odd degrees;
> They froze it in a freezer that was cold as Banished Hope
> And washed it in permanganate with carbolated soap.[4]

Note that while the poor bunny was subjected to rather extreme treatments, they still had primarily to do with temperature (steaming and freezing) and soap, which is washed off with water after lathering. Many commercial products today are designed instead to be left on the surface after wiping. The difference is their solubility in water— soap leaves a scum whereas synthetic detergents, being more soluble, do not. It is an innovation that has entirely changed the way we clean, but because of synthetic detergents' solubility, they initially posed a problem with respect to their packaging.

Low-Density Polyethylene (LDPE) and High-Density Polyethylene (HDPE), while related, are not the same. HDPE sprang forth from LDPE, the first polyethylene, which was discovered in 1933 in England by Imperial Chemical Institute (ICI). Polyethylene's wartime scaling and use in radar equipment during World War II has been covered elsewhere.[5] HDPE, on the other hand, came along postwar, as a result of basic research by Karl Ziegler at the Max Planck Institut für Kohlenforschung. As I will show below, HDPE was scaled in North America not due to wartime necessity, but instead due to the increasingly sophisticated consumer market of the long boom postwar.

LDPE and HDPE's combined impact on the postwar world of plastics and disposability was huge. Their inventions had three main effects. First, LDPE and HDPE greatly contributed to the feminization, domestication, and infantilization of plastics through Tupperware and the hula hoop. The story of LDPE was intertwined with that of Earl Tupper and his invention Tupperware, which is indelibly associated with domesticity and the highly feminized "Tupperware parties" of the 1950s. HDPE, on the other hand, was successfully scaled almost entirely due to a toy, with the hula hoop's invention in 1958. Next, LDPE and HDPE greatly accelerated the conversion of plastics from durable materials to disposable ones. Until HDPE's invention, the plastics industry was still very focused on using plastic for more durable uses, particularly in the building and automotive industries. The proliferation of HDPE and LDPE provided huge leaps for plastics to become synonymous with packaging. Finally, HDPE and LDPE, because of their materiality as polyolefin plastics, ushered in a new paradigm of chemical cleaning, which undermined the damp-cloth utopianism of earlier thermoset plastics. HDPE was able to contain the synthetic detergents that became popular during this time, as its inertness to caustic materials made it a better candidate for packaging. Synthetic detergents' packaging in HDPE made them widely available, which in turn drove even more polyethylene production, as the synthetic detergents were effective in cleaning polyolefins where water was not.

Thermoplastics, upon their birth, were very different from the dominant thermoset materials of the interwar years. They were not so much materials birthed to fulfill a specific need, but materials created to soak up the excess capacity of crude oil and ethylene during the Depression and then massively scaled during World War II. They were materials for which markets were later created, especially postwar.[6] The birth of HDPE was one of the final death knells of plastics as durable materials. Unlike the high regard that polystyrene, polyvinyl chloride, and LDPE held during the war, HDPE did not have a positive wartime utility to shed. As I will show, the hula hoop fad played a key role in the path that plastics took toward disposability, as their explosive proliferation scaled HDPE's production across North America. The entanglements between hygiene, disposability, the polyethylenes, and synthetic detergents worked together to create the story that follows.

Polyethylene before HDPE: Tupperware, femininity, and changing ideals of hygiene

When DuPont entered into licensing agreements with ICI to produce Polythene in 1943, Earl Tupper was well situated to take advantage of the material. Having previously worked as a sample maker for two years under contract to DuPont, Tupper had started his own company in 1939. Based on the connections he had made while working at Doyle Works, his company received several lucrative commissions to manufacture war materiel, which gave him access to polyethylene before other custom molders were able to get their hands on the coveted material. As a sample maker, Tupper created prototypes for the consumer market, which, if successful, were then manufactured by custom molders. Their role was to "field changes in consumer tastes and introduce unfamiliar articles to reticent retailers."[7] DuPont's ad hoc relationship with Tupper meant that he had unrestricted access to "machinery, methods and materials" for his experiments. This access was "reputedly the basis of his first polyethylene experiments."[8] Tupper was therefore able to invent the very first incarnation of what would become a defining feature of domesticity in postwar America: Tupperware.

To understand Tupperware, we must appreciate that "on the one hand, Tupperware taught thrift and containment; on the other, excess and abundance. These contradictions [represented] a historical shift from the Depression economy to a postwar boom."[9] Earl Tupper's status as a self-taught gentleman inventor meant that he imbued his product with all of the machine-age utopianism of the 1930s, and believed wholeheartedly that plastics were the answer to a futurist land of plenty. Alison Clarke writes that "Tupper envisioned a world utterly transformed through the appropriate application of polyethylene" and that he particularly wanted to transform the lives of women through his products, as his personal diaries were rife with inventions designed for them.[10]

However, Tupper's ability to understand, let alone market to, women left something to be desired. For example, an advertisement run by the Tupper Corporation in 1948 appearing in both *Modern Plastics* and *Modern Packaging* takes its title, "The Colonel's Lady and Judy O'Grady" from a Rudyard Kipling poem called "The Ladies," which is a fairly risqué recounting of a British army man's sexual conquests around the world. It contained within such gems as:

Now I aren't no 'and with the ladies,
For, takin' 'em all along,

You never can say till you've tried 'em,
An' then you are like to be wrong.[11]

The "Colonel's Lady" advertisement was a 1948 pitch to make Tupperware into a premium gift, as it had suffered from poor sales in department stores; similarly clumsy attempts were the hallmark of early Tupperware campaigns. Tupper was therefore attempting to be "chummy" with those (overwhelmingly male) executives who might make decisions about premiums to offer with their products. Tupperware might have been lost to the mists of history had not Brownie Wise stepped into the picture and changed the way that Tupperware was sold forever through the Tupperware Party. Brownie Wise was a single mother from Detroit who piqued Tupper's interest with her astronomical sales figures. That interest translated into Tupper appointing her the vice-president of the Tupperware Home Parties division, based out of Orlando, Florida.[12] The "soaring increase in household expenditure, women doing their own housework, and . . . homebound mothers eager to earn extra income and thwart social isolation became enthusiastic organizers of the . . . Tupperware party" and spawned an entire genre of sales through conviviality, socializing, and the exploitation of women's friendships and networks.[13]

We would be hard pressed to find a postwar consumer object more thoroughly feminized than Tupperware. Since it was sold entirely through the Tupperware party, the wares allowed a generation of women who were otherwise cut off from the workplace to earn a living in a way that remained socially acceptable in the conservative 1950s. Clarke writes that "women have stood at the forefront of changes in capitalist consumer society, [as] their social roles and cultural identities have been inscribed with the moral contentions and meanings of consumption."[14] Those changes reflected a fundamental shift in the competing moral values of thrift and hygiene. Gavin Lucas points out that "before disposability as a concept emerged, all waste was effectively regarded as inefficient and arising through improper management."[15] He argues that this idea fundamentally changed in the twentieth century, when the moral value of thrift was pushed out in favour of the moral value of hygiene. Tupperware was a microcosm of this shift, as it marketed effectively to both moral codes. It functioned as a way to keep food fresher for longer and save leftovers for later consumption, so it appealed to thrift. It was also marketed as a "sanitary, easy to clean" object that "protected pies and cakes against insects and dust,"[14] therefore it appealed to hygiene and as a safeguard against the threat of contamination.[16]

This shift had its roots in the rise of home economics in North America. As the germ theory of disease became widely accepted, cleanliness became not only an indication of moral goodness and class status, but also a bulwark against the spread of disease. As Elizabeth Shove points out, germ theory bore a new responsibility for homemakers: "If germs cause disease and if they can be killed by scrupulous hygiene, it is reasonable to interpret the visitation of illness not as an accident of fate but an indication of domestic failure and lax standards . . . [to this day] cleanliness is still used as an index of domestic responsibility and care."[17]

As the idea of hygiene became the norm, notions of disposability were increasingly tied to it. Through this moral lens, we see the rise of throwaway packaging as a way to both market and disseminate consumer goods. Well into mid-century, plastic containers were rarely seen as something that one could simply throw away, as was evidenced by the explosive popularity of the comparatively expensive and decidedly not disposable Tupperware. But increasingly, disposability was specifically marketed as sanitary, as "every time a product was purchased off the shelf, the consumer was assured that the package was new and therefore clean, and this assurance was strengthened precisely because they threw away the package after using the product." While Lucas primarily looks at the ways in which card and paper packaging played this role at the turn of the twentieth century, he also acknowledges the ways that plastics have become important mediators of disposable packages.[18] However, he seems to assume that plastic, in packaging and otherwise, is *prima facie* disposable. But there is a specific and traceable history to plastic becoming associated with packaging, and plastic packaging becoming associated with disposability, which were both mediated through its domestication.

While Tupperware largely escaped this shift to disposability, many containers that were made of LDPE and HDPE did not. HDPE's birth and massive growth closed the door on many different paths plastics could have taken—from frankly plastic architecture, to plastic's utility as a space-aged, easy-care material which required no more than a damp cloth to clean. Its trajectory—from its invention by Karl Ziegler in 1953 to its proliferation as packaging for synthetic detergents—follows.

Exotic materials with no industrial significance whatsoever: the Ziegler process

At the same time that Tupperware was being disseminated via women's networks into the homes of American housewives, a different, but related, plastic was beginning to take shape in Germany. Karl Ziegler of the Max Planck Institut für Kohlenforschung (coal research) had been working since the early 1940s on the effect that organometallic catalysts have on organic chemical reactions, and in 1953 he perfected the catalytic method for producing polyethylene at far lower pressures than the extant polyethylene on the market. By comparison, LDPE required pressures exceeding 300 atmospheres and temperatures exceeding 300° C (572° F), whereas HDPE could be achieved at pressures as low as atmospheric pressure and temperatures no higher than 90° C (194° F).

Ziegler was one of the driving forces behind the dominance of the German chemical industry before, during, and especially after World War II. For his work with the polymerization of butadiene, essential in Germany's lead in synthetic rubber manufacture (Buna Rubber), he was awarded the civilian "war merit cross" in 1940. As a condition of his appointment at the Max Planck Institute in 1943, he "insisted that [he] must be given complete freedom to pursue the entire field of compounds of carbon (organic chemistry) irrespective of whether a clear relationship could be recognized between [his] work and coal, or not."[19] Postwar, he was fundamental in rebuilding the chemical industry, from a "shortage of chemicals, glassware, and equipment; recent foreign books and periodicals seldom available; and heat, water, and current often interrupted."[20] Because he was such an important figure in the German chemical industry, he was given complete freedom to do as he pleased.

The only time that Ziegler ever seemed to question himself with respect to the utility of his basic research was, ironically, in the year immediately preceding the discovery that would both win him the Nobel Prize in 1963 and change the face of the world as we knew it forever, asking himself in 1952 "whether, in the very difficult period after the war, one could really justify continuing investigations ... into the properties of exotic materials with no industrial significance whatsoever?"[21] He continued his research despite his doubts, and in 1953 patented his method for using organometallic catalysts in polyethylene production.

Being able to use organometallic catalysts to create polymers shifted the plastics paradigm in a way not seen since Staudinger's discovery of their macromolecular structure.[22] While polymers were, after thirty years of refinement, able to be created in a relatively uniform manner by manufacturing techniques alone, catalysts meant the reaction could be made safer and more predictable. The reaction could therefore be more precisely controlled, and polymer chain lengths could be customized to suit specific needs. The 1963 Nobel speech lauded the fact that

"Ziegler catalysts . . . have simplified and rationalized polymerization processes and given us new and better synthetic materials."[23] Their elucidation made manufacturing easier, but it also moved their perfection from the realm of manufacturers into the realm of organic chemists, which was reflected in the rhetorical shifts to come. Regardless of what it was called, however, the plastics industry embraced the material excitedly, as it was the "next big thing"[24] in plastics.

Polyethylene grabs the spotlight

The plastics industry knew that HDPE was a remarkable material from the beginning. In the first of two articles about the new polyethylenes run in *Modern Plastics* in September and October of 1955, the lead photograph exhibited how HDPE held up in comparison to LDPE when "bent and soaked in detergent for several days."[25] Showing several test strips of both materials bent 180 degrees and wedged into a gutter, all five of the low-density samples had broken at the stress point, whereas none of the high-density samples had done so. The further examples pictured on the following pages had a distinctly industrial bent to them, showing gloved hands pouring acidic liquids into oversized funnels or men in white lab coats and safety glasses submitting the new plastic to stress tests. When speaking of packaging, the emphasis here was on "bottles and carboys that will have chemical inertness, rigidity and impact strength . . . bottle molders are particularly smitten with the possibilities of low pressure polyethylenes."[26]

The second article, however, focused on LDPE instead of HDPE, and immediately acknowledged the material's domestication. Rather than leading with industrial uses of LDPE, they instead chose a picture of a housewife buying domestic objects heavily resembling Tupperware. In its "Growth Measurements" section, it was quickly becoming the most used plastic of all, even without additional HDPE plants coming online. One of the largest uses of LDPE today, in single-use grocery bags, was first proposed here. They wrote that "odd uses in the packaging field crop up almost every week . . . A newspaper publisher delivered a special edition in re-usable waterproof polyethylene bags."[27] While these uses are quite normal today, that was not the case in 1955, emphasizing as they did the newspaper bag's reusability. In a later section on the potential market for polyethylenes in blow molding bottles, it is assumed that any polyethylene that would replace glass soda or milk bottles would be reusable, sterilizable, and last longer than the average glass bottle's six trips, which was a reason for using it in spite of its far higher cost in comparison to glass.[28]

The "Molding Material" cheered LDPE's domestication. It stated that the material's biggest use was "for housewares. That outlet has been so big that molders haven't taken much time to develop other markets ... so far, molded polyethylene has been largely used for housewares and toys."[29] With hundreds of Tupperware parties taking place each week by 1955, Brownie Wise had turned the brand into a household name. She was featured in *McCall's* and *Business Week* magazines and newspapers at the time, with headers like "'Just a Housewife' Builds 30 000$ Business on Faith," and "The Soft Sell with the Social Service."[30] The relatively high price and prestige of Tupperware, though, meant that while LDPE was certainly feminized and domesticated, it was still valued as a durable, beautiful good.

The Seventh National Plastics Exposition: an anticipated and unanticipated turning point

The National Plastics Exposition (NPE) took place every two years and showcased current innovations and developments in the plastics industry. Given its status as the cutting edge, it is a good gauge of where the industry's aspirations were at the time. At the expositions leading up to 1958, the trend clearly pointed toward glass-reinforced polyesters for use in larger and more permanent applications. In 1954, the marquee exhibit at the Sixth NPE was the Corvette, "North America's first all plastic bodied car."[31] The Corvette, while iconic, was never widely produced; production capacity at GM started at three and peaked at only thirty vehicles per day, compared to over 7,500 cars/day being produced of steel. But the larger innovation came at the seventh NPE, in 1956. There, the "Monsanto House of the Future" was revealed. Like the body of the Corvette, it was built out of Fiberglas. Constructed at MIT with the purpose of "search[ing] for a way to use plastics to the fullest extent possible," and funded by Monsanto, the House of the Future was a "stressed skin" design.[32] Modularity was used to ensure that thermal expansion and contraction could occur without causing stress fractures. A central column with a base of concrete held all the functions of the house—plumbing, electrical, and storage rose from there. The 'wings' of the house attached to the central structure and floated five feet above the ground. Each wing was internally designed to serve a different purpose. At a cost of $1M to create, it was not a low-cost alternative to conventional building, despite assurances that it could be scaled to a point where the cost could be brought down to approximately twenty thousand dollars.

The Monsanto house was the end result of over twenty years of work that the plastics industry had been doing in order to align itself with the building industry. Starting in 1935, in an interview with celebrated modernist architect Harvey Wiley Corbett, it was asserted that "social and scientific trends must logically arrive at the pre-fabricated house." His argument was that there was a "potential market of efficient housing for the masses comparable to that of supplying them with automobiles" and that "mass production of identical units . . . [can enable] housing for the masses to be made profitable by employing such a method, and employing those materials which are most suitable for it."[33] He continued to argue that if the pre-fabricated house no longer suited one's needs, one could either "turn in the old model and get a new one; or [one] can alter the old one adding to it for additional members of the family or to indicate greater affluence."[34] These ideas of a modular architecture would continue to inform the all-plastics house push for the next twenty years.

While design became more pragmatic than the high modernist ideals of the 1930s and 1940s, the plastics industry still had a difficult time cracking the building industry's code. Part of the problem was that architects and engineers continued to fundamentally not understand the material. The first editorial about better building codes appeared in 1954, admitting that "it is easy to understand why architects and builders are loath to accept 'new' materials. First, a material must be tried and tested over a period of years before it may be recommended for a building application . . . second, the historic skills in the construction field do not lend themselves readily to the use of new materials."[35] Nonetheless, it ends on an optimistic note regarding "polyester-fibrous glass sheeting" (aka Fiberglas), knowing that the Monsanto house was already in development.[36]

In the lead up to unveiling the Monsanto House of the Future at the seventh NPE in 1956, the push to standardize and include structural plastics in building codes was increasingly fever pitched. Editorial after editorial throughout 1954 and 1955 covered progress in this realm. When the first sketches and floor plans of the Monsanto house were finally revealed in December 1955, it was described as "flinging aside tradition-bound concepts of architectural engineering . . . an experimental 'house of tomorrow' takes full advantage of the inherent design potential of plastics as applied to the field of building and construction."[37] With the exposition featuring the Monsanto house, as well as "Building in Plastics" conferences occurring and the concerted push by the plastics industry to get building codes changed, this was a moment in which it seemed as if the industry's move into structural building forms was inevitable.

But while the plastics industry continued to attempt to break into the building industry for the next several years, a letter to the editor in

November 1957 points to some of the issues that would ultimately be insurmountable. The president of "Architectural Plastics Corp." wrote that "the plastics industry has created ample interest in its materials. It is now most appropriate for it to direct its attention to a comprehensive educational program . . . I find only a small number of Schools of Architecture prepared to offer materials courses in plastics."[38] Billie Faircloth makes a useful distinction in talking about plastics in architecture, in that while plastics-in-building was made legible, all-plastics architecture was not, and claims that "its inability to define itself [was] because of its ceaseless redefinition, commoditization, branding, and its sources."[39] While the material object (doorknob, insulation, siding, piping) could be easily interpreted, stress tested, and perfected, all-plastics architecture was not. Sadly, the House of the Future never arrived. While no one can deny that there is currently a lot of plastic in construction, most of it is hidden within walls as plumbing or electrical work, or as coatings on other materials such as paint or laminate. This failed foray into an all-plastic architecture illustrates the crossroads to which the plastics industry came in the following two years, the divergent path that it took, and how quickly it happened.

Bonanza for extruders

In the spring of 1958, a toy company called Wham-O released a simple toy. Based on a traditional toy that had existed for millennia, the partners from Wham-O decided to manufacture theirs from Marlex, the new high-density polyethylene from Phillips Petroleum. In contrast to the sixth and seventh NPE, whose marquee exhibits had been the Corvette and the Monsanto House of the Future respectively, the eighth NPE, taking place in 1958, featured the hula hoop as its main attraction.[40] In fifteen short months, an estimated 100 million hoops were sold worldwide.[41] Both the toy industry and the plastics industry were taken completely by surprise by the fad. It took

> a still-new plastic material which was undergoing careful and methodical sales development and overnight pushed it into big-volume extrusion. By the time the fever was at its height, scores of extruders had been educated in the use of high-density polyethylene and were running it on a routine basis. Thousands of laymen had been made aware of the new plastic. Hoops greatly speeded the wide scale debut of high-density polyethylene, and most sales estimates for 1958 have had to be revised upwards.[42]

The original hoops used ten ounces of material and were "handsome, two-color products, made by feeding two extruders through the same die." There were also cut-rate or discount producers looking to capitalize on the fad, and much to the chagrin of an industry that still worried greatly about its image, "some hoop producers started shaving away at price and quality. Some hoops produced weighed as little as 6 oz . . . which reduced the life of the toy."[43]

The complaints that the plastics industry had about skimping on materials in the toy industry were not new. Articles about the use of plastics in toys were nearly always accompanied by reminders that the molds had to be appropriately engineered, particularly with sufficient wall strength. In 1954, coinciding with the annual Toy Fair in New York City, an editorial came out that pointedly targeted poorly made toys as being bad for the sale of other plastics, particularly "toys for children under 5 . . . here were offered most of the poor-quality, badly engineered, thin-walled, weak toys." The editorial continues to warn that "mothers' opinions on plastics toys effect [sic] mothers' opinions on plastics in housewares, furnishings, and other things. For the sake of the plastics industry, plastics toys . . . had better be better this year."[44]

Polystyrene was the most commonly used plastic in toys at the time. In 1953, 17 percent of all polystyrene manufactured went into "toys and games."[45] Polystyrene was brittle, hence the emphasis on wall strength. With high-density polyethylene, however, wall strength became a drastically different concept. HDPE had greater flexibility and strength, which meant it could be molded far more thinly than polystyrene without suffering the same breakage as polystyrene did. Although Dow continued to market its high-impact Styron 475 for use in toys, polyethylene enabled plastics toy manufacturers to cut wall strength considerably and not suffer breakage. While there were still complaints about skimping on hula hoops, they did not break, but bent instead. The hoop became misshapen but not "broken" in the traditional sense of the word, and therefore began to occupy the now familiar liminal space between trash and not trash. It is in this strange liminality that most plastic objects now sit. While the bent hoop can no longer hula, it is still defined as a hula hoop, not a broken toy. Because of the odd space it occupied, it created a continuity that we think *should* be a binary: trash/not-trash, disposable/durable. It's into that continuity's forced fissure that we now step.

Disposables and expendables

Hula hoops were introduced at the tail end of a decade that had profoundly changed American toy manufacturing. The early 1950s were mired in the Korean War, and steel availability again came under threat. The Korean War sped the (already fast) shift from metal to plastic toys, as previous experience in die-cast toys translated easily into injection molding. As long as toy molders could afford the cost of an injection molding machine, they could use the thermoplastics that were a glut on the market. Then, in the mid-1950s, television advertising came into its own. Licensing agreements became emphasized in the American toy industry, and because those licensing agreements were expensive, the larger toy companies sought less expensive manufacturing. Throughout the late 1950s, most toy manufacturing moved offshore. Domestic molders therefore had a problem. Toys in the baby boom era had been big business, trebling to a $1.5 billion/annum market in the decade from 1947–57.[46] Their exit to offshore manufacturing meant that molders were left looking for new markets.

They found them in plastic packaging. In 1956, the reaction to Lloyd Stouffer saying to an SPI conference that the future of plastics was "in the trash can" was dismay. A mere seven years later, he stated that "it is a measure of progress in packaging ... that this remark will no longer raise any eyebrows."[47] The rest of the paragraph, read from 2020's perspective of horrifying ocean pollution and mountains of plastic waste, is chilling: "you are filling the trash cans, the rubbish dumps and the incinerators with literally billions of plastic bottles, plastics jugs, plastics tubes, blisters, and skin packs, plastics bags and films and sheet packages—and now, even plastics cans. The happy day has arrived when nobody any longer considers the plastics package too good to throw away."[48] The "happy day" Stouffer references was surprisingly hard fought. The industry had to pivot its understanding of its own materials almost entirely in less than ten years. In 1951, for example, when discussing the potential for attractive packaging on television, the main point was that the packaging would be able to be reused—in that "no other materials known to man offer such fabulous combinations of beauty, durability, and cleanability." It was extremely important to ensure that "sleazy film, thin sections of molded parts, poor assembly, bad finishing, misapplication of materials" would not exist, as it "will show up over color television ... it is only an opportunity for the finest plastic products."[49] These themes repeated themselves over and over during the 1950s, trying to emphasize that plastic needed to be held in higher regard.

But even as these admonishments continued, disposability began to creep into the narrative. In April of 1956, just a few months before the Monsanto House of the Future had its debut at the seventh NPE, the first editorial about plastics for disposables was published. The idea of disposables was uncommon enough that the editor felt it necessary to define what disposables even were: "the tin can, the non-return bottle, the paper package, the cellophane wrapper are examples." He continued that "most of our disposables are paper, largely because of cost but also because of the ease with which they may be made to disappear after use."[50] Here he also pointed out that "not long ago, plastics were so expensive that deliberate expendability, except for specialty packaging . . . was not to be considered."[51] The industry was beginning to realize the pull that plastic packaging had on consumers. One case study had paper packages of ice cream removed from sixteen Los Angeles area grocery stores and replaced with "I-C paks" made of transparent molded polystyrene. They pointed out that "even with an increase of $0.05 per pint retail to defray the increased cost of the plastic package, sales in all test stores increased from 200 to 400%, with one recording an *800% increase* during the one-month test period."[52]

One year later, in April of 1957—not even a year after the debut of the Monsanto House of the Future—the industry came to grips with the full potential that the packaging industry seemed to offer. The article began by admitting that

> a major problem with the marketing of disposable and expendable merchandise made of plastics . . . is the disinclination of consumers to accept the fact that such merchandise has been designed to be . . . discardable and destroyable. A decade ago, when thermoplastics materials were more expensive . . . [and] materials for thin section products were few and weak, the plastics industries were at some pains to stress durability and re-use value of plastics products, including packaging, and to educate consumers in the proper care of plastics merchandise. Since plastics products, disposable or not, are pleasant to the eye and touch, and inevitably have some durability, consumer habits of saving plastics items for re-use . . . hang on.[53]

Thin-walled sections were as important to *produce* in packaging as they were important to *avoid* in toys. The same property for which toys were admonished over and over again—insufficient wall strength—was seen as a positive in packaging. HDPE's materiality solved a problem not by sturdy design, but by its status as a polyethylene. It bent, but it did not break.

New champ of detergent bottles: HDPE

Once disposable polyethylenes were scaled via hula hoops, they found an immediate market in packaging for synthetic detergents. Plastics development and chemical cleaning development closely paralleled each other, as both are products of research and development corporate cultures in the interwar period amongst the major chemical companies. After World War I, there was a concerted push for the American chemical companies to first catch up to, and eventually wrest domination away from, the German firms.[54] Many of the household cleaning products that are synonymous with hygiene today were developed during this period, with the first synthetic detergent developed in 1916 in Germany, lacking as they were in a natural source of tallow. For example, the development of Tide at Procter & Gamble commenced in 1931, after their chemists made a visit to Germany's IG Farben Research Laboratories. The same research and development cultures that created many of the thermoplastics of the 1930s were also used to develop a whole host of synthetics: from dyes to pharmaceuticals to pesticides to explosives, chemists used base materials—first coal tar, and then petroleum—and created new domestic products. When HDPE was so quickly scaled due to the hula hoop, it seemed natural to immediately use it for the synthetic detergents which were very quickly becoming dominant on the market. Chemists already knew polyethylene as a sort of wunderkind, as it was the only material that made hydrofluoric acid easily transportable due to its non-reactivity to strong acids and bases.[55] Detergent is harsh and difficult to package, as one has to account for its ability to dissolve both hydrophobic and hydrophilic agents. In August 1959, a mere six months post–hula hoop craze, HDPE was lauded as the "new champ of detergent bottles." At the time, "the material against which [HDPE] is competing here, is principally, steel. Almost all light-duty liquid detergents today are packaged in coated steel cans."[56]

One of the major players in this early adoption of HDPE for detergent was the Hercules Powder Co. Under the subheading "Who is in the picture?" Hercules was stated as the main company to take the initiative along with three bottle blowers, based on the "Ziegler type Hi-fax 1600-E" which, they state, "proved to be the first material with enough stress crack resistance for the job." Here, you see the other side of the "material undergoing careful development" statement in the hula hoop article, where "market testing was a long drawn-out affair because the market loomed so large that no one could afford to take any chances on failure. Containers were redesigned time and again, molds were built and rebuilt, blowing techniques were revised, cartoning had to be re-engineered, filling equipment had to be developed and built."[57] Three months later, a two page "info-tisement" run by Hercules titled "The hoops have had it . . . now what?" specifically points toward the "Hi-Fax" bottles as where the next boom is; however, it takes care to emphasize that "housewares, appliances, sporting goods and toys are among the many fields where product planners are now finding ways to whet customer buying appetites with these new plastics."[58]

Hercules Powder Company was formed in 1912 as a result of an anti-trust ruling against DuPont and mostly dealt with "smokeless powder" (i.e., nitrocellulose or guncotton and other ordnance), particularly throughout World War I. As it grew and diversified, it began to incorporate various other facets of chemical manufacturing throughout the interwar period. One of those facets was its 1920 acquisition of a company from Brunswick, Georgia which had developed a method to extract wood rosin from pine stumps. A product which gained considerable consumer traction was pine oil, especially with the postwar ascendance of the popular floor cleaner Pine-Sol. While Hercules did not make Pine-Sol, they made a competitor called Yarmor Pine Oil, which they heavily advertised in trade magazines such as *Soap* and *Soap & Sanitary Chemicals*.[59] Those advertisements were fascinating harbingers of the schmear paradigm of chemical cleaning, in that they referenced cleanliness "as not only a virtue, [but] a necessity." An exemplary advertisement pushes Yarmor Pine Oil as "when made soluble by disinfectant manufacturers, plays an important part in safeguarding health."[60]

The chemical conglomerates that created most consumables postwar were an entangled process of becoming—where the production of one thing begets a waste product that becomes another product through chemistry; where the vertical integration from crude oil to finished object is specifically designed to induce demand. While there are three petrochemical areas that Hercules scaled in the years 1955–61, "crystalline (i.e. high-density) polyethylene . . . was the most complex of Hercules' initial efforts and the last to come to fruition."[61] But by the end of 1957, a mere three years after obtaining the license from Ziegler, Hercules began to commercially produce high-density polyethylene. It was marketed "initially as a substitute for [LDPE] in many existing applications, such as housewares, coated wire, detergent dispensers, fibers, bottles, toys and chemical-ware. In 1958, the product received significant boost from the craze for Hula-Hoops, which consumed 2.5 million pounds of polymer." But "investing heavily in development work and manipulating the characteristics of the polymer . . . helped it gain a secure foothold in the market for blow-molded bottles such as those used to contain liquid detergents and bleach."[62] However, the problems that Hercules encountered with respect to obtaining the chemical feedstock ethylene to produce HDPE pushed it to the point where they temporarily converted the plant over to making polypropylene by analogous process, as the plant designed for HDPE sat idle while Hercules waited on delayed shipments of ethylene from Enjay Company (a predecessor of Exxon Chemicals). The delay showed the need for Hercules to secure its own supply of raw feedstock, which in turn allowed "new raw materials base to supplement what appeared to be limited opportunities for growth in the company's existing business."[63]

While chemical cleaning became ascendant starting in the 1930s with pine oil, for the largest population of people—children—who were affected by the polio outbreaks, it was of limited efficacy, as it was too harsh to be used for children's toys. The issue with toys and childhood disease was that, pre-plastic, most of a child's toys had to be burned or thrown away in order to prevent contagion. Metal toys were not easily washed without causing rust, and it was difficult to ensure complete sterilization because of the joints and rolls used to prevent sharp corners. Stuffed toys were nearly impossible to clean without (as yet non-existent) high-heat dryers; wood is naturally porous, so could not be easily sterilized without boiling and possibly ruining any adhesives used in its assembly. In the immediate postwar years, where polio was a greatly feared and severe virus without a cure, only plastic was easily cleaned, with the "damp cloth" of damp-cloth utopianism. Advertisement after advertisement in *Playthings* magazine shows plastic toys being broadcast as "Safe! Sanitary! Washable!" and every other combination of those words.[64]

As Emily Martin points out, the polio outbreak during the damp-cloth utopian era represented the last major outbreak before changing ideas around the body's natural and internal immunity became the dominant medical paradigm.[65] Advertisements at the time for cleaning compounds "stressed the presence of deadly disease germs by the millions that may lurk unseen in ordinary house-dust."[66] She writes of being tripped up more than once in her own assumptions when reading about various measures that were taken; assuming, for example, that the suggestion that children not get teeth or tonsils removed during summer polio months was to prevent a hit to the immune system; until reading, chagrined, that it was instead reasoned that the open wound would be considered an entry point for the polio. She writes that the "external threat" rhetoric surrounding disease narratives bore remarkable similarities to the threats of cold war annihilation, where keeping one's own house in order (through domestic or political hygiene) was essential to keep the threat at bay. While pine oil was wonderful for floors and furniture, things that would go into children's mouths, such as toys, required less harsh treatments, which is why plastic was the ideal candidate for them. Throughout the 1950s, however, more and more products came onto the market that were based in synthetic detergents, with the introduction of now common cleaning products such as Mr. Clean (1958).

But the damp-cloth utopianism that existed with thermoset materials like Bakelite or melamine was also undermined by the material composition of the new polyethylenes. Unlike the smooth and hard surfaces of many of the older thermoset plastics like Bakelite and urea formaldehyde, polyethylene has a "waxy" or "oily" surface feel to it. The technical term for polyethylenes is "polyolefins"—olefin being derived from the mid-nineteenth-century French descriptor of ethylene dichloride "oléfiant" or, chemically, the "oil-forming" tendency of alkenes in reaction with chlorine gas. Consequently a damp cloth is no longer sufficient to make a polyethylene surface feel as if it is clean, because the surface still feels greasy (as anyone who has washed a Tupperware container more than once to ensure its cleanliness knows). Polyethylenes therefore undermined the idea that one could simply wipe the material with a damp cloth for cleanliness, at the same time as providing a reliable container for what would indeed get it clean—the liquid soap detergent.

As the reader can see, the way that we arrive at a packaged-in-plastic planet comes from a constellation of circumstances. Far from the common-sense story of plastic being used in packaging due to its cheapness, there are competing trajectories that plastics were approached with in the 1950s, and a child's toy—the hula hoop—tipped the balance in giving one of the plastics that are now synonymous with packaging a huge boost in the late 1950s. The 1950s plastics industry had, for the most part, been moving its "marquee" type projects into more durable, permanent applications. The 1954 Corvette and the 1956 Monsanto House of the Future were the apex of a thirty-year push by the plastics industry to create both an "all-plastics" car and an "all-plastics" house, ensuring their dominance over two of the largest industries in postwar America. But while this was the focus in much of the plastics industry nearly from the moment of its birth, it is not its current reality. Instead, the dominance of plastics in packaging became something that would define their use over the subsequent time period all the way to the present.

That boost, which contributed to HDPE's increasing ubiquity, as well as the "greasy," visceral properties of that plastic and the proliferation of Wham-O's remarkable toy, all facilitated the changing perception of plastics from one of utopian usefulness and durability to a paradigm of a more quickly discardable material. This process culminated in HDPE's evolution into consumer packaging, to contain the synthetic products that the same industrial concerns that made the plastic itself were bringing to market. The intelligibility of plastic was undermined repeatedly by its ever-changing formulations, compositions, and branding, which ultimately required the building industry to be far more nimble than it could be in order to adopt the new materials. However, consumers were much easier to educate and take from the practice of reusing packaging to throwing it away after a single use, a process that was facilitated by the proliferation of plastic toys. The 1950s are where the realm of plastic pollution we today see in every environment—from the Mariana Trench to the tip of Mount Everest—began its inexorable growth.

Notes

1. Mirja Salkinoja-Salonen, "Cleaning Can Also Be a Source of Indoor Air Problems" (translated from Finnish), *Occupational Health and Safety*, *TTT* (blog), October 11, 2017.
2. Raimo Mikkola et al., "20-Residue and 11-Residue Peptaibols from the Fungus Trichoderma Longibrachiatum Are Synergistic in Forming Na+/K+-Permeable Channels and Adverse Action towards Mammalian Cells," *The FEBS Journal* 279, no. 22 (2012): 4172–90.
3. Dorzar Toy Studios, Inc., "The Antiseptic Baby," in *Playthings: The National Magazine of the Toy Trade* (January 1945).
4. Arthur Guiterman, "Strictly Germ-Proof," in *Modern American Poetry: An Introduction*, ed. Louis Untermeyer (New York: Harcourt, Brace and Howe, 1919), 170, https://www.bartleby.com/104/50.html.
5. M. Kaufman, *The First Century of Plastics: Celluloid and Its Sequel* (London: The Plastics and Rubber Institute, 1963).
6. Jeffrey L. Meikle, *American Plastic: A Cultural History* (New Brunswick: Rutgers University Press, 1995).
7. Alison J. Clarke, *Tupperware: The Promise of Plastic in 1950s America* (Washington, DC: Smithsonian Books, 2014), 27.
8. Ibid. 28.
9. Ibid. 3.
10. Ibid. 10.
11. Rudyard Kipling, *Rudyard Kipling's Verse*, inclusive edition, 1885–1918 (Garden City: Doubleday, Page & Co., 1922), www.bartleby.com/364/.
12. Bob Kealing, *Tupperware Unsealed: Brownie Wise, Earl Tupper, and the Home Party Pioneers* (Gainesville: University Press of Florida, 2008).
13. Clarke, *Tupperware*, 100.
14. Ibid. 63.
15. Gavin Lucas, "Disposability and Dispossession in the Twentieth Century," *Journal of Material Culture* 7, no. 1 (January 3, 2002): 6.
16. Clarke, *Tupperware*, 114.
17. Elizabeth Shove, *Comfort, Cleanliness, and Convenience: The Social Organization of Normality*, New Technologies/New Cultures Series (Virginia: Berg Publishers, 2003), 87.
18. Lucas, "Disposability and Dispossession," 12.
19. Ziegler, 1956, qtd. in Günther Wilke, "Fifty Years of Ziegler Catalysts: Consequences and Development of an Invention," *Angewandte Chemie International Edition* 42, no. 41 (October 27, 2003): 5000–8, https://doi .org/10.1002/anie.200330056.

20. Ralph E. Oesper, "Karl Ziegler," *Journal of Chemical Education* 25, no. 9 (September 1, 1948): 510; qtd. in Wilke, "Fifty Years of Ziegler Catalysts."

21. Wilke, "Fifty Years of Ziegler Catalysts."

22. Bernadette Bensaude-Vincent and Isabelle Stengers, *A History of Chemistry*, trans. Deborah van Dam (Cambridge, MA and London: Harvard University Press, 1996), 202.

23. Nobel Lectures Chemistry, 1972, qtd. in Wilke, "Fifty Years of Ziegler Catalysts."

24. "Polyethylene Grabs the Spotlight Part 2," *Modern Plastics* (October 1955): n.p.

25. "Polyethylene Grabs the Spotlight," *Modern Plastics* (September 1955): 85.

26. Ibid. 90.

27. "Polyethylene Grabs the Spotlight Part 2," *Modern Plastics* (October 1955): 104.

28. Ibid. 233.

29. Ibid. 226.

30. Clarke, *Tupperware*, 189.

31. Glenn L. Beall, "The Evolution of Plastics in America: As Seen Through the National Plastics Expositions," Occasional Paper (London: Plastics Historical Society, 2005).

32. Billie Faircloth, *Plastics Now: On Architecture's Relationship to a Continuously Emerging Material* (Abingdon: Routledge, 2015), 205.

33. Dock Curtis, "Seeing the Ready Made House with Harvey Wiley Corbett," *Modern Plastics* (July 1935): 18.

34. Ibid. 52.

35. Hiram McCann, "Editorial: Better Building Codes," *Modern Plastics* (January 1954): n.p.

36. Ibid.

37. "New Architectural Concepts in Plastics House," *Modern Plastics* (December 1955): n.p.

38. George R. Hermach, "The Platform," *Modern Plastics* (November 1957): n.p.

39. Faircloth, *Plastics Now*, 25.

40. Beall, "The Evolution of Plastics in America."

41. Tim Walsh, *Wham-O Super Book: Celebrating 60 Years Inside the Fun Factory* (San Francisco: Chronicle Books, 2007).

42. "Bonanza for Extruders," *Modern Plastics* (November 1958): n.p.

43. Ibid.

44. Hiram McCann, "Toys Had Better Be Better," *Modern Plastics* (March 1955): n.p.

45. "Highlights of the Plastics Picture," *Modern Plastics* (January 1954): 78.

46. "Plastics' Stake in Toys," *Modern Plastics* (March 1958): n.p.

47. Lloyd Stouffer, "Plastics Packaging: Today and Tomorrow," vol. Section

6-A (1963 National Plastics Conference, Chicago: The Society of the Plastics Industry, Inc., 1963), 1–3.

48. Ibid. 1–3.

49. Editorial, "Color TV: An Opportunity and a Warning," *Modern Plastics* (August 1951): n.p.

50. Editorial, "Plastics for Disposables," *Modern Plastics* (April 1956): n.p.

51. Ibid.

52. "Polystyrene," *Modern Plastics* (November 1955): n.p.

53. "Plastics in Disposables and Expendables," *Modern Plastics* (April 1957): n.p.

54. Kathryn Steen, *The American Synthetic Organic Chemicals Industry: War and Politics, 1910–1930* (Chapel Hill: University of North Carolina Press, 2014), 150.

55. In an absolutely unforgettable scene in the popular show *Breaking Bad* (2008–13), Jesse learns about HF acid's ability to corrode ceramic the hard way while trying to dissolve a body in a bathtub. Adam Bernstein, *Cat's in the Bag . . .*, Crime, Drama, Thriller (High Bridge Productions, Gran Via Productions, Sony Pictures Television, 2008).

56. "New Champ of Detergent Bottles: High Density PE," *Modern Plastics* (August 1959): n.p.

57. Ibid.

58. Hercules Powder Company, "The Hoops Have Had It . . . Now What?" *Modern Plastics* (March 1959): n.p.

59. Hercules Incorporated, Russell H. Dunham, and Karl Ziegler, eds., *Records of Hercules Incorporated*, n.d.

60. Hercules Incorporated, "Cleanliness Is Not Only a Virtue . . . It Is a Necessity," *1930 Hercules Advertisements*, Records of Hercules Incorporated, 1930 (May 1930).

61. Davis Dyer and David B. Sicilia, *Labors of a Modern Hercules: The Evolution of a Chemical Company* (Boston: Harvard Business School Press, 1990), 296.

62. Ibid. 302.

63. Ibid. 307.

64. For an extremely small representative sampling of what was in virtually every advertisement for plastic toys immediately postwar, see Ideal Toy Company's advertisements of the plastic telephone (appearing in every issue of the 1945 *Playthings* magazine); Pliotoys advertisements for the "Plio Blocks," appearing in the same issues; Knickerbocker Plastic Co.'s advertisements for their "California Designed Plastic Toys and Infant Novelties," appearing on page 87 of March 1947's *Playthings* magazine; and so on.

65. Emily Martin, *Flexible Bodies: The Role of Immunity in American Culture from the Days of Polio to the Age of AIDS* (Boston: Beacon Press, 1995).

66. Ibid. 24.

3

The Anti-Plastic City: Local Governments, Plastic Waste, and the Undoing of the Weak Recycling Waste Regime in the United States

Lily Baum Pollans

Over the first two decades of the twenty-first century, plastic waste has gone from being a fringe environmental concern to a global environmental priority. Like climate change, its existence is largely attributable to the fossil fuel industry. But unlike the greenhouse gas emissions responsible for altering the earth's atmosphere, which are regulated through a patchwork of federal regulations, plastic waste is managed almost exclusively by municipal governments as part of their responsibility for municipal solid waste management.

In the US context, municipal waste management has been a core municipal service since the late nineteenth century.[1] Between the 1880s and 2020, municipal waste management evolved as a result of occasional crises brought on by changes in consumption patterns and changing public perceptions of waste and wasting.[2] Municipal waste management decisions are made within the context of the United States' "weak recycling waste regime." This waste regime prioritizes production and consumption, and sanctions municipally organized recycling of a few key packaging materials. Producer and waste management industries work together through direct lobbying and a series of industry-funded think tanks and non-profit organizations to maintain the stability of this regime, to keep the costs of production and waste management on individuals and municipal governments, and to ensure that materials can continue to flow one way through the economy.

Two key changes have put increasing pressure on municipal waste systems over the 2010s and now, I contend, are thrusting the weak recycling waste regime into crisis. The first was China's National Sword policy, which placed new restrictions on the types of material and level of contamination that China would accept for recycling. The National Sword policy altered global recycling markets, making recycling more difficult and costly for American cities.[3] Secondly, the decade saw increases in plastic production as natural gas producers sought outputs for surplus gas. Plastic production has become a key investment area for the fossil fuel industry as energy systems, domestically and globally, target more renewables.[4]

At the national level, the American federal government has been very hesitant to regulate plastics production, and national environmental laws provide almost no regulatory guidance for managing plastics in the waste stream. This has left municipalities to cope with increasing volumes of plastic wastes on their own. As volumes and costs have risen, some cities have sought to "turn off the tap" by regulating the consumption of single-use plastic products like lightweight plastic shopping bags, straws, and polystyrene foam takeout containers. Plastic waste regulations take a number of forms, from outright bans, to taxes or fees, to some combination of both, but all share the goal of limiting the amount of plastic waste that is generated, thus reducing the municipality's responsibility for plastic disposal.

Municipal programs like plastic bag and straw bans have been controversial, even among environmentalists. One frequently cited study indicated that paper bags—the most common alternative to plastic—could be more carbon-intensive than plastic.[5] Others have argued that single-use plastic bans are band-aids that cannot make a dent in global environmental catastrophes.[6] But it is the strong and organized attempts of producer-industry interests to discipline municipalities' attempts that signals an incipient crisis. Municipal regulations to reduce plastic usage and plastic waste are a radical departure from sanctioned management strategies available within the weak recycling waste regime, and thus are a potentially transformative move by city governments.

To build the argument that the municipal regulation of plastic consumption is a radical move within the current waste regime, I'll first review the context in which local governments make decisions about waste management. This context—America's weak recycling waste regime—is an assemblage of waste and producer industries that has historically defined the options for waste disposal and set the cultural definitions and practices for waste and wasting. The chapter will then review the current landscape of municipal and state plastic regulations. Finally, I'll show how industry has mobilized a multi-scalar campaign to prevent regulation, but with only limited success. The fact that reducing plastic consumption remains an option in public discourse, despite the all-out campaign by industry, signals that the weak recycling waste regime is on the verge of transformation.

Waste regimes

Every city in the United States makes relatively independent decisions about how to collect and manage waste materials produced
by everyday resident and commercial activities. Though some states,
like California, have relatively strict state-level requirements, most
states do not.[7] The federal government regulates hazardous waste and
municipal waste disposal infrastructure, but leaves management decisions entirely up to state and local governments. Despite this ostensi-
ble regulatory independence, however, local governments are highly
constrained. There are only a few management options: waste may be
disposed in landfills or through incineration with energy recovery; a
few materials can be recycled.

Waste regimes, a theory first articulated by the sociologist Zsuzsa
Gille, describe national-scale structures of waste production. In the
US context, waste regime theory provides a clear explanation for why
cities in a relatively loose regulatory environment face such a consistent landscape of management options and constraints.[8] Gille's waste
regimes are constituted through constant negotiations around the
representation, production, and politics of waste. "Representation"
captures how waste and the larger systems of production and consumption are understood or perceived by the public. This includes
the values that the public associates with certain waste management
practices such as landfilling or recycling, and how they interpret
waste-related phenomena. The "politics" of waste refers to how it
is governed and by whom. Waste politics are structured by local and
national policy, tools for regulating and governing waste, the key
institutions responsible for handling it, and importantly, the groups
or institutions who wield the power to decide how waste in general
is produced, handled, and disposed. "Production" refers to the actors
and processes that generate waste, as well as the markets that guide
the end-of-life of wastes, especially domestic and international scrap
markets. It also includes practices of consumption and modes of waste
disposal, including landfilling, incineration, and recycling. Within a
particular waste regime, these three dimensions constitute a common
definition of waste and how it should be handled.[9]

The United States is currently characterized by a weak recycling
waste regime, in which coordinated coalitions of waste management
and manufacturing industries actively promote recycling as a solution to ever-increasing material consumption and disposability. The
weak recycling waste regime was formed over the 1970s and '80s in
response to shifting perceptions of waste and wasting in the United
States. As disposable products became increasingly available in the

postwar era, litter became a widespread public nuisance. Across the country, citizens mobilized against waste.

Riding momentum and rhetoric from the growing environmental movement, Americans increasingly defined solid waste as the "third pollution," alongside air and water pollution.[10] The framing of garbage as pollution put industry on the defensive. For a moment, it seemed that infinite consumption of disposable products could not be sustained. But over the 1970s and '80s, beverage manufacturers and producers of all manner of newly disposable products created messaging campaigns and "environmental" non-profits to reframe the public perception of the solid waste problem. Through these campaigns, which included the infamous "crying Indian" imagery, a host of catchy slogans like "Don't be a litter bug," and much, much more, manufacturers successfully reframed the waste problem in America. These campaigns redirected public and regulatory attention away from production and corporate producers. Americans internalized an understanding of litter as a failure of individuals to dispose of their waste properly. Eventually, municipal governments assumed responsibility for designing and implementing recycling programs to salve consumer consciences.[11]

The weak recycling waste regime was built through a concerted industry effort to control the public understanding of solid wastes and the possibilities available for management. Within this regime, municipal recycling programs emphasize only a few common packaging materials—glass, paper, aluminum, and plastics—the materials that were the most problematic for industry during the mid-twentieth century.[12] But these materials are not necessarily the most economically or environmentally efficacious materials to recycle. As Samantha MacBride and others have shown, recycling textiles and organics would be more environmentally valuable, and yet without the support of either the waste management or producer industries, only a small handful of cities have any organics or textile recycling programs.[13]

The weak recycling waste regime relies on "sacrifice zones"[14] for disposal and resource extraction; it relies on cheap labor throughout the supply and disposal chains. It ensures that waste is removed efficiently from urban space, and that just enough is recycled—or appears to be recycled—that consumers don't get anxious about consumption. In short: it ensures that consumption can continue without limit. The success of the weak recycling waste regime is evidenced by the fact that, by 2020, the United States consumed and disposed of more plastic than almost any other country on earth.[15]

Not all cities participate equally in the unmitigated consumption of the weak recycling waste regime. Most US cities support disposability through efficient waste collection and regime-sanctioned recycling. These cities bear the costs of disposability and offload the burden of materials management onto individuals and waste workers. But others resist the waste regime by finding ways to redirect costs back onto producers, either through extended producer responsibility programs or by reducing upstream material flows. This chapter is primarily concerned with various efforts to reduce the consumption of problematic materials through novel approaches to waste management practice. These cities have "defiant wasteways": processes of waste management and waste decision-making that resist disposability and industry interference, and introduce a broader variety of knowledge and values in decisions than more conventional systems.[16]

Municipal and state regulation of plastic has become a key arena for expressing resistance to the unsustainable material flows protected by the weak recycling waste regime. However, it has become increasingly controversial as more cities have tried to stem the rising tide of plastic waste. By examining recent attempts to regulate plastic consumption, we can see how plastic producers are attempting to protect the weak recycling waste regime and, simultaneously, their own ability to offload the costs of plastic pollution onto the public sector, individuals, and the environment. Plastic lobbying in the context of the coronavirus pandemic displayed acute characteristics of disaster capitalism as industry exploited public anxieties in order to control policy and public discourse.

The state of municipal plastics regulation

According to PlasticsEurope, an association of plastics manufacturers based in the European Union, plastics are a "global success story." As an industry, plastic has exhibited "continuous growth for more than 50 years."[17] In material terms, global plastic production increased from roughly 2 Mt annually in 1950 to 380 Mt annually in 2015.[18] In total, nearly 7.8 billion tons of plastic have been produced since 1950.[19]

As plastic production has increased, so has plastic waste. As a consumer of plastic and generator of plastic waste, few countries rival the United States. According to the US Environmental Protection Agency (EPA), plastic waste increased nearly a hundredfold between 1960 and 2018. In 1960, 390 tons of plastic waste were generated in the US. By 2018, that number had grown to over 35,000 tons. Less than 10 percent of that volume was recycled. Though some was combusted with energy recovery, the vast majority—over 27,000 tons—of that waste was landfilled.[20] Per capita, the US is the fourth largest generator of plastic waste in the world, and the second largest waste generator overall.[21]

Since the 1980s, manufacturers, distributors, and retailers have touted recycling as a solution to the proliferation of disposable products, especially plastics.[22] But industry defines recycling very narrowly: recycling of a few visible packaging materials, sorted and disposed by consumers, and processed and paid for by municipal governments. Most corporate producers do not invest in their own recycling systems, utilize recycled feedstocks, or manufacture products that are technically recyclable. These failures leave municipal governments and individuals with a costly and impossible task. Many of the plastic products in the waste stream are a real problem for city recycling programs. Some forms of plastic, like plastic bags, are nuisances for cities. They are technically recyclable, but they gum up standard Material Recovery Facility (MRF) equipment, and therefore most cities do not accept them as part of curbside recycling programs.[23] Many other forms of plastic are accepted for municipal collection—like the catch-all #7 plastics—but are rarely actually recycled because of limited back-end market applications, technical difficulties, and cost.[24] This means that most post-consumer plastic waste is bound either for disposal or, worse, the ocean. Since plastic first entered the municipal waste stream in large quantities in the 1970s, only 9 percent of plastics have been recycled globally, meaning that almost all plastic ever made still exists on earth in its original form.[25]

As manufacturers and environmentalists coalesced around recycling as a suitable solution for plastic waste, the percentage of recycled plastic did increase slowly between 1980 and 2018. China absorbed plastic waste from all over the world, recycling, downcycling, and disposing of much of the world's post-consumer plastic. In 2018, however, China's National Sword Policy upended global recycling markets and exacerbated preexisting challenges with plastic recycling. China's plastic imports decreased 99 percent over 2018.[26] Even as other East Asian countries ramped up recycling capacity, the value of recyclables fell precipitously. By 2019, the revenues from the sale of municipal recyclables had fallen by 50 percent.[27] Between 2018 and 2020, at least sixty municipalities across the US canceled their recycling programs altogether.[28] Many others altered or reduced recycling services, a trend that was accelerated due to pandemic-related budget crises and falling oil prices, which reduced the price of virgin plastics.[29]

The National Sword, however, was only the most recent challenge to municipal recycling. Many critics have noted that plastic recycling has never been an economical investment, for many reasons. Most plastics degrade in quality with each remanufacture, meaning that each "recycling" is really a "downcycling." Because virgin plastic is cheap and of reliable quality, most manufacturers prefer it. Recycling plastic is also a toxic and energy-intensive endeavor.[30] A key reason that municipalities continue to invest in plastic recycling is because intensive public messaging from industry-funded interest groups, like the Recycling Partnership and Keep America Beautiful, has built a strong public constituency for recycling. Oil, retail, and manufacturing industries have worked hard to market recycling as the solution to waste. In so doing, they have successfully salved consumer worries and built political support for municipal recycling programs. It is hard to cancel the popular programs.[31]

All of this, however, only applies to materials that are currently accepted for recycling in municipal recycling programs. Some plastic items, like single-use plastic shopping bags, straws, and polystyrene foam—though technically recyclable—are not compatible with the technologies used to sort and process other curbside recyclables. City governments are therefore forced to pay for their disposal. New York City alone estimated that it cost $12 million annually to dispose of the 10 billion single-use plastic bags thrown away every single year in the city prior to a statewide ban.[32]

Industry's attempts to prop up the weak recycling waste regime

As recycling fails to offset increasing disposal costs, more and more cities are moving to "turn off the tap" of plastic wastes. Over the 2010s, cities and states targeted the most-difficult-to-recycle plastic items, like bags and polystyrene foam, and experimented with different techniques for reducing or eliminating local distribution and consumption. As of 2020, eight states had statewide policies mandating bans or fees on plastic bags. Hundreds more cities had enacted similar policies, within and outside of states with statewide plastics regulations.[33]

Early analysis of individual bans suggests that they can effectively cut down plastic bag usage and plastic pollution in local waterways.[34] However, in many states, cities have lost the ability to regulate plastic wastes. As a result of concerted industry lobbying, by 2019 fifteen state governments had enacted laws prohibiting local governments from regulating single-use plastics, forcing dozens of cities across the country to roll back plastic regulations. Another nine states had preemption regulation pending.[35]

States with reliably Republican legislatures and large fossil fuel industry footprints were among the first to enact preemption laws, or to encourage interpretation of existing laws so as to prevent municipal plastic regulation. Texas was among the first states to preempt municipal plastics regulation. Several Texas municipalities had passed municipal plastic bag bans and fees in the early 2010s. In Laredo, the local retail association sued the city, arguing that the bag ban violated an obscure provision in a 1993 solid waste disposal law. After a series of appeals, the Texas Supreme Court ultimately ruled for the plaintiff, forcing the repeal of about a dozen plastic bag bans in Texas municipalities.[36]

The Texas case was notable for the unusual coalition of interests that came together to support municipal bans on plastic bags. In classic defiant-wasteway form, cotton ginners, bass fishermen, and cattle ranchers joined with more conventional environmentalists and municipal officials in supporting local government's right to regulate plastic consumption.[37] A smaller—but evidently more powerful—coalition of interests opposed that right. The only organization to file a brief in support of the Laredo retailers was the BCCA Appeal Group, Inc., a coalition of industrial facility owners that included ExxonMobil and Dow Chemical, two corporations with vested interests in the unmitigated production of plastic.[38]

The fingerprints of the fossil fuel and plastics industries can usually be identified in attempts to reduce the regulation of single-use plastics. Operating through a variety of lobbying and marketing entities, plastic and fossil fuel interests are omnipresent in local and state-level plastic regulation processes. For instance, New York City passed a plastic bag ban in 2016, with hearty support from both environmental and environmental justice advocacy organizations.[39] In short order, the New York State legislature imposed an indefinite delay on the measure, citing concerns about an inequitable financial burden imposed by the proposed five-cent fee.[40] It was a surprising argument ideologically, since conservative lawmakers usually show little concern about regressive taxation. But this line of argumentation was likely a direct reflection of industry lobbying. The American Recyclable Plastic Bag Alliance (ARPBA, formerly the American Progressive Bag Alliance, a lobbying group that bills itself as "frontline defense against plastic bag bans and taxes nationwide")[41] released a statement arguing that the City's proposed fee would "disproportionately impact those who can least afford it—without providing any meaningful benefit to the environment or New Yorkers."[42] Ultimately, New York's delay was only temporary. The legislature was widely criticized for its veto, and in 2019 passed a statewide plastic bag law.[43] New York's law, however, was not a simple defeat for industry. The state ban included key compromise provisions that environmentalists felt weakened the effectiveness of the law.[44]

The ARPBA, its subsidiary project "Bag the Ban," its lobbying partner, the Plastics Industry Association, and an array of plastic and fossil-fuel representatives like BCCA in Laredo, have used every tool to weaken, delay, and preempt municipal attempts to regulate plastic consumption. The COVID-19 pandemic offered a useful crisis for the plastics industry and its allies. Industry interests actively exploited public anxiety and uncertainty about disease vectors to undermine public trust in reusable products and to coax policymakers to suspend, roll back, or delay implementation of plastic consumption regulations.

Disaster capitalism: industry use of COVID-19 to promote plastic regulation rollbacks

Plastics and allied industries already had a well-worn playbook by the time the novel coronavirus began to spread in the United States in early 2020. But the pandemic offered a new opportunity for lobbyists. Across the US, plastics lobbyists amplified specious claims about the ability of reusable bags to spread infection and successfully instigated policy suspensions and rollbacks across the country.

On March 20, 2020, Tony Radoszewski, CEO of the Plastics Industry Association, issued a statement requesting that all local and state policymakers include plastic producers on their lists of essential workers to ensure uninterrupted plastic production. Plastic was essential to fight the virus, he argued, and not only in the form of protective equipment for medical workers. "Single-use plastic bags provide a sanitary and convenient way to carry our groceries home while protecting supermarket employees and customers from whatever is lurking on reusable bags."[45] Just five days later, the health commissioner in Massachusetts announced that reusable bags were prohibited in grocery stores and pharmacies until further notice.[46] New York State postponed implementation of its new bag law.[47] Over the month of March, hundreds of municipalities followed suit, temporarily outlawing the use of reusable shopping bags, out of an "abundance of caution."[48]

There was no clear evidence at the time that reusable shopping bags were a likely vehicle for virus transmission. In fact, early evidence indicated that the virus remained active longer on plastic surfaces than on paper, and there was not yet any research about cloth.[49] But there was nevertheless a concerted effort by right-wing think tanks and plastic industry lobbyists to exploit public anxiety and promote fear around reusable bags. Over the spring and summer of 2020, the Competitive Enterprise Institute (CEI) republished dozens

of opinion pieces from local newspapers across the country expressing concerns—largely without evidence—that reusable bags were
contributing to the uncontrolled spread of the coronavirus.[50] The
CEI published its own study arguing that bag bans were endangering
the public; but this study simply reported on American Chemistry
Council-funded research from 2011 showing that unwashed reusable
bags may contain a host of bacteria.[51] The Pacific Research Institute
wrote that reusable bags were "carriers of a simmering noxious
stew," in an attempt to shame California lawmakers into repealing
the statewide bag ban.[52]

While writing and reprinting a steady stream of opinion pieces
about the dangers of reusable items, industry representatives also
directly lobbied policymakers at national, state, and local levels to
postpone or permanently roll back plastics regulations. The Plastics
Industry Association, for example, sent a letter to the US Department
of Health and Human Services urging the agency to broadcast the
dangers of reusable products and to "make a public statement on the
health and safety benefits seen in single-use plastics."[53] Meanwhile,
news outlets across the political spectrum observed industry lobbyists
working furiously to overturn or postpone plastics regulations across
the country.[54]

Limits to plastic growth?

Though the plastics and allied industries worked hard at all scales to
prevent plastics regulation, and though they fully exploited the coronavirus pandemic to advance their policy objectives, and though they
have had some success in state-level preemption, plastics deregulation
efforts have not been entirely successful. Lobbyists deployed a well-
worn playbook to promote their anti-regulatory agenda and maintain
the weak recycling waste regime. But their attempts have not lessened
public anxiety about plastic waste. Across the country, states and cities
have reinstated plastic waste regulations. Massachusetts began allowing reusable bags again in July of 2020.[55] After delays, New York
State implemented its statewide ban in October 2020.[56] In Texas, a
coalition of ranchers and municipal leaders ramped up a messaging
campaign highlighting the impact of plastic bag waste on the state's
cows and horses after the municipal bans were repealed.[57] The first
ever national plastics law, the Break Free From Plastic Pollution Act,
was introduced in Congress in 2020.[58] Simultaneously, science journals and mainstream media outlets began to report on a new torrent
of single-use plastic waste in the form of coronavirus PPE.[59] Not only
did industry fail to create permanent policy change by exploiting the

pandemic, but it unleashed a new wave of plastic waste that only magnified public awareness of the world's growing problem of plastic accumulation.[60]

As of this writing, the plastic industry was girding for the possibility of more interest in regulating plastics at the federal level as the new Democratic administration prepared to assume power. One plastics industry newsletter wrote in November 2020 that "there's been a lot of bipartisan interest in the topic" of marine plastics. The newsletter also reported that some "plastics industry groups and business associations representing consumer brand companies have recently endorsed things like packaging fees to fund recycling."[61] This is a minor pivot, firmly within the plastic industry's neoliberal ideological tradition, that would not fundamentally change the relations of the weak recycling waste regime. It would continue to allow unabated production and consumption, maintain emphasis on recycling, and divert responsibility for plastic waste to individual consumers and public recycling programs. However, even this minor shift signals that industry senses a changing political landscape.

Plastic may be the undoing of the weak recycling waste regime. The municipal and state governments that have already banned the most problematic plastics, circumventing industry lobbyists and targeting production and distribution, may indicate a broader willingness to change the terms of the regime for the first time in generations. Even as the case of municipal plastic regulation shows the power of industry to subvert local regulatory agendas, it also demonstrates the power of municipal governments to meaningfully alter material consumption practices. Ultimately, to truly "turn off the tap," a consistent landscape of state regulations, or federal regulation, will be necessary. But as of now, cities are playing a crucial role, and proving that with proper support, citizens and retailers are perfectly capable of thriving with less plastic.

Notes

1. Martin V. Melosi, *Garbage in the Cities: Refuse, Reform, and the Environment*, revised edition (Pittsburgh: University of Pittsburgh Press, 2005).
2. Lily Baum Pollans, *Resisting Garbage* (Austin: University of Texas Press, 2021); Zsuzsa Gille, *From the Cult of Waste to the Trash Heap of History: The Politics of Waste in Socialist and Postsocialist Hungary* (Bloomington: Indiana University Press, 2007).
3. Nicole Javorsky, "How American Recycling Is Changing after China's National Sword," *Bloomberg City Lab*, April 1, 2019, https://www.

bloomberg.com/news/articles/2019-04-01/how-china-s-policy-shift-is-changing-u-s-recycling.

4. Michael Corkery, "A Giant Factory Rises to Make a Product Filling Up the World: Plastic (Published 2019)," *New York Times*, August 12, 2019, https://www.nytimes.com/2019/08/12/business/energy-environment/plastics-shell-pennsylvania-plant.html; Reid Frazier, "The US Natural Gas Boom Is Fueling a Global Plastics Boom," *NPR.org*, November 15, 2019, https://www.npr.org/2019/11/15/778665357/the-u-s-natural-gas-boom-is-fueling-a-global-plastics-boom; Beth Gardiner, "The Plastics Pipeline: A Surge of New Production Is on the Way," *Yale E360* (blog), December 19, 2019, https://e360.yale.edu/features/the-plastics-pipeline-a-surge-of-new-production-is-on-the-way.

5. Subramanian S. Muthu et al., "An Exploratory Comparative Study on Eco-Impact of Paper and Plastic Bags," *Journal of Fiber Bioengineering and Informatics* 1, no. 4 (February 2009): 307–20.

6. Björn Lomborg, "Opinion: Sorry, Banning Plastic Bags Won't Save Our Planet," *The Globe and Mail*, June 17, 2019, https://www.theglobeandmail.com/opinion/article-sorry-banning-plastic-bags-wont-save-our-planet/; Mathy Stanislaus, "Banning Straws and Bags Won't Solve Our Plastic Problem," *World Resources Institute* (blog), August 16, 2018, https://www.wri.org/blog/2018/08/banning-straws-and-bags-wont-solve-our-plastic-problem.

7. NERC, "Disposal Bans and Mandatory Recycling in the United States" (Brattleboro, VT: Northeast Recycling Council, May 1, 2017), https://nerc.org/documents/disposal_bans_mandatory_recycling_united_states.pdf; NERC, "Compendium of State Disposal Bans & Mandatory Recycling Laws Updated" (NERC, July 9, 2020), https://nerc.org/documents/nercnews/compendium%20of%20state%20disposal%20bans%20_%20mandatory%20recycling%20laws%20updated.pdf.

8. Gille, *From the Cult of Waste*.

9. Discussion of Gille's argument adapted from Green, Pollans, and Krones, in preparation.

10. William E. Small, *Third Pollution: The National Problem of Solid Waste Disposal* (New York: Praeger Publishers, 1971); US EPA, *The Third Pollution*, documentary, 1972, https://nepis.epa.gov/.

11. Finis Dunaway, *Seeing Green: The Use and Abuse of Environmental Images* (Chicago: University of Chicago Press, 2015); Bartow J. Elmore, "The American Beverage Industry and the Development of Curbside Recycling Programs, 1950–2000," *Business History Review* 86, no. 03 (2012): 477–501, https://doi.org/10.1017/S0007680512000785; Samantha MacBride, *Recycling Reconsidered: The Present Failure and Future Promise of Environmental Action in the United States* (Cambridge, MA; London, England: MIT Press, 2012); Pollans, *Resisting Garbage*.

12. MacBride, *Recycling Reconsidered*.

13. Frank Ackerman, *Why Do We Recycle? Markets, Values, and Public Policy* (Washington, DC: Island Press, 1996); MacBride, *Recycling*

Reconsidered; Brenda Platt and Nora Goldstein, "State Of Composting In The US," *BioCycle*, July 2014; Pollans, *Resisting Garbage*; Lily Baum Pollans, Jonathan S. Krones, and Eran Ben-Joseph, "Patterns in Municipal Food Scrap Management in Mid-Sized US Cities," *Resources, Conservation and Recycling* 125 (October 2017): 308–14.

14. Steve Lerner, *Sacrifice Zones: The Front Lines of Toxic Chemical Exposure in the United States*, reprint edition (Cambridge, MA: MIT Press, 2012).

15. Hannah Ritchie and Max Roser, "Plastic Pollution," *Our World in Data*, September 2018, https://ourworldindata.org/plastic-pollution.

16. Pollans, *Resisting Garbage*.

17. PlasticsEurope, "World Plastics Production 1950–2015," August 2016, https://committee.iso.org/files/live/sites/tc61/files/The%20Plastic%20Industry%20Berlin%20Aug%202016%20-%20Copy.pdf.

18. Roland Geyer, Jenna R. Jambeck, and Kara Lavender Law, "Production, Use, and Fate of All Plastics Ever Made," *Science Advances* 3, no. 7 (July 1, 2017): e1700782, https://doi.org/10.1126/sciadv.1700782.

19. Ritchie and Roser, "Plastic Pollution."

20. US EPA, "Plastics: Material-Specific Data," Collections and Lists, US EPA, September 10, 2020, https://www.epa.gov/facts-and-figures-about-materials-waste-and-recycling/plastics-material-specific-data.

21. Ritchie and Roser, "Plastic Pollution."

22. Elmore, "The American Beverage Industry and the Development of Curbside Recycling Programs"; Pollans, *Resisting Garbage*.

23. Alexia Elejalde-Ruiz, "Plastic Bags a Headache for Recyclers," *Chicago Tribune*, July 30, 2015, https://www.chicagotribune.com/opinion/commentary/ct-plastic-bag-ban-recycling-0731-biz-20150730-story.html.

24. National Geographic Society Newsroom, "7 Things You Didn't Know About Plastic (and Recycling)," *National Geographic*, April 4, 2018, https://blog.nationalgeographic.org/2018/04/04/7-things-you-didnt-know-about-plastic-and-recycling/.

25. Geyer, Jambeck, and Law, "Production, Use, and Fate of All Plastics Ever Made." Though plastics break into smaller and smaller pieces and leach chemicals into surrounding environments, they have not been materially or chemically transformed.

26. Colin Staub, "China: Plastic Imports down 99 Percent, Paper down a Third," *Resource Recycling News* (blog), January 29, 2019, https://resource-recycling.com/recycling/2019/01/29/china-plastic-imports-down-99-percent-paper-down-a-third/.

27. Waste360 Staff, "SWANA Report: National Sword Impact and Solutions," *Waste360*, September 11, 2019, https://www.waste360.com/business/swana-report-national-sword-impact-and-solutions.

28. E. A. Crundun and Cole Rosengren, "How Many Curbside Recycling Programs Have Been Cut?" *Waste Dive*, October 7, 2020, https://www.wastedive.com/news/curbside-recycling-cancellation-tracker/569250/.

29. Michael Bermish and Alexandra Tennant, "Has COVID-19 Changed the

Economics of Plastics Recycling?" *Wood Mackenzie* (blog), August 25, 2020, https://www.woodmac.com/news/opinion/has-covid-19-changed -the-economics-of-plastics-recycling/; Colin Staub, "Budget Shortfalls Threaten Local Recycling Programs," *Resource Recycling News* (blog), May 27, 2020, https://resource-recycling.com/recycling/2020/05/27/bud get-shortfalls-threaten-local-recycling-programs/.

30. National Geographic Society Newsroom, "7 Things You Didn't Know."

31. Elmore, "The American Beverage Industry and the Development of Curbside Recycling Programs"; MacBride, *Recycling Reconsidered*; Pollans, *Resisting Garbage*.

32. New York Department of Sanitation, "Carryout Bags," 2020, https:// www1.nyc.gov/assets/dsny/site/our-work/zero-waste/carryout-bags.

33. "In Your State," *Bag the Ban*, 2019, https://www.bagtheban.com/in-yo ur-state/; Trevor Nace, "Here's A List Of Every City In The US To Ban Plastic Bags, Will Your City Be Next?" *Forbes*, September 20, 2018, https://www.forbes.com/sites/trevornace/2018/09/20/heres-a-list-of-eve ry-city-in-the-us-to-ban-plastic-bags-will-your-city-be-next/.

34. John Hite, "The Truth about Plastic Bag Bans," *Conservation Law Foundation* (blog), June 16, 2020, https://www.clf.org/blog/the-truth -about-plastic-bag-bans/; "Why Bag Laws Work: A Summary of Plastic Bag Law Effectiveness," *Surfrider Foundation*, 2020, https://www.sur frider.org/coastal-blog/entry/why-bag-bans-work-a-summary-of-plastic -bag-law-effectiveness.

35. "Preemption," *PlasticBagLaws.org*, https://www.plasticbaglaws.org/pre emption.

36. Emma Platoff, "Texas Supreme Court Strikes down Laredo's Plastic Bag Ban, Likely Ending Others," *Texas Tribune*, June 22, 2018, https://www .texastribune.org/2018/06/22/texas-supreme-court-rules-bag-bans/; Isabelle Taft, "Laredo's Bag Ban Becomes Flashpoint in Debate Over Local Control," *Texas Tribune*, June 28, 2016, https://www.texastri bune.org/2016/06/28/laredos-bag-ban-becomes-flashpoint-debate-over -loc/.

37. Billy Joe Easter, "Brief of Amici Curiae: Texas Cotton Ginner's Association," August 23, 2017, http://www.search.txcourts.gov/Search Media.aspx?MediaVersionID=437be899-e4f3-40b6-b8c0-411604fdc65 e&coa=cossup&DT=BRIEFS&MediaID=5395cb60-f1c5-4128-ab55-a9 3d8ab6af23; Edward Parten et al., "Brief of Amici Curiae: Texax Black Bass Unlimited," January 4, 2018, http://www.search.txcourts.gov/Sea rchMedia.aspx?MediaVersionID=a9858390-252f-4727-a621-a6511fd6 7b29&coa=cossup&DT=BRIEFS&MediaID=ef8d2bf0-89c6-40fb-80af -ef345a867a43; Jennie Romer, "Supreme Court of Texas Strikes down Laredo's Bag Ban—Why We Should All Be Alarmed," July 20, 2018, https://www.leonardodicaprio.org/supreme-court-of-texas-strikes-down -laredos-bag-ban-why-we-should-all-be-alarmed/.

38. "City of Laredo, Texas, V. Laredo Merchants Association," *PlasticB*

agLaws.org, https://www.plasticbaglaws.org/949904626842; Michael Kurlya, Evan A. Young, and Ellen Springer, "Brief of Amicus Curiae BCCA Appeal Group, Inc.," March 20, 2018, http://www.search.txcour ts.gov/SearchMedia.aspx?MediaVersionID=5db1664a-02ec-4121-920e -e14e5d7bc085&coa=cossup&DT=BRIEFS&MediaID=55434d89-2d 81-4207-874a-504a11d1eef4.

39. Ben Adler, "Banning Plastic Bags Is Great for the World, Right?" *Grist*, June 2, 2016, https://grist.org/climate-energy/are-plastic-bag-ba ns-good-for-the-climate/; J. David Goodman, "5¢ Fee on Plastic Bags Is Approved by New York City Council (Published 2016)," *New York Times*, May 5, 2016, sec. New York, https://www.nytimes.com/2016 /05/06/nyregion/new-york-city-council-backs-5-cent-fee-on-plastic-bags .html.

40. William Neuman, "Bag Fee for New York City Is Delayed by State Legislature (Published 2017)," *New York Times*, February 4, 2017, sec. New York, https://www.nytimes.com/2017/02/03/nyregion/new-york-plastic-bag-fee.html.

41. "American Recyclable Plastic Bag Alliance," *American Recyclable Plastic Bag Alliance*, https://bagalliance.org/.

42. Statement quoted in: Sarah Maslin Nir, "State Senate Takes Aim at Plastic Bag Fee in New York City (Published 2017)," *New York Times*, January 18, 2017, sec. New York, https://www.nytimes.com/2017/01 /17/nyregion/plastic-bags-new-york.html.

43. Anne Barnard, "Get Ready, New York: The Plastic Bag Ban Is Starting," *New York Times*, February 28, 2020, sec. New York, https://www.nyt imes.com/2020/02/28/nyregion/new-york-state-ban-plastic-bags.html.

44. Zach Williams, "Everyone's Got a Problem with the Plastic Bag Ban," *City & State* NY, February 11, 2020, https://www.cityandstateny.com/ articles/policy/energy-environment/everyones-got-problem-plastic-bag-ban.html.

45. Tony Radoszewski, "Plastics Industry 'Essential' As First Line of Defense With Products To Fight Coronavirus," Text, Plastics Industry Association, March 20, 2020, https://www.plasticsindustry.org/article /plastics-industry-essential-first-line-defense-products-fight-coronavirus.

46. Monica Bharel, "Order of the Commissioner of Public Health" (Massachusetts Executive Office of Health and Human Services, March 25, 2020), https://www.mass.gov/doc/march-25-2020-pharmacy-grocery -order.

47. Sophia Chang, "Enforcement of New York's Plastic Bag Ban Delayed to June," *Gothamist*, April 18, 2020, sec. Food, https://gothamist.com/fo od/enforcement-new-yorks-plastic-bag-ban-delayed-june.

48. Paul Tuthill, "Plastic Bag Bans Suspended Over Coronavirus Fears," *WAMC*, April 1, 2020, https://www.wamc.org/post/plastic-bag-bans-suspended-over-coronavirus-fears.

49. Neeltje van Doremalen et al., "Aerosol and Surface Stability of SARS-CoV-2 as Compared with SARS-CoV-1," *New England Journal of*

Medicine 382, no. 16 (April 16, 2020): 1564–7, https://doi.org/10.1056/NEJMc2004973.

50. "CEI Newsroom," *Competitive Enterprise Institute*, https://cei.org/newsroom/.

51. Angela Logomasini, "Lift #NeverNeeded Plastics Regulations and Bans—Permanently," *Competitive Enterprise Institute*, June 2, 2020; David L. Williams et al., "Assessment of the Potential for Cross-Contamination of Food Products by Reusable Shopping Bags," *Food Protection Trends* 31, no. 8 (August 2011): 508–13.

52. Kerry Jackson, "Not Even the Threat of Spreading COVID-19 Can Change California's Plastic Bag Ban," *Pacific Research Institute*, April 2, 2020, https://www.pacificresearch.org/not-even-the-threat-of-spreading-covid-19-can-change-californias-plastic-bag-ban/.

53. As quoted in: Ivy Schlegel, "How the Plastic Industry Is Exploiting Anxiety about COVID-19," *Greenpeace USA* (blog), March 26, 2020, https://www.greenpeace.org/usa/how-the-plastic-industry-is-exploiting-anxiety-about-covid-19/; but also reported here: Associated Press, "Coronavirus Pandemic Deals Blow to Plastic Bag Bans," *Fox Business News* (*Fox Business*, April 8, 2020), https://www.foxbusiness.com/lifestyle/coronavirus-plastic-bag-ban.

54. Associated Press, "Coronavirus Pandemic Deals Blow to Plastic Bag Bans"; Justine Calma, "Plastic Bags Are Making a Comeback Because of COVID-19," *The Verge*, April 2, 2020, https://www.theverge.com/2020/4/2/21204094/plastic-bag-ban-reusable-grocery-coronavirus-covid-19; Perry Wheeler, "Plastics Industry Using Coronavirus to Demonize Reusable Bags," *Baltimore Sun*, May 6, 2020, sec. Op-ed, Opinion, https://www.baltimoresun.com/opinion/op-ed/bs-ed-op-0507-plastic-bags-coronavirus-20200506-4vznrwoqovbgtlahtkjgpj3y5i-story.html.

55. John Hilliard, "Mass. Department of Public Health Lifts Ban on Reusable Bags," *Boston Globe*, July 11, 2020, https://www.bostonglobe.com/2020/07/11/metro/mass-department-public-health-lifts-ban-reusable-bags/.

56. NYS Department of Environmental Conservation, "DEC Announces Enforcement of New York's Plastic Bag Ban to Start on Oct. 19, 2020" (NYS DEC, September 18, 2020), https://www.dec.ny.gov/press/121415.html.

57. Lara Korte, "Plastic Bags Are Killing Horses and Cows across the State. What's Texas to Do?" *Texas Tribune*, August 14, 2019, https://www.texastribune.org/2019/08/14/texas-wont-approve-bans-plastic-bags-which-can-be-fatal-livestock/.

58. Allan S. Lowenthal, "H.R. 5845—Break Free From Plastic Pollution Act of 2020," Pub. L. No. H.R. 5845 (2020).

59. Ashifa Kassam, "'More masks than jellyfish': Coronavirus Waste Ends up in Ocean," *Guardian*, June 8, 2020, https://www.theguardian.com/environment/2020/jun/08/more-masks-than-jellyfish-coronavirus-waste-ends-up-in-ocean; Tanveer M. Adyel, "Accumulation of Plastic Waste

during COVID-19," *Science* 369, no. 6509 (September 11, 2020): 1314–15, https://doi.org/10.1126/science.abd9925; Carly Fletcher, "What Happens to Waste PPE during the Coronavirus Pandemic?" *The Conversation*, May 12, 2020, http://theconversation.com/what-happens -to-waste-ppe-during-the-coronavirus-pandemic-137632.

60. Dave Ford, "COVID-19 Has Worsened the Ocean Plastic Pollution Problem," *Scientific American*, August 17, 2020, https://www.scientifica merican.com/article/covid-19-has-worsened-the-ocean-plastic-pollution -problem/; Parija Kavilanz, "Used Masks and Gloves Are Showing up on Beaches and in Oceans," *CNN*, October 28, 2020, sec. Business, https:// www.cnn.com/2020/10/28/business/ppe-ocean-waste/index.html.

61. Steve Toloken, "Plastics Industry Ponders What Biden Administration Will Mean for Legislation," *Plastics Industry News*, November 8, 2020, https://www.plasticsnews.com/news/plastics-industry-ponders-what- biden-administration-will-mean-legislation.

Part II

Plastic Proliferation

4

From Plasticity to the Aesthesis of Queer Toxicity

Amanda Boetzkes and Dana Feldman

Philosopher Catherine Malabou argues that plasticity is a motor-scheme of contemporary thinking.[1] It is both a neoliberal ideal of the subject's infinite flexibility, and a philosophical capability by which ideas and recalcitrant matter co-actualize and co-mind one another. Ultimately, she suggests, because plasticity rests at the axis of ideology and philosophical ideal, we must take hold of plasticity for contemporary thinking. But, what would the taking hold of plasticity entail in relation to plastic's recalcitrant materiality?

This chapter considers how plastic produces reaction formations—schemas, behaviors, environmental topologies, and ultimately a queer futurity. Plastic's topological spread is a challenge to Malabou's call to take hold of philosophical plasticity, for it suggests a materialist dilemma in addition to a metaphysical struggle. Plastic as such predetermines the material conditioning of performative behavior. More than standing as an ideal neoliberal subject, plasticity has produced toxic entanglements. Plastic matters above and beyond a philosophical ethic by which it might come to grips with the temptations of neoliberalism. In materialist terms, plastic replicates the capitalist logic that brought it into being, and this very logic continues to reproduce itself as environmental contamination that retroactively affirms the supremacy of capitalism. In other words, plastic materiality entrenches its ideological motor-scheme *as* a form of toxic capitalism.

What counts as a plastic resistance to plastic, then? We argue that in retrieving plastic's antithesis, its material recalcitrance and dysfunctionality, we might arrive at the limit of the plastic paradigm and thereby imagine its unlimitation in the queering of bodies. Drawing from Malabou's paradigm, as well as the scholarship of Gay Hawkins, Heather Davis, Myra Bird, Nicole Seymour, and Mel Chen, we consider how an ethics of queer coexistence with plastic emerges from a reflection on its *aesthesis*. We therefore address Davis and Seymour's suggestion that we might read plastic as a queer material, while nevertheless maintaining an insistence that its toxicity is not a measure of its adaptation to queer subjectivity (as failure, dysmorphia, or dysfunctional normativity). We suggest that plastic's toxicity is a signal to its queering of ontology altogether. We therefore differentiate plastic's queering of ontology through toxicity from other forms of subjective distortion, diffraction, and play with normativity. Through a discussion of three works of contemporary art by Sophia Oppel, Tuula Närhinen, and Harley Morman, we consider plastic's consumption of subjectivity, bodies, and futurity. In its absorption of life and the future, plastic's toxicity presents a challenge to biotic life in totalizing terms. Its toxic effects queer the body in a broader process of planetary wasting. A queer ethics thus does not seek to exacerbate toxic difference but rather to seize futurity as a means to preserve the queer life of bodies.

The material limit of plasticity's futurity

Contemporary plasticity is an inversion of its historical connotations in Hegelian philosophy. Plasticity is the term Hegel uses for the dialectical movement of thought, as substance and Idea collide and shape one another, metamorphosing over time.[2] In this regard, the "plastic arts," from architecture to sculpture and painting, occupy a privileged position in exemplifying this philosophical movement over the course of history from one era to the next. Plasticity in this sense is a tensile figuration of time itself. It encompasses the figurative impulse of the Idea as it seizes, molds, and shapes the future, while equally remaining malleable and receptive to form as it comes into being.

Catherine Malabou argues that plasticity is essentially the anticipatory structure of the dialectic; it *is* the future. As she reads the Hegelian version of plasticity, the future occurs in a "philosophical face-to-face between two temporal modalities." Plasticity is the co-shaping of teleological circularity and representational linearity; of what is actual and what is potential; of the retrospective and the prospective.[3] Thus, thinking itself in the Hegelian vein, is written in a tense that waits for what is to come (according to a linear and representational thinking), while presupposing that the outcome has already arrived. In short, plasticity is the dynamic temporal system in which a time ushers in its future, a future that configures its history, that imagines its past as the future, and its future as coming to pass.

However, in her account of neuroplasticity, Malabou redeploys Hegelian plasticity with a view to distinguishing it from its ideological forms. Plasticity has come to refer to the specific molding of the brain in and through consciousness, rather than the Idea and substance co-shaping one another in the abstract. Consciousness—the consciousness of our very plasticity—has been put under pressure by a bad version of itself. Plasticity has begun to enable the restrictions of the economy by encouraging a flexible subject in a system that neurologically maximizes desirable behavior and a general "positivity." Neuroplasticity reweights the discourse of plasticity to put an emphasis on adaptability and a "feedback model" of subjectivity, enabling the latest form of global capitalism as a decentered and networked organization reliant on a pliable neoliberal subject.[4]

Important to Malabou's reclaiming of plasticity through Hegel is that it is also a philosophical disposition, a speculative attitude to the possible configurations of the future coming to pass, and to the unknowability of the specific materialization of the event in the future. Plasticity is heterogeneous, and cannot be contained by its particularity at any given moment in history.[5] In this regard, Malabou recovers the association of plasticity with plastic explosives. The dialectic might effectively take shape through the explosion of given forms. In fact, that is precisely the result of the polarizing energies of dialectical oppositions.

The questions remain, however, can plasticity be rescued from its ideological double? How can plasticity be recuperated in and against the material counterpart to its ideological form, plastic itself? Further, how might we reconsider plasticity in relation to the economic basis of plastic? For insofar as plastic appears as a purely artificial material, one that has effaced its earthly source, we might be hard-pressed to make the connection between plastics and global oil. Whereas plastic is used for its transposability, and therefore its ontological pliability, its toxicity is nevertheless an extension of the logic of the oil economy that conceived it. It persists and accumulates as a form of waste that is unaccounted for by the economy that imagined it as a disposable material. Where oil is scarce, undoubtedly earthen (extracted only by extreme measures), it is desired, consumed, and promises plenitude and wealth, plastic is everywhere, readily available, primed to be discarded. In a sense, plastic is the dreamwork of the oil economy: a petroleum-based commodity that inserts itself into social relations, cultivates dependencies, and is designed to be wasted. In the oil imaginary, plasticity has become a myth of eternal and limitless transformation of the material world. In this paradigm plastic becomes both a signifier and an agent of the oil economy.

The philosopher Michael Marder argues that the schema at work in capitalism's current energy imaginary can be characterized by the desire to liquidate potential energy from the static, inward dwelling energy of all beings and things.[6] In its quest to harvest energy and keep it in perpetual expansion through exchange, the oil economy flattens ontological differences and displaces all boundaries. Under the spell of these energy dreams, the conceptual and spatial framing of the very concept of nature has collapsed. Further, the economy has expanded precisely through a topological procession through planetary systems.

Plastic as economic expression

Plastic pervades environments and mediates how we see them. Toronto-based artist Sophia Oppel captures this effect in her 2018 site-specific sculpture *Figures of Distance*, which appeared at QueenSpecific, a window exhibition space in the core of a busy downtown Toronto neighborhood (see fig. 4.1). Made of Plexiglas, mirrors, and other reflective materials, Oppel's sculpture captures the visual effects of gentrification and the schemas of capitalist space that drive industrial materials into the fabric of urban environments, reducing their complexity to agglomerations of useless commodities.

Plastic objects such as a disposable fork, a rubber band, a small heart-shaped toy, a ripped Ziploc bag, and other miscellaneous items appear trapped between the sheets of Plexiglas. The artist laser-cut barely legible letters, words, and phrases onto the surface. The result was a misshapen totality, a sculpture bent by patches of heavy melting and industrial litter. The sculpture consolidates its flotsam and jetsam into one fluid sheet of plasticized paper, wrinkled up and tossed away with garbage fused inside it. Blending street garbage with reflections of the street itself as a work of art presented in a shop window, *Figures of Distance* captures the collision of industrial material, commodity, street garbage, and capitalist space, in its simulacral visuality. It conflates what is for sale with the window that displays it, using a material that is continuous with the urban environment.

Whereas in pre-modern historical eras waste was conceptualized and managed in a spatial paradigm by which it could be expelled from human habitats and social visibility to marginal zones, over the course of the twentieth century, the boundaries between interiority and exteriority along with the false binary of human/nature have collapsed. Plastic is an exemplary form of contemporary waste in this regard. On the one hand, it was designed from its very production to be *disposable*—designed to be wasted. But it has now so permeated planetary ecosystems that microplastics can be found accumulating in Arctic sea ice and inducing genetic defects in animals. There was no natural exterior to which it could be permanently relegated. Plastic is

Figure 4.1 *Sophia Oppel,* Figures of Distance, *2018. Laser-cut mirrored acrylic, resin, acrylic medium and found waste. Installed at *queenspecific, curated by The Shell Projects.*

now integrated into the living fabric of the planet, while at the same time causing its systemic necrosis.

Over the course of the late twentieth century, plastic became an increasingly informed material whose chemical make-up was thickened and enriched as it was designed to fulfill the demand to be cheap, available, sanitary, light, durable, and expendable.[7] And it is exactly this informational richness that destines it to penetrate into all environments. As it accumulates in oceans, rivers, open-air landfills, in animal bodies, it transforms them all into postindustrial environments. Plastic's force as a geocultural agent is fully revealed in the way it exposes the unthought of the oil economy—its total limitation as material intractability—in its recursion with that economy. In other words, plastic waste evolved in and through a paradoxical movement of disposability into the environment and eternal return as pollution.

Australian cultural theorist Gay Hawkins argues that plastic must be understood through its patterns of emergent causation.[8] Rather than analyzing how external structures situate plastic in time and space, instead she reads the ways that plastic evolved through processes of chemical refinement, entrenching itself in social and material relations as it moved through global and planetary systems. Plastic generated a vital co-evolution of dense chemical programming, economic flexibility, and the earth's material substrate, which exposed the limitations of waste management infrastructure. It now behaves as a geochemical agent and a geocultural schema. It shapes social parameters through its transposability, while it induces chemical reactions in the environment, bodily malfunctions, hormonal imbalances, and death. Plastic reactions subsume vitalities, beings and objects into a common topology. The living earth is flattened as it is consumed by plastic.

Insofar as plastic is an economic agent, then, it interpenetrates social relations and subjectivities as it is exchanged. It is in this sense that plastic can also be thought as an economic *expression*; it was chemically refined in accordance with a market demand for disposability. But this very economic expression effected an ecological trajectory in much the manner of the genetic expression of a DNA code. As plastic has been wasted in waterways, atmosphere, and dump sites, it effects chemical reactions that intervene on living beings, drawing them into a common topology. The economic criteria that dictated the chemical design of plastic destined it to a material reversal, whereby it became a virulent form of waste in its uselessness, persistence, and recalcitrance. Its viability inverted, and it turned into its own economic antithesis. As it became an increasingly mobile and liquid medium that integrated into the biosphere on all possible levels, plastic transformed into a toxic element, an inhibitor of life itself. The spaces that once contained waste have collapsed entirely, and we now confront its return as a vital force that denaturalizes planetary ecologies and overturns the foundational concepts and material basis of biological life in its wake.

The aesthesis *of plastic as the displacement of life*

As much as plastic has evolved to become more chemically complex, Roland Barthes's insightful analysis from 1957 still holds: "It is impregnated throughout with this wonder: it is less a thing than the trace of a movement."[9] It is through the economy's transformation of

environments into topologies of plastic waste that we see the full fruition of its contradictory logic. Plastic expresses the economy aesthetically in the sense of expression as we commonly think of it, as a spatial
movement from an interiority to an exteriority, from the economic
system outward to the planet as a communication of that very system.
But the fallacy of this spatial logic reveals itself in the way that plastic
inverts its expression. It lures the living with its stimulating appearance, but once it enters living systems it integrates itself, accumulates,
and poisons them. Thus we can consider the aesthesis of plastic waste
as both an operation of economic expression and planetary recursion,
as it permeates living ecologies and collapses the boundaries between
organic and inorganic; living and non-living; vitality and artificiality;
biological evolution and toxic spillover.

In order to consider the topological movement of plastic, we also
consider its aesthetic efficacy: how it moves through systems based on
the interpretation (or perhaps misinterpretation) of its appearance.
Thus we can consider the expressivity of plastic as an operation of
aesthesis: a virulent movement across the planet in patterns of expression, reaction, and recursion. As part and parcel of our approach to
plastic's movement as aesthetic per se, we also consider plastic's indifference to the living as integral to its aesthetic movement. It relies on
its appeal to living beings, while it pollutes indifferently.

It is precisely because plastic appears to be a vital organic material
that it is able to penetrate environments and organisms. Its appearance is deceptive, and it is easily mistaken for organic matter that
is consumed by animals and microorganisms. This aesthetic deceit
enables its passage from an economic system into living ecologies. But
while its topological movement from the economy to the planetary can
be considered an expression of sorts, it is also the dark antithesis of
expression in terms of a meaningful communication. Plastic expresses
as pure biochemical reaction with no meaningful content, and no
defined spatio-temporal destination. It spreads only as a necrotic
force. As in the case of a chemical reaction, the catalytic movement
is not so much intentional and directed as it is technical and irreversible. Whereas an expression might be subject to interpretation—"what
does this mean?"—the reaction to plastic occurs at the level of pure
stimulus-response. Living ecologies cannot interpret plastic as different or foreign to it, and due to the incapacity to interpret its difference,
become vulnerable to contamination.

We might consider Chris Jordan's film and photograph series
Midway: Message from the Gyre (2009—) which documents the
effects of plastic accumulation in the North Pacific Ocean. Jordan has
been documenting albatrosses on the island of Midway through film

Figure 4.2 Chris Jordan, Midway: Message from the Gyre, 2011.
© *Chris Jordan.*

and photography for over ten years. The problem of plastic pollution in the ocean has come to roost in the albatross ecosystem, as the birds are hopelessly attracted to the colorful objects when they are fishing in the water, ingest them, and then die from the obstruction (see fig. 4.2). Jordan's photographs reveal plastic as an uncanny presence that remains still vibrant in the albatross's exposed digestive tract. Listed as an Endangered Species under the US Fish and Wildlife Service, the albatross is one of millions of species that are threatened by anthropogenic change. His film *Albatross* (2017) tracks the trajectory of plastic as it enters the albatross's lifecycle and activates intense struggle, pain, and death. Jordan's images of albatross bodies expose plastic's knotting of economy, ecology, and extinction. Such a reflection might easily induce a depressive paralysis in the place of responsibility.[10] Plastic's aesthesis is so wholly destructive that it cannot be contained by either knowledge of economy or science. It marks a complete lapse in meaningful interpretation of living beings and environments. Plastic marks the displacement of life with human economic expression. It therefore behooves us to address the force of this displacement.

While the analysis of plastic from a purely socioeconomic perspective might lead us to understand the paradoxes of capitalism (its systemic drive toward planetary collapse), our focus on the aesthetic expression of plastic leads us to imagine how its materiality out-strips its economic ideal and propels it into a future where its very usefulness—its disposability—has left it with no living being to dispose of it. In its wasted form, plastic conjures a future in which it is the only species left. Plastic's reactivity propels it outside of the phenomenological experience of space and time and into a future in which it has supplemented life out of existence altogether. Once it has exited economic circulation, it wanders the planet, lonely for want of being used and disposed. Here is where plastic's recalcitrance betrays its dependency on social connection to ensure its circulation. It *wants* to insinuate itself into relations. Yet it also expresses environmental catastrophe: it is the harbinger of the catastrophe it brings. Waste is a material problem, the symptom of the problem, and its ecological extension all in one. Its economic expressivity absorbs its social efficacy, leaving it behind as a quasi-life without a human society in which to dwell. In this recalcitrant state, its futural form, it becomes indifferent to living beings altogether.

Such a futural form in which plastic has become indistinguishable from other beings is captured in *Baltic Sea Plastique* (2013) by Helsinki-based artist Tuula Närhinen (see fig. 4.3 and fig. 4.4). Närhinen created nine marine "creatures" comprised of plastic waste that had washed ashore on Harakka Island. She produced videos of the creatures floating underwater amidst other forms of marine life and debris. Like drifting jellyfish, the creatures appear self-generated and

Figure 4.3 Tuula Närhinen, Baltic Sea Plastique, *2013.*

Figure 4.4 Tuula Närhinen, Baltic Sea Plastique, *2013.*

unproblematic. There is no cause or explanation for their appearance in these waters; they seem to inhabit it naturally. Närhinen then displayed the creatures in cylindrical vessels filled with water, as though they were specimens in an aquarium. She paired the display with drawings of invented organisms, also made of beach waste, ensuring that each being had been thought through and living through synthetic relations.[11] The project replicates the fluidity of marine life using plastic, and posits this plasticity as central to the ecological relations of the site.

Sociologist Myra Hird argues that waste can exist as an ontological contradiction, as both waste and non-waste. She denounces the disappointing statistics that inform readers of the amounts of garbage produced annually.[12] Such statistics which focus on quantity have no fixed or defining qualities: "Waste is an inherently ambiguous linguistic signifier: anything and everything can become waste, and things can simultaneously be and not be waste, depending on the perceiver."[13] Plastic disappears in such statistics. But it exerts itself in its obstruction of ecologies. For example, Hird notes that Canada's waste is particularly "wicked" for its paradoxical complexity: solving one aspect of (any) waste crisis simply results in the introduction of newer, more complex issues down the road. Even by improving recycling technology, for instance, different challenges tend to arise—like increased toxic runoffs from new systems of recycling.[14] Plastic produces an inescapable predicament in just this way. It not only embodies its primary economic value—its disposability—it develops this quality so thoroughly that it has begun to express the economy as a self-consuming force. In other words, plastic is an expression of the economy's self-eradication through its displacement of life.

Plastic toxicity toward a queer future

Plastic materializes through performative repetitions of the economy, as it is chemically designed for disposability, sold, bought, used, and discarded *ad infinitum*. These acts guarantee its dispersion into a planetary topology. In this materialization, plastic becomes an autonomous geological agent that conditions social relations and ecologies. As agent, it inverts the primary condition of planetary life, turning it into toxic non-life. Indeed, it alters the spatio-temporal paradigm by which we understand the earth as *physis*, a fundamental basis from which life and growth proceeds. But if there is a plastic movement to be taken hold of, and a quality of its topological spread worth reconsidering, it is that this toxic reactivity is a process of differencing, and indeed queering, bodies. For plastic's toxicity also points toward the queer ontology of planetary life itself. How then can a queer politics express itself at its juncture with toxic ontology? Or rather, how can queer ontology resist a collapse into a toxic politics? What distinguishes these domains in a uniformly plastic field?

The aesthesis of plastic—its economic movement as disposability and recursion as planetary toxicity—occurs through its patterning onto the performativity of gender and sexuality. While plastic attains its characteristic traits through regular practices of use and disposal, it integrates itself into the social systems that define the partitioning of normative identities and their excesses. Plastic aggregates precisely at the junctures of gender and sexuality, those hierarchical qualities of identity that determine the social value that certain bodies hold and the power they can exert. But while gender binarism and heteronormativity are systems of control unlinked to biology but instead dictating the way bodies function in a social totality, plastic locates the cracks in normative identities and relations, becoming toxic at their point of excess. Though it has become integral to social functioning, it overflows into non-normative bodies, exacerbating their difference. The dependency on plastic's performative requirement—its demand to be used, disposed of, and reingested as toxic waste—has procured a debt that, as Heather Davis argues, will be paid "of the flesh."[15]

On the one hand, then, plastic vanishes into biotic life as it enacts its flexibility and disposability. As Gay Hawkins explains, plastic does not accumulate its value based on its structure or molecular composition. Rather, plastic is "enacted" upon and receptive, performing the qualities we necessitate most.[16] Because we manufacture plastic to be a material that *is* malleable, cost-efficient, and useful, it becomes defined by those critical traits upon which we depend most. It begins as the ideal material we want it to be. But the full integration of plastic into bodies, elements, and the earth itself is unavoidable. Along with the acknowledgment of its total integration into materiality itself comes an imperative to live in and through its toxicity.

Plastic toxins queer bodies by altering the heteronormative ability of affected bodies to biologically (sexually) reproduce. As Davis notes, since the mid-twentieth century scientists have been observing that plastic toxins have the effect of procuring defective (or diseased) sexual organs, behavioral abnormalities, and impaired fertility amongst humans and animals across the world. Owing to the association between queerness and non-reproductivity, that correlation has been extended to plastic toxicity as well. Plastic effects a queering of bodies, whilst simultaneously establishing an additional relationality between queerness and the petrochemical industry.[17] Plastic toxins have also frequently been reported to cause behavioral abnormalities: the femininization of males and increased masculinity of girls and/or women. Geographer and plastics specialist Max Liboiron probes the ethics and harm such a queering plastic's toxicity procures. Ultimately, they state that neither the LGBTQ+ community nor chemical industries consider veering away from gender-normalcy to be problematic. Plastic therefore actualizes itself in unanticipated ways: it both determines and exists in queer futurity.

While there are obviously despairing associations with toxicity—the environmental wasting of Indigenous lands, the ghettoization of racialized groups in toxic urban areas, and the displacement of waste to developing nations—queer theory suggests a toxic productivity might be recuperated from the conditioned effects that we (must) learn to live with in our plastic future. Harley Morman (previously Megan Morman) is a queer artist based in Lethbridge, Canada, working pri-

Figure 4.5 Harley Mormon, Art Party, *2012. Courtesy of the artist.*

marily with craft media such as needlepoint and plastic beads.[18] His ongoing series *Art Party* is an installation displaying life-sized queer performance artists that Morman has rendered out of fusible plastic beads (see fig. 4.5). As poet Lucas Crawford wrote in his review of Morman's *Art Party*, "Morman's work reminds you that your self and body are bricolage. The terrifying and beautiful truth is revealed: you may conceive of yourself as a whole, but you are (in) pieces."[19] This is particularly relevant to the consideration of sympathetic intoxication and preparing for a futurity queered by plastic. As Crawford explains, Morman's work is the visual summation of "the many molecular possibilities that underlie any one person's attempt to fixedly cohere as a molar human individual."[20] The dissolution of the confining aspects of bodily norms into atomized possibilities is therefore thematized in Mormon's representation of the body in pieces. He suggests that normativity fixes us within our bodies—indeed normativity is a suffocating totality—so that we forget about the possibilities of reorganizing partitions of the body or subjectivity.

Morman composes his portraits from thousands of small plastic beads that the artist fuses together. This approach to portraiture calls to mind the conventional molecular composition of all bodies, while reminding us that bodies are nevertheless able to divide, change, and even break down, all while remaining whole. His work becomes the representation of hopefulness, demonstrative of the multifarious capabilities of bodies to integrate within our new and ever-growing plastic surroundings. By composing portraits of exclusively queer artists, Morman maintains the emergent ontological connection between queerness and plastic. As well, his use of plastic beads, rather than sheets of plastic, for instance, is indicative of his focus on the commonality of plastic and the body: both operate and change at the molecular level.

A kind of play on ethics, Morman's use of plastic beads, perhaps in response to the challenge of microbeads polluting bodies of water, is also informed by the sexual subject matter of the installation. *Art Party*, as Crawford explains, is actually a "coded story about transgressive sex acts."[21] The plastic diamonds that accompany Morman's portraits gesture to the folded handkerchiefs that queers have historically used to signify their sexual desires. In this way, the plastic medium carries forward a queer history into a potentially plastic futurity. The work becomes a kind of "queer utopia" for Morman, enabling queer bodies, sexualities, and living with, or in this case *as*, plastic. Crawford ends his review of Morman's exhibition with a warning: "You may fall apart. It might feel like a relief."[22]

Morman's exhibition thus captures a queer ethic which Heather Davis describes as a coinvestment in ecological advancements, environmental recuperation, and queer values, by learning to live with such toxicity.[23] In a similar vein, cultural theorist Mel Y. Chen considers queer bodies *as* toxic assets.[24] Therefore, it is increasingly critical to learn from queer theory's proposed futurities, to understand how to move forward into a non-reproducing "plastisphere." We therefore seek to establish a distinction between the corporeal effects of plastic toxicity on queer bodies, and the queer ontology of bodies that we might embrace as a futural form.

It is critical to make a clear distinction between toxic plasticity, which effects a toxic totality, and queer futurity, which entails care for the differences emerging in toxic bodies. This queer ethic implies a distinction between plasticity and toxicity at the level of ontology. The plastic future—as ecologically taxing as it may be—is effecting the disassembling of gender (confines) and heteronormative sexuality, whether this is desired or not. Davis suggests that the phenomenological correlation between plastic futurity and queer methodology contributes to developing a "queer technobacterial future."[25] An ethics of queer futurity is our best strategy for living alongside plastic toxicity while anticipating the continued emergence of newer and more harmful chemicals. We must have a strategy in place for defense against these threats of extinction, low reproductivity, and the destabilization of social structures via indecipherable differences of gender and sexuality. Davis thus argues for embracing what Lee Edelman calls a "no future" in the service of queer theory.[26] This is the break-down and reassembling of a new social order, for queer systems that secure the possibilities of relations of care in the future. Such a reas-sembly acknowledges the eternal return of plastic while, in a sense, riding its toxic effects as a pretense for taking hold of the future.

From the common origin of plasticity and its queer toxicity, it becomes possible to imagine a praxis of living with toxicity. Cultural theorist Nicole Seymour advocates for learning to improve our rela-tionship with the environment by learning to live with toxicity in the future. She argues that environmental discourse is implicitly heter-onormative, since ecological discourses on toxins are often used to mobilize a fear of abnormal bodies.[27] Seymour suggests that anti-toxicity is overtly synonymous with campaigns of homophobia and transphobia, since to be actively living with plastic toxins requires fighting the destruction of systems of categorization. But taking into account the lived experience of non-normative bodies, bodies that do not comply to social standards and as such reject the necessity of social standards such as gender or sexuality, deepens our appre-ciation of the plastic predicament. Queer theories reveal social con-structs and the political agenda of maintaining such confining norms as nothing more than systems of hierarchy between different kinds of bodies.

Mel Y. Chen suggests that we must think *with* toxicity to con-sider the relationality between affective toxicity and queerness.[28] They compare the capitalist values that inform the standardized fear of toxicity, or desire for immunity, precisely because of the invalidity affected bodies are allied with. Chen argues that given the rapidly reproducing use of toxin-procuring materials like plastic and lead, there is a certain "persistent allure" to re-signifying the ontological connotation of toxicity by crediting queer bodies for already living with the attributes most associated with toxins: non-normative traits of identity (like gender or sexuality) and non-biological reproductiv-ity.[29] Toxins, Chen explains, threaten to alter our realms of comfort and current sociocultural norms. Considering the political and capi-talist agendas of governing structures to racialize certain toxins, or toxify certain races, sexualities, and bodies, demonstrates the social, political, and economic complexities of disassociating toxicity from fear and abnormality.

Chen looks specifically at the racialized fear associated with lead toxicity, constructing racialized bodies as toxins akin with the chemi-cal pollutants produced by lead. They write, "which assets have gone toxic (lead), which assets are considered toxic (bodies of color), which assets must be prevented from becoming toxic?"[30] Toxic assets, Chen describes, become the materialized extensions of toxic bodies. They explain that if toxins literally and ontologically threaten to alter our interior composition, then everything becomes toxic. Ultimately, Chen aims to promote that if everything is threatened to change then nothing will be abnormal in futurity; we must be urged to invite a sympathetic intoxication, and value past lived experiences as demonstrative of how to function *with* toxicity.[31] Chen writes that queer bodies are often treated as toxic, following the segregation of queer bodies and the biopolitical privileging of heteronormative bodies still evident in the twenty-first century. Of course, since "toxicity straddles boundaries of 'life' and 'nonlife,'" it must be considered that toxicity perhaps releases capacities of life from affected bodies. But to see toxicity as a new but attainable challenge of generating life means to remove toxicity from the realm of death.

To remove fear from connotations of toxicity means to pull from queer studies and adapt ecological considerations of heteronormative environmentalism. Chen suggests looking at toxic futurity under the guise of animacy. "We can then ask not 'who is alive, or dead,' but 'what is animate, or inanimate, or less animate'; relationally, we can ask about the possibilities of the interobjective, above and beyond the intersubjective."[32] Non-reproductivity is the most immanent threat of our toxic future, however, losing the capacity to biologically reproduce should not render futurity nonhuman, or "dead." Rather, a new, queer existence must be considered as the result of toxic exposure to bodies and the environment at large. If we are to consider animacy a false, performed confine, as Chen suggests, then we must also render heteronormativity *performed* (and thus false). Again, this brings toxic-procuring substances, like plastic, as functioning within the present but from the future. In this respect, Chen reveals that plasticity, whether as philosophical ideal or neoliberalist antithesis, can be shifted toward a new concept of futurity, one that is not condemned to the distinction of life or death in the biopolitical regime. Instead, plasticity can be embraced for its queer effects as it renders bodies toxic.

Conclusion

In its original production, plastic was designed to be disposable. Yet it has now so permeated planetary ecologies that its disposability must be understood, paradoxically, as the very impossibility of waste. Plastic cannot be disposed of and is instead an all-encompassing material and a toxic condition. Plastic has thus become a vital synthetic agent, co-evolving with the earth's material substrate and with other living beings. Its force as a geocultural agent is therefore fully revealed in the way it exposes the unthought of the economy: the reingestion of the economy's waste as plastic's material recursion into an environmental totality. In other words, plastic expresses the self-eradication of the economy in and through the planetary ecologies it has produced. But in this movement, one that is so deeply defined by the aesthetic appearance of plastic and the performative acts by which its economic logic is integrated into all beings and their relations—in *aesthesis* itself—a queer difference can be recuperated that alters the framing and philosophical dilemma of plasticity. The future consciousness to be taken hold of is discovered in plastic's toxic assets, the movement by which it queers bodies and those same bodies present their toxicity to normativity itself. As the plastic condition forecloses the future, it comes to rest on a new understanding of "plasticity" as its very queer toxicity.

Notes

1. Catherine Malabou, *What Should We Do with Our Brain?* (New York: Fordham University Press, 2008).
2. G. W. F. Hegel, *Phenomenology of Spirit*, trans. Adrian V. Miller (Oxford: Clarendon, 1977), 39.
3. Catherine Malabou, *The Future of Hegel: Plasticity, Temporality and Dialectic* (London: Routledge, 2005), 17.
4. Malabou, *What Should We Do*, 43.
5. Jacques Derrida, "From Restricted to General Economy: A Hegelianism Without Reserve," in *Bataille: A Critical Reader*, ed. Fred Botting and Scott Wilson (Malden: Blackwell, 1998), 102–38.
6. Michael Marder, *Energy Dreams of Actuality* (New York: Columbia University Press, 2017), 56.
7. Gay Hawkins, "Made to Be Wasted: PET and Topologies of Disposability," in *Accumulation: The Material Politics of Plastic*, eds. Jennifer Gabrys, Gay Hawkins, and Mike Michael (London: Routledge, 2013), 55.
8. Ibid. 49–67.
9. Roland Barthes, "Plastic," in *Mythologies*, trans. Annette Lavers (New York: Noonday Press, 1972), 97.
10. For a longer discussion on ecology and depressive paralysis see Amanda Boetzkes, *Plastic Capitalism: Contemporary Art and the Drive to Waste* (Cambridge, MA: MIT Press, 2019).
11. Tuula Närhinen, "Baltic Sea Plastique: Installation Views," Tuula Närhinen, http://www.tuulanarhinen.net/artworks/baltplast/plastinstallation.html.
12. Myra J. Hird, "Knowing Waste: Towards an Inhuman Epistemology," *Social Epistemology* 26, no. 3–4 (2012): 454.

13. Ibid. 454.
14. Myra Hird, "Biography," *Myra Hird*, https://www.myrahird.com/biog raphy.
15. Heather Davis, "Toxic Progeny: The Plastisphere and Other Queer Futures," *PhiloSOPHIA* 5, no. 2 (2015): 234.
16. Heather Davis, "Life and Death in the Anthropocene: A Short History of Plastic," ed. Etienne Turpin, *Art in the Anthropocene: Encounters Among Aesthetics, Politics, Environments and Epistemologies*, 2015, http://heathermdavis.com/wp-content/uploads/2014/08/Life-and-Death-in-the-Anthropocene.pdf, 349.
17. Davis, "Toxic Progeny," 237.
18. Megan Morman, "Artist Statement," *Megan Morman*, June 12, 2016, https://www.populust.ca/?page_id=13.
19. Lucas Crawford, "Art Party—Megan Morman: Stride Gallery," Stride Gallery (Stride Gallery, February 21, 2018), http://www.stride.ab.ca/art -party-megan-morman/.
20. Ibid.
21. Ibid.
22. Ibid.
23. Davis, "Toxic Progeny," 232.
24. Mel Y. Chen, "Toxic Animacies, Inanimate Affections," *GLQ: A Journal of Lesbian and Gay Studies* 17, no. 2–3 (2011): 273.
25. Heather Davis, "Imperceptibility and Accumulation: Political Strategies of Plastic," *Camera Obscura: Feminism, Culture, and Media Studies* 31, no. 2 92 (2016): 188–9.
26. *Heather Davis: The Queer Futurity of Plastic* (Sonic Acts, 2017), https:// www.youtube.com/watch?v=CwR3hfqz-58&ab_channel=SonicActs.
27. Nicole Seymour, *Bad Environmentalism: Irony and Irreverence in the Ecological Age* (Minneapolis: University of Minnesota Press, 2018).
28. Chen, "Toxic Animacies, Inanimate Affections," 265.
29. Ibid. 266.
30. Ibid. 270.
31. Ibid. 273.
32. Ibid. 280.

5

Microplastics in Arctic Sea Ice:
A Petromodern Archive Fever

Chantelle Mitchell and Jaxon Waterhouse

This makes capitalism very much like the Thing in John Carpenter's film of the same name: a monstrous, infinitely plastic entity, capable of metabolizing and absorbing anything with which it comes into contact.

Mark Fisher, *Capitalist Realism*

Reflecting upon the perceived limitlessness of the ocean in 1809, French naturalist Jean Baptiste de Lamarck remarked, ". . . animals living in the water, especially the sea waters, are protected from the destruction of their species by Man. Their multiplication is so rapid and their means of evading pursuit or traps are so great that there is no likelihood of his being able to destroy the entire species of any of these animals."[1] In the two centuries since Lamarck's comments, the oceanic surfaces and depths have been permanently altered by human impacts, as have those species dwelling within the watery deep. Once perceived as a wellspring of watery abundance, the seas have been exhausted by overfishing and overrun by waste. An alarm sounds over the surface of these waters, one hastened as a result of rising temperatures: it warns us that by 2050, plastics in the ocean will outnumber fish, and that even the most remote shores on Earth already have plastic castaways washing ashore. Reflecting on this contemporary reality, not only has de Lamarck's thesis been disproved, but his dictum has been replaced—animal becomes a cipher for plastic, as it spreads and embeds itself within ocean ecologies. The limitlessness of life within seawater is no longer, brought into stark relief by the limitlessness of plastic. Mobilizing the pervasive and insidious presence of microplastics within sea ice, we follow its origins from hydrocarbon to Anthropocene marker embedded within the natural. In recognition of the reanimation of this carbon, we become attentive to spectral and supernatural connotations. In doing so, we align our reading of microplastics with the work of John Carpenter and H. P. Lovecraft to further grasp the material and affective horrors of the Anthropocene. Entering these speculative and spectral sites, we encounter a complex temporal entanglement, as reanimated carbon-as-plastic extends beyond human lifespans into increasingly uncertain futures.

In recognition of the archival and agentic resonances of microplastics, we refigure their presence as part of a broader understanding of circulations of matter within New Materialist and contemporary capitalist frameworks. Our investigation enters speculative territory as we locate microplastics within a capitalist realist ontology, one in which capital/ism is totalizing, endless, and seemingly sentient. Moving through these speculative frames against the backdrop of accelerating climate change, we apprehend the entrapment of microplastics as a form of self-archiving within the soft stratigraphies of the body and the unstable stratigraphies of sea ice.

Plastic from Anthropocene to anthrop-ocean

The global production of plastics has rapidly increased from 2 million tons per year in 1950 to 360 million tons as of 2018—down from over 380 million tons in 2015.[2] The lives of these plastics, including polyvinyl chloride, polyethylene, cellulose, and polyester, are not solely locked into a relationship between production and use. They linger, extending their presence as they shift out of these frames and into those of waste and persistence. This extension beyond production and consumption is where the persistence of this matter, and the severity of this persistence, comes into critical focus. The posthumous persistence of plastics dwarfs the short lifetimes of the plastic object. Within the frame of the Anthropocene, these afterlives are "the longest part of their lives."[3]

The Anthropocene is a pervasive but debated epoch; a framework in which earth systems and stratigraphy coalesce with human history, in seeking to delineate a period whereby the central role of humankind in geology and ecology is recognized.[4] There is a sonorous terrestriality to the Anthropocene, predicated upon geologic stratigraphy, extraction, and above-ground industrial activity, the imposing nature of the problem and the lexicon that has developed around it. Millions of miles of roads, farmlands, and cities press up against disappearing forests and melting permafrosts, merging with fears about the slow creep of rising sea levels over low-lying sites. Whilst there is an inherent terrestriality to the Anthropocene, the permeation of human activity that demarcates it as such pervades beyond the hard edge of the land, spilling over into the fluid space of the ocean. What occurs here, in this passage toward and into the volumetric space of the sea, is an expansion of the Anthropocene, towards an *anthrop-ocean*.

We see this anthrop-ocean manifest further, in the rapid increase of overfishing and the depletion of fish populations by 35 percent, studies remarking on the devastating effects of bottom trawling, of a new "global gold rush" in rapid expansions of deep-sea mining.[5] The statistics surrounding the presence of plastics within the ocean further emphasize the bleed between Anthropocene terrestriality, the emergent Fourth Industrial Revolution, and the theorization of a Blue Economy.[6] These are the characteristics that point to the existence of an anthrop-ocean; a wet ontological framework which moves beyond the terrestrial into a consideration of fluvial and oceanic consequences of human impacts upon that which covers more than 70 percent of the globe. As explored by Elspeth Probyn, this mercurial anthrop-ocean, in its watery formulation, lacks the solidity or fixed stratigraphy of the terrestrial. The little fixity the sea can offer us, in the form of sea ice, disappears as the shimmer of heat mixes with the salty spray cast up as it calves. The Heraclitean river runs to the sea; troubling temporalities as sea life, waste, refuse, and detritus mix in together.

As Colin Waters from the British Geological Survey states: "We have become accustomed to living amongst plastic refuse, but it is the 'unseen' contribution of plastic microbeads from cosmetics and toothpaste or the artificial fibers washed from our clothes that are increasingly accumulating on sea and lake beds and perhaps have the greatest potential for leaving a lasting legacy in the geological record."[7] When a piece of plastic becomes waste, it moves in passage toward the ocean. Significantly, in this process it accumulates oily chemicals, "such as the pesticides DDT and HCH, the chemical coolant PCB, flame retardants like PBDEs, and surfactants (waterproofing material) like PFOA," becoming a conglomerate.[8] When this combines with the natural, it forms what Corcoran et al. have deemed a "plastiglomerate": an object that extends past scientific curiosity, becoming instead an "object of power that invokes an emotional reaction," a "symptom," and a "consequence."[9] These plastiglomerates, as conglomerations of refuse, natural matter, and detritus, resonate with a significance that surpasses those of its individual constituents. As these individual components mix, they retain traces of industry and nature through their passage. Once formed, they travel and traverse fluvial sites, with the plastiglomerate holding within it the memory of what these materials once were, becoming simultaneously archive and archived. There are ghosts in these archives—both within the plastiglomerate, haunted by the long-dead organic matter that constitutes it, but further; the plastiglomerate is itself a peculiarly Anthropocene phenomenon, haunting in what it represents.

Plastic pervasiveness in petromodern times

Jenna Jambeck et al. estimated that 4.8 to 12.7 million tons of plastic entered the ocean in 2010, with 80 percent of this plastic entering the ocean from terrestrial sources.[10] In their traversal of global waters, these ocean plastics enter into the global commons as they slip across territorial borders and markers of place—becoming a collective problem, as they leave no site unvisited by their petrochemical presence.[11] Drawn from fossil hydrocarbons, plastics proliferate exponentially within the accelerated contemporary. From containers, plastic wrap, and packaging to incomprehensible amounts of synthetic fibers developed for easy consumption, these plastics spread across a backdrop of slicks and spills of oil. As petrocultural detritus, they herald the advent of petromodernity. Oil's centrality within the 20th century, alongside the precarity of immediate futures inferred by ongoing speculation regarding peak oil, is an "inverse relationship." This relationship is one in which the causal entanglements of oil and geopolitics, neoliberalism and the temporal delineation of the Anthropocene, is at odds with the relative absence of oil in "cultural and social imaginaries," until recently.[12] Oil, with its "multiform liquidity and imbrication in networks of power," is all-pervasive materially and economically.[13]

For Rob Nixon, this pervasiveness is a marker of contemporary life—but life lived on "borrowed time."[14] This borrowed time is both deep geological time and the accelerated temporalities of future collapse. In recognition of modes of living built upon petroleum industries and enframed by the Anthropocene, this period is presented as a time of petromodernity. In the words of Stephanie LeMenager, this time is one in which we are experiencing "every day in oil, living within oil, breathing it and registering it with our senses."[15] The notion of "living oil" is understood as inhabiting economic, cultural, social, and environmental structures oriented around oil itself. However, this "living oil" reflects also the realities of living with the presence of oil as inscribed within the body. The impact of "oil-based catastrophes is not confined to an initial event but unfolds in space and time in often imperceptible ways."[16] For example, we see this not just in the initial flow of oil and mineral sands from the 2010 *Deepwater Horizon* disaster, but the subsequent destruction wreaked upon the Louisiana coastline and the slow, catastrophic violence of that suspended indefinitely within the water column of the Gulf of Mexico; the impacts upon those whom the Gulf sustained, economically and in other ways; and extending even to the affective resonance of the images that circulated following the disaster, of sea life caked in thick, black sludge, and their worldwide reach.

But the presence of oil seeps deeper, impacting the human beyond the emotional response. A 2019 study found that humans ingest upwards of 50,000 particles of microplastics every year.[17] These oily, petrochemical presences linger within the body, triggering immune reactions as they embed in tissues. As paleontologist Neil Schubin states, "If you know how to look, our body becomes a time capsule that, when opened, tells of critical moments in the history of our planet and of a distant past in ancient oceans, streams, and forests."[18] This time capsule of the body is a register of present trace markers; the plastics and toxins, the registers of industry and capitalist outputs, but subsequently also a record of the passage and disruption of time. It is not only in the soft registers of the body that these microplastics are found, but the far reaches of ecology and natural systems as well.

Fragmentation and storage: from plastic to microplastic

Microplastics are defined as plastic particles up to 5mm in diameter, either primary (microbeads, fibers, or "nurdles," lentil-sized balls used to make plastic objects) or secondary (the fragmentation of larger pieces of plastic through wear, damage, or UV impact). Microplastics, in their form, ability to permeate, and spread, challenge boundaries of self and other, environment and invader, organic and inorganic, whilst troubling temporal framings due to their ability to outlast. Their time-scales are wracked by an uncertain longevity, but one which extends beyond finite human lifespans. Some studies suggest this can range from approximately 100 years to 5,000 years, depending on the form the plastic takes.[19] The persistence of microplastics within the human body speaks to—however protracted—a geologic presence within the body, but the body is also a product of watery origins and evolution. We understand life as emerging from the sea, and, following the words of Rachel Carson, "the sea's first children lived on the organic substances then present in the ocean waters, or, like the iron and sulphur bacteria that exist today, lived directly on inorganic food."[20] What once seemed to be relatively clear evolutionary and temporal pathways become troubled and unclear; organic substances that once nourished become increasingly constituted, through the bioaccumulation of plastic, by their inorganic counterparts. The human, envisioned as the top of the food chain, brings back into itself the waste it has sent out into the world through the consumption of seafood, and indirectly communes with the earliest of lifeforms, those bacteria that lived on the inorganic.

Through the accumulation of microplastics in the body, we enter here into a consideration of the depth of Anthropocentric/anthropoceanic activity's penetration into the ecologic, enframed by our petromodern enclosure. In speaking to the holding of time that this accumulation represents, we see the body acting as archive. This is a notion that extends beyond the human, however, to encompass the natural world at large. Within the frame of the Anthropocene, the notion of the archive is expanded; the ur-marker of time, the stratigraphic layers, becoming forever changed by direct and forceful human intervention. Through activities including but not limited to resource extraction, we have become the first lifeform to "contemporaneously communicate with geological time": not only extending the extractivist arm of capital and industry into the earth, but leaving traces of human activity within geologic strata via the embedding of plastic waste within the layers.[21] Indeed, as Zalasiewicz et al. state, plastics are now starting to be used as "stratigraphic markers" in the conduct of archaeological field practice, because of the relative dating information their presence offers.[22] As we look to the ocean, however, we encounter a volumetric space that lacks this fixed stratigraphy. There is a temporal flattening that occurs despite the ocean's depth, as objects and marine life—all registers of being—are thrown together; a gathering of matter that is exacerbated as rising temperatures trigger the rapid transformation of the solid into the fluvial. As we traverse these full waters, however, we move through the ocean as a temporally complicated vessel, arriving at a rapidly destabilized register of fluvial history and present—sea ice.

The unstable stratigraphy of ice

The rise of global temperatures has unquestionable impacts upon environments and ecosystems. A hyper-visible casualty of these growing increases are conglomerations of sea ice in the Arctic and Antarctic. Permanent or multi-year sea ice is thicker and stronger than seasonal sea ice, which freezes and melts in response to climatic shifts between winter and summer months. More than 70 percent of current sea ice is now seasonal, susceptible to summer melts. Permanent sea ice has reduced by 50 percent in the years between 2002 and 2017, drastically reshaping Arctic and Antarctic environments, a reduction directly linked to the effects of climate change.[23] Recent studies have recorded a decrease in ice sheet thickness by two meters, or some 66 percent occurring with each successive melt season, demonstrated over a period of six decades.[24] In light of this, sea ice becomes a troubled stratigraphic site, positioned as an increasingly unstable time capsule as its melting triggers the release of that which was captured within—most commonly CO_2 gases, but increasingly microplastics as well.[25] In a study conducted between 2011 and 2013, scientists tracked the movement of plastics through Arctic waters by taking core samples. Through their mapping, they were able to posit that certain areas of sea ice in the Arctic operate as a type of sink, findings which were supported by further studies in 2013/14, 2016, and 2018.[26] Studies conducted in the Antarctic suggest that this is not a problem limited to the Northern Hemisphere, however, with microplastics being found in sea ice samples from as early as 2009.[27] Rather than being a remote site, far removed from the impacts of human waste, these locations operate as containers for this waste through the flows of currents from the North Atlantic to the Greenland and Barents Seas.[28] Whilst the sea ice here operates as a sink, it is an ephemeral one, moving from oceanic fluviality to stratigraphic register as it freezes, before becoming fluid once more. We encounter here also a collision of temporalities, made visible through this capture of microplastics in sea ice. The temporal register of stratigraphy collides with the longue durée of plastic, set against the ever-accelerating rate of melt within sea ice. Whilst this is a process that occurred prior to the advent of the Anthropocene, the rapidity with which melting occurs now aligns with what Yusoff terms the "forceful" use of time amidst the Anthropocene, and the bearing down of climate change as the actual, the real, and the present.[29]

Despite the increasing destabilization of sea ice by rapidly rising temperatures, its unstable stratigraphy is a readable archival marker within the present. This readability, and the archival impulse in relation to the presence of microplastics within sea ice, can be mapped across multiple trajectories. One of these trajectories relates to the inherent *plasticity* of plastic objects allowing them to take many forms; a human-industrial composition of matter, which, as it enters into geologic, hydrologic, and bodily strata, becomes not fossil—but technofossil.[30] These technofossils express a durability beyond the organic, imbued with the ability to persist within the strata as Anthropocene trace markers, in which the Plastic Age may last, fossilized within the earth, for millennia.[31] Technofossils form part of the "planetary archive," a register of human activity as inscription upon and within the Earth and its systems. Every fragment of plastic contains the geologic memory of the planet—a reflection of Derridean archive fever in the extent to which they transcribe the present into the past, retaining the traces for unstable but insistent and looming futures.[32]

Archive fevers in the floes

This notion of the archive, between the fluvial and geologic, is no new thing—rather a way of thinking through the Earth and strata that has persisted for hundreds of years.[33] The stratigraphic layers of the Earth are markers of Deep Time, in which histories may be "'read' in the succession of fossils embedded in the strata."[34] This stratigraphy is troubled by the Anthropocene, and troubled too as stratigraphy is read through increasingly unstable ice registers. The Anthropocene force at work, and its extension via entanglements of plastics and petromodernity, is representative in some ways of the Derridean concept of the "archiviolithic"; the death drive is at play within the archive's own structures, destroying the stratigraphy of the sea ice at the same time as it seeks to embed them within.[35] But it is not truly archiviolithic—there is a slight divorce from the "lithic" suffix, which speaks also to lithic or geologic matter, as we read plastics in sea ice. It is instead a troubled form of this archiviolithic impulse, as the true death drive seeks to leave no trace, no monument nor marker. But looking to the global proliferation of microplastics, these traces may be read instead as a monument dispersed. When Derrida writes that "It is the future which is at issue here," we see within the frames of our accelerated contemporary that it is the future which is at *stake*. When Derrida then locates the archive as an "irreducible experience of th[is] future," we read this as living amidst the irreducible experience of the future's uncertainty.[36]

A key experience of the Anthropocene is that of solastalgia, or petromelancholia. These contemporary theories explore the impact that collapsing climates have on the witness, and the affective impacts the horrors of the speculative future have on the human psyche.[37] There is a resonance within these theories, of the ghostly life cycles of plastics; emerging from oil, embedding in the body, and finding life in oceanic strata, reinforcing Laurel Peacock's recognition that one of the great difficulties of the Anthropocene is the struggle to "continue to distinguish human actor from passive environment."[38] As much as human activity intervenes into the environment, the environment also, quite literally, intervenes into the body. When we consider the physiological and affective dimensions of this, we recognize the world of wounds to which Aldo Leopold speaks when he delineates it as "one of the penalties of an ecological education."[39] Recognizing the temporal dimensions of this wounding as emerging uniquely through an Anthropocene context, we return to the archive. This irreducible experience of the future to which Derrida speaks is not simply an archive fever, as translated from the French into English, but *mal d'archive*—in which *mal* is etymologically rooted in an ill. Our ills here, manifest through the archives of the Earth's strata, the ice and the body, and our irreducible and uncertain experience of the future, operate as both literal and affective wound, as a sickness or illness which strikes both the human and the other than human. Indeed, as Heather Houser notes, it is through affect that we are able to arrive at an understanding of the "troubled interdependence of the individual body and large-scale environmental change."[40]

When we consider the present in relation to the past, we accept the trajectory that has led the world toward this place of crisis. As Colebrook acknowledges, the Anthropocene operates similarly to capitalism, wherein it "makes sense of its prehistory, enabling a universal and all-encompassing narrative that explains every other social form as its precursor."[41] Casting our gaze backward, we see all trajectories having led us into the Anthropocene, despite its relatively recent recognition. Indeed, as Timothy Morton suggests, "the end of the world has already occurred and it was [in] April [of] 1784, when James Watt patented the steam engine."[42] In acknowledging this epoch, we accept the futures wrought by these trajectories which color the past with urgency, accountability, and the blinding error of having taken a path which writes within it our own undoing. In his now ubiquitous adage, Frederic Jameson tells us that "it is easier to imagine the end of the world than to imagine the end of capitalism."[43] Within our Anthropocene context, the inseparability of these two shapes our future, colors our present, and reshapes our pasts.

Entering the mountains of madness

In laying out the foundations for the interrelations of plastic and sea ice, petromodernity and climate crisis, we enter into territory here resembling the strange and speculative fictions of H. P. Lovecraft. In perhaps his most famous novella, *At the Mountains of Madness*, Lovecraft recounts the tale of an ill-fated scientific expedition to Antarctica undertaken in 1930 in which the narrator, geologist William Dyer, and the expedition party hoped to obtain "as great as possible a variety of the upper fossiliferous rocks, since the primal life-history of this bleak realm of ice and death is of the highest importance to our knowledge of the earth's past."[44] As the party settles into their work routine, however, they find the first sign of the horrors that are to unfold: "a queer triangular, striated marking about a foot in greatest diameter," which the party puzzles over, this "radically unclassifiable organism of advanced evolution," unable to comprehend its presence within rock that is potentially pre-Cambrian.[45] The intrigue grows as, some time later, the party locates a cave within which are a range of specimens, some dating from as long as 300 million years ago and others as recent as 30 million years ago. Alongside those fossils that are recognizable, however, are strange remains that the party "cannot yet assign positively to animal or vegetable kingdom."[46] It is this discovery that leads to the evisceration of a large number of the expedition party, and to Dyer's entry into a "hideously amplified world of lurking horrors which nothing can erase from our emotions."[47] From reading on, we know that these are ancient creatures and that Dyer's journey into that world marks his entry into a somewhat spectral realm, the ancient and eldritch remains of a city created by creatures that are eternities old, but simultaneously far more advanced than humanity.

In apprehending Lovecraft's fiction here, we not only recognize the similarity between Dyer's "hideously amplified world" and the world of wounds to which Leopold spoke, but attempt to tease out further the affective aspects of the Anthropocene.[48] But further, we deploy this fiction to draw a parallel between the discovery of fantastic and horrifying entities that disrupt temporalities within the Antarctic ice and the lurking presence of microplastics. We will come to the limitless overwhelm of climate change, but here textual references and pop culture become a way to comprehend and to grasp something amidst the oily slick of looming catastrophe. Whilst there is an esotericism to Lovecraft, the lexicon of horror aligns with the apocalyptic nature of climate change discourse. It becomes a useful tool through which we can mediate emotional and cultural reverberations of the unstable futures ahead—particularly as we come to realize the monsters of the present and the future are ones that we have created. Horror has a contemporary resurgence, "as a narrative discourse through which to map the metamorphoses of present reality."[49] Indeed, as the novella draws to an end, Dyer posits the question: "Could one be sure of what might or might not linger even to this day in the lightless and unplumbed abysses of Earth's deepest waters? Those things had seemingly been able to withstand any amount of pressure—and men of the sea have fished up curious objects at times."[50] The sad thing is that we know what exists at those deepest of depths: plastic bags.[51] It is hard to imagine anything more horrifying, saddening, or shameful than that.

Lovecraft's horror can be aligned with John Carpenter's *The Thing* (1982). The Thing which haunts this film is an alien organism which, when released from Antarctic ice, changes its form to imitate and then assimilate human and other-than-human life in an act of destructive consumption—an enactment of the multiform liquidity we see in oil, particularly as it becomes plastic. Fisher provides to us not only a framing of this plastic *Thing*, as a "monstrous, infinitely plastic entity, capable of metabolizing and absorbing anything with which it comes into contact" akin to capital, but a hauntological lens through which we can read the presence of microplastics within sea ice.[52]

This framing of plastics enacts its oil-based liquidity; the petroculture snaking through water, geology, human body and other-than-human body, respectively. Drawing from Derrida, Fisher tells us that the future is "always experienced as a haunting; as a virtuality that already impinges on the present."[53] Our climate futures impinge hauntologically also, through projections of collapse, extinction, and an increasingly uninhabitable world. The haunting is a failed mourning, a specter which "will not allow us to settle into/for the mediocre satisfactions one can glean in a world governed by capitalist realism" and continually reminds us of the horrors ahead.[54]

Plastics from object to abject

As we struggle to comprehend the scale of climate crisis present and future, we engage with the frame of the hyperobject, as presented by Morton. Stretching into this future are plastics, with their long afterlives extending centuries beyond our own lifetimes. Viewed through waste, the hyperobject distends, becoming hyper*abject*—"a planetary infrastructure of waste . . . [defined by] the clogging of economic and ecological circulations."[55] Through the specters of the hyperobject and hyperabject, the present is shadowed over by the looming specter of a future of collapse and catastrophe. As Kristeva explains it, the abject delineates a "topology of catastrophe," one which "notifies us of the limits of the universe."[56] We are haunted by the encroachment of these specters, which Andreas Malm describes as "the rolling invasion of the past into the present," but our pasts are no longer just the end of the world which was wrought by the invention of the steam engine, but the petromodern present within which we are now enframed, and the futures of catastrophe which await us.[57]

Malm is speaking to the slow cancellation of the future and the shrinking of temporalities as they bleed into each other—the horror of inescapable pasts and impossible futures, as they emerge through Fisher's key concept of capitalist realism.[58] For Fisher, capitalist realism is, in its very base sense, the inability of the world to imagine ways of being outside of the grasp of capital, drawing from the work of Jameson. The "Real" of this capitalism is a "traumatic void," an opening through which reality breaks to reveal the facade of capitalism as a so-called "fantasy structure"—a curtain obscuring the consequences of unbridled capitalist expansion upon the finite resources of the Earth.[59] But further, both metaphorically and materially significant is the plasticity of capital, its ability to reshape and reform in order to persist eternally, without possibility of escape from its stranglehold upon life, industry, and being.

Capital, particularly as framed in the work of Fisher, spills out and over the everyday: "an abstract parasite, an insatiable vampire and zombie-maker."[60] In apprehending the political and affective impacts of this state of capitalist realism, Fisher furthers Deleuze and Guattari's framing of schizophrenia as a condition marking the edges of capitalism, and presents depression as a collective experience of this "abstract parasite."[61] For Fisher, we are zombies, deadened and necrotized by the structures of capital that disable normative affective responses to the everyday. In his work, capital wears down the individual—in essence, the life gets sucked out of them, leaving them beholden to the forces of capitalism. These psychological-political framings are an affective force, a Derridean hauntology, which Fisher extends toward a "hauntological melancholia" in which the effects of capitalist realism, deeply entwined with the specters of future, "will not allow us to settle into/for the mediocre satisfactions one can glean in a world governed by capitalist realism."[62] Instead, these specters haunt, their depressive forces pressing down on the present. The zombies and specters emerge in plastic forms too, as reanimated fossil fuels and as the slick, multiform liquidity of petrochemicals. This plastic infection courses into waterways, once-pristine landscapes, and flesh—with nothing seemingly able to escape its grasp. A 2019 oceanographic expedition, the Northwest Passage Project, found that core samples taken from melting sea ice in the Barrow Strait in Canada's Arctic waters contained colored markings—revealed to be plastic threads, nurdles, and microbeads. In reflecting upon this discovery, a member of the team remarked, "You'd assume that the Arctic is pristine and unaffected by the pollution that is occurring in other parts of the world. But that is clearly not the case."[63] As the true pervasiveness of microplastics is uncovered, their embeddedness reveals an agential global network of commodities—a spread beyond that ever thought possible. It is through sampling these stratigraphies of sea ice that we recognize the reach of this agentic matter.

Initially, plastics were naturally derived and not heavily processed; it was the treatment of cotton fiber cellulose with camphor that produced the first, in 1869. The first "modern" and fully synthetic plastic, Bakelite, wasn't produced until 1907. From here, however, we see the growth of plastics occurring alongside the development and spread of capitalism as these new materials enabled rapid and expansive production, providing a cheaper alternative to natural materials such as wood, paper, or glass. As Peter Ryan demonstrates in his brief survey of marine litter research, an inference can be drawn that links the post-World War II economic boom to the discovery of plastics within the ocean.[64] From the late 1950s, records emerged of turtles ingesting plastic bags, a trend which continued throughout the 1960s as studies emerged from all corners of the globe detailing the ingestion of plastics by marine life: plastics were found in birds' stomachs, seals were found entangled in nets, and fish were found trapped in rubber bands.[65] As the global economy grew, the proliferation of plastics enabled this growth, becoming the physical manifestation of the invisible networks and supply chains that crisscrossed the globe: markers of production, consumption, and the rapidity with which these occurred.

This current context, in which the mapping of these trace markers of activity becomes necessary for the comprehension of the true impact of human activity, leads us toward reflections of the intra-action of matter between the boundaries of human and other-than-human. This is a New Materialist framework in which the vibrancy of matter, its agency and vitality, is apprehended as a means of countering a purely anthropocentric world view. The liveliness that Fisher ascribes to capital by comparing it to Carpenter's "Thing" has been recognized as a key to apprehensions of the nature of matter itself. Indeed, Jane Bennett recognizes this in her interrogations of the vibrancy of other-than-human matter as she imbues it with "thing-power," a power which "gestures toward the strange ability of ordinary, man-made items to exceed their status as objects and to manifest traces of independence of aliveness, constituting the outside of our own experience."[66] In resurrecting the fossil of fossil fuels and reanimating it through the capitalist machine, we give it new life as plastic and must accept our responsibility for the dormant forces we have awoken.

Anthropocene archives: self-archiving systems of capital

We mobilize here an example—a polyester fleece sweater, cheap and readily available from a department store—to draw in the further contours of our argument. Studies indicate that fibers shed from clothing in the washing process are one of the largest sources of microplastics found in the ocean, with polyester fleece being a particularly notorious source.[67] This sweater is a good, one that circulates through the systems of capital. Manufactured cheaply somewhere in Bangladesh, or China, or Vietnam, these items are produced in large volumes, flooding shelves with clothing available at affordable prices. When this sweater is purchased, taken home, and worn, it then needs to be washed. As plastic microfibers unravel from the sweater in the wash, they filter down drains, disappearing into the ocean—gathering in sinks and traveling to the far reaches of the world, whereby they become embedded. They, as markers of pervasive capital, engage in a line of flight across the ocean, slipping through the gaps of any net that may be drawn across the seas.

Following our line of inquiry through Bennett and Fisher, we posit that having been manufactured and processed, they become engulfed within the Carpentarian *Thing*—this metastasizing system of capital—although the fibers shed in the washing process have escaped, in essence, from the cycle of capital, production, and commodification. However, by virtue of entering into it, they have become subject to and eternally changed by the zombification process Fisher writes of; they are now forever markers of capital. Adrift at sea, the microplastics are lost: no longer natural, yet beholden to the broader forces that order the natural world. Like the zombie driven by base impulse, wandering mindlessly looking for food, microplastics drift until some are eventually encased within sea ice. As we recognize increasing awareness of the ocean-as-destination for many microplastics, we suggest that we can read this process as a way of capital archiving itself within the natural world.

To provide another example of this archiving of capital, we return here to the idea of the technofossil. As Pam Longobardi writes, "Plastic objects are the cultural archaeology of our time, a future storehouse of oil, and the future fossils of the Anthropocene . . . at one end, raw, telluric matter, at the other, the finished, human object."[68] Within this Anthropocene framing, we see the full commoditization of the globe by capital—the market no longer stretching over the globe, to engulf it, but to penetrate it, as well. The strata is plumbed, only to be replaced by a Baudrillardian simulacrum of what it once was. No stone is unturned nor unplasticized. This transformation of matter from organic to human-machinic-composite in plastics positions them as plastiglomerate, which, while now a curiosity and a worrying indicator, will one day be all that is left behind, waiting to be discovered. Even before they undergo the process of becoming refossilized and transform back into carbon, plastics will exist amidst a timescale far beyond human existence—and, as in *At the Mountains of Madness*, their discovery will speak to an advanced and ancient civilization, but one steeped within unimaginable horrors. Like the Elder Ones creating the shoggoths as a servitor race, only to have them rebel against them, we have created plastic to aid in the development of our civilization, only to have it escape from our control.

Conclusion

Donna Haraway and others position the Anthropocene as a boundary event and threshold that we are passing through on our way into the future and whatever comes next—whether it be new kinship structures which recognize and revere the other-than-human, or the terraforming of Earth toward more habitable structures.[69] The reality of our current state, however, is that certain effects of the Anthropocene are irreversible, among which we include the proliferation of microplastics. As we draw our nets across the ocean, microplastics slip through the gaps, and the fibers of the net too unravel and mix in with the water—ferried toward sea ice structures which form unstable archives of petromodern pasts, presents, and futures. Plastics, in their spread across the globe, do, and will continue to, rewrite and replace the organic. Through this chapter, we have sought to outline a speculative engagement with microplastic presences in sea ice. Employing Fisher's notion of capitalist realism, we posit that metaphorically and materially significant is the plasticity of capital, its ability to reshape and reform in order to persist eternally, as rendered visible through the endless and dizzying array of plastics that exist solely through machinations of capital. Reading the twinned development of plastics and capital, we suggest that perhaps more appropriate and disturbing is the *microplasticity* of capital, given scientific findings about the presence of plastics within organic material and the most distant of sea ice—its ability to permeate and become embedded within organic material as it breaks up and traverses the globe. Following Fisher's line of Carpentarian suggestion, we view these plastics through agentic and New Materialist lenses to see an Anthropocene archival presence within sea ice; the technofossils amidst unstable stratigraphies revealing to us the tangled futures of capital and nature, beyond human presence.

Notes

1. Carl Safina, "The World's Imperiled Fish," *Scientific American* (November 1995): 59.
2. Roland Geyer, Jenna R. Jambeck, and Kara Lavender Law, "Production, Use, and Fate of All Plastics Ever Made," *Science Advances* 3, no. 7 (2017): 1; Plastics Europe, "Plastics—the Facts 2019," information sheet, https://www.plasticseurope.org/application/files/9715/7129/9584 /FINAL_web_version_Plastics_the_facts2019_14102019.pdf.
3. Max Liboiron, "Redefining Pollution and Action: The Matter of Plastics," *Journal of Material Culture* 21, no. 1 (2016): 103.
4. Paul J. Crutzen and Eugene F. Stoermer, "The Anthropocene," *IGBP* [International Geosphere-Biosphere Programme] *Newsletter* 41 (2000): 17; quoted in Dipesh Chakrabarty, "The Climate of History: Four Theses," *Critical Inquiry* 35, no. 2 (2009): 209, doi: 10.1086/596640.
5. Christopher M. Free et al., "Impacts of Historical Warming on Marine

Fisheries Production," *Science* 363, no. 6430 (2019): 979, https://doi:10.1126/science.aau1758; Julie Hunter, Pradeep Singh, and Julian Aguon, "Broadening Cultural Heritage: Addressing Gaps in the Deep Sea Mining Regulatory Regime," *Harvard Environmental Law Review*, last modified April 16, 2018, https://harvardelr.com/2018/04/16/broadening-common-heritage/.

6. In a presentation to the United Nations, the then Chair of the World Economic Forum's Global Agenda Council on Oceans directly links the Fourth Industrial Revolution to the plastic industry; see Nishan Degnarain, "The Fourth Industrial Revolution and the New Plastics Economy" (presentation, repr., UN Consultative Process on Oceans and the Law of the Sea, New York, 2016). The blue economy speaks to an interaction in the world's ocean between economic growth and sustainability; see Jean-Baptiste Jouffray et al., "The Blue Acceleration: the Trajectory of Human Expansion into the Ocean," *One Earth* 2, no. 1 (2020): 43–54, https://doi:10.1016/j.oneear.2019.12.016.

7. University of Leicester, "Human Impact Has Created a 'Plastic Planet', Research Shows," 2017, https://www2.le.ac.uk/offices/press/press-releases/2016/january/human-impact-has-created-a-2018plastic-planet2019-research-shows.

8. Liboiron, "Redefining Pollution and Action," 100; see also Yukie Mato et al., "Plastic Resin Pellets as a Transport Medium for Toxic Chemicals in the Marine Environment," *Environmental Science & Technology* 35, no. 2 (2001): 318–24, https://doi:10.1021/es0010498; Yuko Ogata, "International Pellet Watch: Global Monitoring of Persistent Organic Pollutants (Pops) In Coastal Water," *Marine Pollution Bulletin* 58 (2009): 1437–46, https://doi:10.1016/j.marpolbul.2009.06.014; Christiane Zarfl, Eberhard Karls, and Michael Matthies, "Are Marine Plastic Particles Transport Vectors for Organic Pollutants to the Arctic?" *Marine Pollution Bulletin* (2010):1810–14, https://doi:10.1016/j.marpolbul.2010.05.026.

9. Patricia L. Corcoran, Charles J. Moore, and Kelly Jazvac, "An Anthropogenic Marker Horizon in the Future Rock Record," *GSA Today* 24, no. 6 (June 2014): 4–8, https://doi:10.1130/GSAT-G198A.1; Patricia L. Corcoran and Kelly Jazvac, "The Consequence That Is Plastiglomerate," *Nature Reviews Earth and Environment* 1 (2020): 6–7, https://doi:10.1038/s43017-019-0010-9.

10. Jenna R. Jambeck et al., "Plastic Waste Inputs from Land into the Ocean," *Science* 347, no. 6223 (2015): 768, https://doi:10.1126/science.1260352.

11. Kate O'Neill, *Waste* (Cambridge: Polity, 2019), 127.

12. Imre Szeman, Sheena Wilson, and Adam Carlson, "On Petrocultures: Or, Why We Need to Understand Oil to Understand Everything Else," in *Petrocultures: Oil, Politics, Culture*, ed. Sheena Wilson, Adam Carlson, and Imre Szeman (Montreal: McGill-Queen's University Press, 2017), 5.

13. Stephanie LeMenager, *Living Oil: Petroleum Culture in the American Century* (New York: Oxford University Press, 2014), 185.
14. Rob Nixon, *Slow Violence and the Environmentalism of the Poor* (Cambridge, MA: Harvard University Press, 2011), 69.
15. Ibid. 6.
16. Ibid. 71.
17. Kieran D. Cox et al., "Human Consumption of Microplastics," *Environmental Science & Technology* 53, no. 12 (2019): 7068, https://doi:10.1021/acs.est.9b01517.
18. Neil Schubin, "Recognizing Your Inner Fish," interview by Andrea Seabrooke, radio (NPR, 2008).
19. Ali Chamas et al., "Degradation Rates of Plastics in the Environment," *ACS Sustainable Chemistry & Engineering* 8, no. 9 (2020): 3494, https://doi:10.1021/acssuschemeng.9b06635.
20. Rachel Carson, *Sea around Us* (New York: Open Road Media, 2011), 117–18.
21. Robin Mackay, "A Brief History of Geotrauma," in *Leper Creativity: Cyclonopedia Symposium*, ed. Edward Keller, Nicola Masciandaro, and Eugene Thacker (Brooklyn: Punctum Books, 2012), 29.
22. Jan Zalasiewicz et al., "The Geological Cycle of Plastics and Their Use as a Stratigraphic Indicator of the Anthropocene," *Anthropocene* 13 (2016): 8, https://doi:10.1016/j.ancene.2016.01.002.
23. Richard Kwok, "Arctic Sea Ice Thickness, Volume, and Multiyear Ice Coverage: Losses and Coupled Variability (1958–2018)," *Environmental Research Letters* 13, no. 10 (2018): n.p., https://doi:10.1088/1748-9326/aae3ec.
24. Ibid.
25. Guillaume Lamarche-Gagnon et al., "Greenland Melt Drives Continuous Export of Methane from the Ice-Sheet Bed," *Nature*, 2019: 565, https://doi:10.1038/s41586-018-0800-0.
26. Nicholas-Xavier Geilfus et al., "Distribution and Impacts of Microplastic Incorporation within Sea Ice," *Marine Pollution Bulletin* 145 (2019): 463–73, https://doi:10.1016/j.marpolbul.2019.06.029; La Daana K. Kanhai et al., "Deep Sea Sediments of the Arctic Central Basin: A Potential Sink for Microplastics," *Deep Sea Research* Part I: Oceanographic Research Papers 145 (2019): 137–42, https://doi:10.1016/j.dsr.2019.03.003; La Daana K. Kanhai et al., "Microplastics in Sea Ice and Seawater beneath Ice Floes from the Arctic Ocean," *Scientific Reports* 10, no. 1 (2020), https://doi:10.1038/s41598-020-61948-6; Ilka Peeken et al., "Arctic Sea Ice is an Important Temporal Sink and Means of Transport for Microplastic," *Nature Communications* 9, no. 1 (2018), https://doi:10.1038/s41467-018-03825-5.
27. Anna Kelly et al., "Microplastic Contamination in East Antarctic Sea Ice," *Marine Pollution Bulletin* 154 (2020), https://doi:10.1016/j.marpolbul.2020.111130.
28. Andrés Cózar et al., "The Arctic Ocean as a Dead End for Floating

Plastics in the North Atlantic Branch of the Thermohaline Circulation," *Science Advances* 3, no. 4 (2017): 1, https://doi:10.1126/sciadv.160 0582.

29. Kathryn Yusoff, "Geologic Life: Prehistory, Climate, Futures in the Anthropocene," *Environment and Planning D: Society and Space* 31, no. 5 (2013): 779, https://doi:10.1068/d11512.

30. Amanda Boetzkes, "Plastic Vision and the Sight of Petroculture," *Petrocultures: Oil, Politics, Culture*, ed. Imre Szeman, Sheena Wilson, and Adam Carlson (Kingston: McGill-Queen's Press: 2017), 239; see also Jan Zalasiewicz et al., "The Technofossil Record of Humans," *The Anthropocene Review* 1, no. 1 (2014): 34–43, https://doi:10.1177/2053 019613514953.

31. Jan Zalasiewicz, "The Geological Cycle of Plastics and Their Use as a Stratigraphic Indicator of the Anthropocene," *Anthropocene* (2016): 4, https://doi:10.1016/j.ancene.2016.01.002.

32. Tom Chadwick and Pieter Vermeulen, "Literature in the New Archival Landscape," *Lit: Literature Interpretation Theory* 31, no. 1 (2020): 3, https://doi:10.1080/10436928.2020.1712793.

33. *Torbern Olof Bergman, Physical Description of the Earth (Uppsala, 1766)*, quoted in David Sepkoski, "The Earth as Archive: Contingency, Narrative, and the History of Life," *Science in the Archives: Pasts, Presents, Futures*, ed. Lorainne Daston (Chicago: University of Chicago Press, 2017), 59.

34. Sepkoski, "The Earth as Archive," 57.

35. Jacques Derrida, *Archive Fever: A Freudian Impression* (Chicago: The University of Chicago Press, 1996), 14.

36. Ibid. 45.

37. *Solastalgia*, as presented by Glenn Albrecht, is a neologism which interrogates the experience of climate change in affective and psychological frames of distress and mourning. Glenn Albrecht et al., "Solastalgia: The Distress Caused by Environmental Change," *Australasian Psychiatry* 15, no. 1 (2007): 95–8, https://doi:10.1080/10398560701701288. *Petromelancholia* emerges from notions of peak oil and petroculture, speculating on the destructive impacts and oil industries, particularly in relation to this becoming-scarce resource. LeMenager, *Living Oil*, 102.

38. Laurel Peacock, "SAD in the Anthropocene: Brenda Hillman's Ecopoetics of Affect," *Environmental Humanities* 1, no. 1 (2012): 86, https://doi.org/10.1215/22011919-3609985.

39. Aldo Leopold, *Writings: Round River* (Oxford: Oxford University Press), 168.

40. Heather Houser, "Wondrous Strange: Eco-Sickness, Emotion, and The Echo Maker," *American Literature* 84, no. 2 (2012): 385, https://doi:10.1215/00029831-1587386.

41. Claire Colebrook, "'A Grandiose Time of Coexistence': Stratigraphy of the Anthropocene," *Deleuze Studies* 10, no. 4 (2016): 441, https://doi:10.3366/dls.2016.0238.

42. Timothy Morton, *Hyperobjects: Philosophy and Ecology after the End of the World* (Minneapolis: University of Minnesota Press, 2013), 7.
43. See Fredric Jameson, *The Seeds of Time* (New York: Columbia University Press, 1994), xii; and "Future City," *New Left Review* 21 (May/June 2003): 76. The original association of the end of capitalism and the end of the world is to be found in H. Bruce Franklin, "What Are We to Make of J. G. Ballard's Apocalypse?" Franklin accuses Ballard of "mistaking the end of capitalism for the end of the world" and asks in conclusion, "What could Ballard create if he were able to envision the end of capitalism as not the end, but the beginning, of a human world?"
44. H. P. Lovecraft, *At the Mountains of Madness* (New York: Norton, 1997), 615–16.
45. Ibid. 620.
46. These fossils are found to be the remains of Elder Ones, an ancient alien civilization that recurs frequently throughout Lovecraft's work. Ibid. 631.
47. Ibid. 638.
48. Ibid. 638.
49. Sarah Dillon extends the work of Veronica Hollinger from science fiction in 2002 to horror in "The Horror of the Anthropocene," *C21 Literature: Journal of 21st-Century Writings* 6, no. 1 (2018): n.p., https://doi.org/10. 16995/c21.38.
50. Lovecraft, *At the Mountains of Madness*, 685.
51. Sarah Gibbens, "Plastic Bag Found at the Bottom of World's Deepest Ocean Trench," *National Geographic*, 2019, https://www.nationalgeog raphic.org/article/plastic-bag-found-bottom-worlds-deepest-ocean-trench/ #:~:text=But%20if%20you%20thought%20the,feet.
52. Mark Fisher, *Ghosts of My Life: Writings on Depression, Hauntology and Lost Futures* (London: Zero Books, 2014), 10.
53. Mark Fisher, "Hauntology," *Film Quarterly* 66, no. 1 (Fall 2012): 16.
54. Fisher, *Ghosts of My Life*, 22.
55. Mikkel Frantzen and Jens Bjering, "Ecology, Capitalism and Waste: From Hyperobject to Hyperabject," *Theory, Culture & Society* (2020): 3, https://doi:10.1177/0263276420925541.
56. Julia Kristeva, *Powers of Horror: An Essay on Abjection* (Columbia: Columbia University Press, 1982), 9–11, quoted in Frantzen and Bjering, "Ecology, Capitalism and Waste," 14.
57. Andreas Malm, *Fossil Capital* (London: Verso Books, 2016), 10.
58. Fisher, *Ghosts of My Life*, 5.
59. Mark Fisher, *Capitalist Realism* (London: Zero Books, 2009), 19.
60. Ibid. 15.
61. Ibid. 35.
62. Fisher, *Ghosts of My Life*, 22.
63. Ed Struzik, "A Northwest Passage Journey Finds Little Ice and Big Changes," *Yale Environment* 360 (2019), https://e360.yale.edu/features /a-northwest-passage-journey-finds-little-ice-and-big-changes.

64. Peter G. Ryan, "A Brief History of Marine Litter Research," in *Marine Anthropogenic Litter*, ed. Melanie Bergmann, Lars Gutow, and Michael Klages (Berlin: Springer Nature, 2015), 2–3.

65. S. H. Cornelius, "Marine Turtle Mortalities along the Pacific Coast of Costa Rica," *Copeia* (1975): 186–7; Ryan, "Marine Litter Research," 4–5.

66. Jane Bennett, *Vibrant Matter: A Political Ecology of Things* (Durham, NC and London: Duke University Press, 2010): 2.

67. Francesca De Falco, Emilia Di Pace, Mariacristina Cocca et al., "The Contribution of Washing Processes of Synthetic Clothes to Microplastic Pollution," *Scientific Reports* 9, 6633 (2019): n.p., https://doi.org/10.10 38/s41598-019-43023-x; Edgar Hernandez, Bernd Nowack, and Denise M. Mitrano, "Polyester Textiles as a Source of Microplastics from Households: A Mechanistic Study to Understand Microfiber Release During Washing," *Environmental Science Technologies* 51, no. 12 (2017): 7036–46, https://doi.org/10.1021/acs.est.7b01750; Imogen E. Napper and Richard C. Thompson, "Release of Synthetic Microplastic Plastic Fibres From Domestic Washing Machines: Effects of Fabric Type and Washing Conditions," *Marine Pollution Bulletin* 112 (2015): 39–45, https://doi.org/10.1016/j.marpolbul.2016.09.025.

68. Pam Longobardi, "The Ocean Gleaner," *Drain (Junk Ocean)* 13, no. 1 (2016): n.p.

69. Donna Haraway, "Anthropocene, Capitalocene, Plantationocene, Chthulucene: Making Kin," *Environmental Humanities* 6 (2015): 160.

6

Jugoplastika: Plastics and Postsocialist Realism

Andrija Filipović

Several times each year, especially during the ever-warmer summer months, there are reports of spontaneous combustion at the landfill site in Vinča. Vinča is the largest landfill in Serbia, located near Belgrade, the capital city. The site was created in 1977, during the era of the Socialist Federal Republic (SFR) of Yugoslavia. The commonest form of waste in Serbia is packaging waste, such as plastic bottles and bags, some 300,000 to 500,000 tons per year.[1] In considering plastic waste in general, one should also count other plastic objects in everyday use, given the ubiquity of plastic in the production of appliances and disposable domestic items. Given that recycling is almost non-existent in Serbia—according to some sources, 95 percent, 97 percent or even 99 percent of waste ends up in legal and illegal landfills or is dumped directly into rivers, lakes, or fields[2]—the short agential non-organic life of plastic objects requires further theoretical attention.

The creation of the Vinča landfill is inextricably linked to the so-called socialist consumerism of twentieth-century Yugoslavia, and today's Belgrade citizens drink, breathe in, and otherwise engage with the consequences of Yugoslav and Serbian production forms and consumer habits. In 2019 the European Bank for Reconstruction and Development (EBRD) approved 72.25 million euros' credit for a waste-to-energy incinerator plant that will heat apartment buildings and produce electricity.[3] Behind this ostensibly innocuous and positive news there lies a discourse of financialization of both natural resources and waste, revealing environmental management techniques based on the free-market economy. This amalgam of the post-Yugoslav transitional state and market liberalization—with its implications for environmental issues—I call postsocialist realism. Moreover, I will show that the ontology of plastics needs to be critically rethought in order to properly understand such an anthropogenic substance and the ways in which it interacts and intra-acts with humans and nonhumans alike.

Plastics need to be thought of as a pluritemporal and multimaterial hyperobject, as a *hyperplastic object*.

The title of this text hangs on a wordplay. Jugoplastika was a factory in Split, Croatia, built in 1952 during the era of SFR Yugoslavia; it produced plastic consumer goods. The factory's name of course also means *Yugoslav plastics*, and so I am using the name to signify Yugoslav and post-Yugoslav production, consumption, and disposal of plastic items. Many factories and products of the era featured this "Jugo-" prefix, which translated to English as "Yugo"—as in Yugo cars. There were also the factories Jugodom (for lumber and wood), Jugoprevoz (transport), the self-explanatory Jugopetrol, Jugoturbina (engines), Jugovinil (plastics, especially PVC), and so on. Therefore "Jugoplastika" is automatically associated with the SFR Yugoslavia. Following this double meaning, this text will look at the continuities and discontinuities of use and disuse of plastic objects, and, as will become apparent, the (dis)continuity of plastic itself is problematic from the ontopolitical point of view. A new ontology is needed for understanding the agency of plastic and the ways in which it acts and is acted upon, the ontopolitics of pluritemporality and multimateriality of the hyperplastic object. This text also calls attention to an omission in the environmental humanities and Anthropocene Studies, namely a paucity of investigations of Eastern Europe and the region's impact on the formation of this new geological epoch in which we supposedly find ourselves. Each country on the eastern side of Iron Curtain—and I would argue Yugoslavia in particular, given its special status after 1948 and Tito's famous "No" to Stalin—had a different effect on the environment and planet in general, via the planned economy, mega-infrastructure projects such as dams and surface coal seams, large-scale agro-industrial activity, etc. This text is, then, a part of Red Anthropocene studies to come, the study of the environmental impacts of historical socialism and communism on local, regional, and planetary scales, and the effects of these impacts we are living with today.

Jugoplastika: socialist consumerism and Yugoslav plastics culture

Jugoplastika, the largest Yugoslav factory producing plastic items for everyday use, was opened in 1952 in Split, Croatia. According to Pavičić, the Communist Party branch of Split decided to separate off part of Jugovinil, the plastics giant of the time, which made goods for mass consumption.[4] Furthermore, "in the 1950s, plastics had just begun to enter the modern world. Until that time, plastic coasters, toys, nylon and synthetic materials in general were still something exotic, something that smelled of the West and brought the aroma of modernization, if not Western decadence." Pavičić claims that the foundation of Jugoplastika met two strategic political goals of the 1950s: the mass employment of women, and economic growth through personal consumption.[5]

As far as the second goal is concerned, the Yugoslav economy prior to 1951 was not a consumerist one. It was based on *petoljetka*, a five-year plan whereby the state and the Party decided what would be produced, imported, and exported. In 1951, the Yugoslav Communist Party partially discarded the concept of the Soviet-type planned economy, and insisted ever more on free pricing: "Companies became more independent, and business tax credits for investment were introduced, making space for market liberalization."[6] These economic reforms, together with political decisions to turn away from the Soviet bloc and open up the federal state, and society in general, toward the West, led to the creation of a consumerist society in Yugoslavia. Both Pavičić and Dimitrijević deem 1958 to be the decisive year in this respect. As Pavičić notes, "[Just a few years later (1958),] the 7th Yugoslav party congress would declare in program directives that 'personal consumption needs to follow economic growth,' and that a goal of the Party is 'better supply of consumable goods.'"[7] Dimitrijević finds that this move away from the Stalinist *doxa* in arguments of the Party ideologues—who claimed that a socialist society is a consumerist society and a society of personal welfare—was a political and economic stance that remained constant throughout the Yugoslav self-governing socialist project.[8] Moreover, the first postwar decade was marked by industrialization and modernization which aimed to respond to "natural" needs for food, shelter, heating, and similar. Dimitrijević argues that the response to consumption needs was later, and that these diverted into two streams, the first being the need for more stylish clothing and the second the need for modern home appliances.[9]

According to Igor Duda's analysis, newspapers were significant in the creation of consumerism. Duda quotes a text from *Vjesnik* where the author claims that "symptoms of more modern spending are reflected in buying less food but more industrial products. We eat a bit less bread and meat, but the consumption of metal goods, vehicles, and electric appliances grew."[10] Newspapers also gave coverage to new products. In 1958, those were the electric blanket, electric sink, slippers with plastic added for durability, the mini-bar and automatic gramophone, and in 1963, sports textiles, spin dryers, imported shaving kits, tastefully packaged dried fruits, enamel saucepans with metal sieves, electric hotplates, plastic garbage bins.[11] These products for mass consumption were sold in supermarkets modeled on American lines. The first supermarket was opened as early as 1956 in Ivanc, near Zagreb, and followed by another one in Zagreb in 1957, and in Belgrade in 1958.[12] The opening of supermarkets was part of US anti-Soviet strategy during the Cold War, but regardless of motivation, the supermarket introduced the Yugoslav public to novel ways of packaging, displaying, and buying goods. The Belgrade supermarket was spread over 600 square meters, with twenty meters of refrigerators of packaged fresh meats, seven refrigerators of frozen food and ten meters of glass-fronted cabinets with fruits and vegetables. It also carried luxury drinks, pots and pans, socks. A particular novelty was individual packages of sugar, coffee, and rice.[13] Now the customers, mostly women, could touch the goods themselves instead of asking staff for the merchandise they wanted. This individualized shopping required shopping baskets, which were imported at first in 1958 from Italy because of the tight deadline for opening the supermarket, and domestic, Yugoslav producers could not develop production technology quickly enough.[14]

Darko Suvin claims that the Party program from 1958, in proclaiming care for everyday needs, strengthened the self-governing aspect of Yugoslav socialism: it marked the transformation of state bodies into self-governing bodies,[15] which is especially important considering Yugoslavia's distinctiveness from other Eastern, as well as Western, forms of government. Given this ideological, political, and economic framework, it is not surprising that by the end of the 1950s, as the buying power of Yugoslavs began to grow, spending also increased. Yugoslavs began to buy consumer goods which had been out of reach during the period of postwar renewal, such as motorcycles (in 1955 there were 12,541 registered, 37,649 in 1958), refrigerators (in 1959 40,000, compared to 15,000 just a year before), and television sets (6,000 in 1958 and 12,000 in 1959).[16]

Besides motorcycles, refrigerators, and televisions, there appeared from 1956 onward, as a result of an economic deal with West Germany, products from brands including Nivea, Schwarzkopf, and Labello.[17] In the late 1960s, others came to be present in the Yugoslav market, including Coca-Cola, Pepsi, Milka, Nestle, Thomy, Dr. Oetker, Tuborg, Converse, Helena Rubinstein, and Dior. Dimitrijević writes that "the new bourgeois class smoked Kent, Marlboro, Winston, Astor and Chesterfield cigarettes, and drank luxury alcoholic beverages—Johnnie Walker or Vat 69 whisky, Martel and Courvoisier cognac, Gordon's gin and Bacardi rum."[18] This new middle class, the so-called "red bourgeoisie," distinguished themselves by imitating the style and everyday habits of the Western middle class. By buying and consuming Western products—some unobtainable for other social groups, in a society that was supposed to be moving toward a classless society—this new middle class marked itself as different, though not by its closeness to the political power.

Mass production and mass consumption of cheaper products boomed in the postwar decades, so much so that a 1958 hit song by Zdenka Vučković had the lyrics: "Daddy buy me a car, bicycle, and romobil [kick scooter]/ buy me a bear and bunny, trolley Jugovinil/ Daddy, buy me cake, bonbons and oranges two/ at least one little baby, I say that's all to you!" Pavičić notes that it was not Jugovinil that produced the trolleys but Jugoplastika. Jugoplastika was replaced to fit the rhyme scheme. However, Jugoplastika was, elsewhere, not so easily replaceable. It was becoming an important player in economic, political, and even cultural areas of Yugoslav society. Besides producing plastic goods for domestic use, and therefore playing its part in the modernization of homes, Jugoplastika worked on cultural development—music, film, and sports—"for the working man to have space to entertain himself after the factory," as Pavičić puts it. The Jugoplastika basketball team (a professional team sponsored by Jugoplastika) won a number of Yugoslav cups in the 1970s, and several Yugoslav and European championships during the 1980s. It is still considered one of the all-time best basketball teams of the former Yugoslav republics. As noted above, among the reasons for opening Jugoplastika was increasing the employment of women. At the peak of production activity the factory employed 13,000 workers, two-thirds of whom were women, who "sewed, tailored, sprayed, managed the machines . . . they were in considerable numbers among designers, managers, directors."[19] Pavičić adds that the perceived gendered nature of work at Jugoplastika was one of the reasons for its demise in post-Yugoslav transitional times.

Jugoplastika not only played an important role in the emancipation of women and in supporting music, film, and sport. It also changed the way Yugoslavs consumed, especially at times when they had increasing amounts of money to spend. Pavičić writes: "At the height of its power, it was impossible to imagine the life of an average Yugoslav without Jugoplastika products. If you were an average resident of that country, you wore a Jugoplastika jacket. Children went to school with Jugoplastika bags, played with Jugoplastika dolls, wore Adidas shoes made under license at Jugoplastika. When they sat down in their Renault 4 or Golf automobile, Yugoslavs drove a car in which the hubcaps, doors, steering wheels and dashboards came from the company in Split's Brodarica neighborhood."[20] Jugoplastika's products were everywhere, and were for everyone. Hence the new middle class distinguished itself by consuming foreign products, while regarding domestic goods as cheap and tasteless. The functioning of this ideological work based on consumerism can be gleaned from Marina Vidas's 2009 short story "Poslanica modernoj ženi." The main character, Maja, recollects her past in the following way: "Maja was perhaps only now aware how jealous girls from school were of her. She wore 'Wranglers,' sweatpants in all colors and sneakers no one in school had. Everybody else wore the same blue sweatpants from Jugoplastika and black textile training shoes. She had a bicycle and a crying baby, and, later, Barbies."[21]

Jugoplastika was not the only producer of plastic toys; there was also the Biserka factory in Zagreb, Croatia. Biserka opened in 1946, initially making textile dolls filled with straw. In 1958 Biserka began the production of PVC dolls. It was the only factory outside the USA that was licensed to produce dolls of Disney characters.[22] Jugoplastika produced various small plastic toys, and in the period from 1981 to 1985, 60 percent of its toys were exported to the West.[23] Up to the late 1950s, Yugoslavs had made their own toys from available materials—wood, straw, wool, pieces of textile, and other discarded domestic materials.[24] These homemade toys of natural or recycled materials were rapidly replaced by plastic and rubber products from Jugoplastika and Biserka. They were cheap and popular, so much so that people today nostalgically remember the smell upon entering Jugoplastika stores, as does Silvija Šesto in her short magazine feature article: "In Ilica, not far from Gundulićeva crossroads, there used to be a small Jugoplastika store which sold the fabled plastic-rubber characters produced by the Biserka toy factory in Zagreb. I loved it when my mother took me there. The smell of rubber and plastic made me happy. Mom always bought me something because it was afford-able, even though our family never splurged money on toys."[25]

This nostalgia for Jugoplastika's products, the toys' affective charge, lies partly in the never-fulfilled promise of a socialist self-governing society of consumers that almost materialized during the late 1960s and 1970s. Edmund Stillman wrote in 1964 that Yugoslavia was a pleasant shock because it represented the first example of a Marxist society of plenty: "It is the first socialist state that enacts the politics of promoting material and immaterial pleasures of individuals more than just the politics of collective welfare."[26] A US citizen of Yugoslav extraction, Duško Doder, wrote about his bafflement when faced with the hybrid society that was Yugoslav self-governing socialism. He wrote that it was best understood as consumerist socialism, or *il socialismo borghese*, bourgeois socialism, when the new middle class was taken into account.[27] However, the promise was never fulfilled, and when Yugoslavia fell apart violently at the beginning of the 1990s, all the plastic waste from socialist consumerism had to go somewhere.

Post-Jugoplastika: plastic waste and postsocialist realism

In spring 2017, two events in Serbia highlighted the pervasive environmental legacy of SFR Yugoslavia, including plastic waste. First, there was an incident of spontaneous combustion at the aforementioned Vinča landfill site, several kilometers southeastward down the river Danube from Belgrade. Residents of several Belgrade neighborhoods voiced their complaints on social media and television news, about smoke entering their homes and even causing breathing problems. Second, a new chapter of negotiations for Serbian accession to the European Union, dealing with environmental preservation, was announced. The environmental legacy of the socialist Yugoslav state underlies both—and one novel way to examine that legacy is via the sense of smell. Analogous to the study of urban landscape and soundscape, is the urban *olfactoscape*. Belgrade's olfactoscape shows a whole assemblage of ways in which smells form contemporary urban life, and they can reveal its underpinnings. In this case, the smell of burning landfill reveals a crumbling infrastructure built during the socialist era and never properly looked after since. Much of the city's life is performed amid the ruins of Yugoslavia and its project to modernize and industrialize its republics—and amid the decades-old effects of socialist consumerism.

The Vinča landfill was opened at the peak of Yugoslavia's self-governing socialist society of plenty. It spreads over 68 hectares, with 2,700 tons of waste deposited daily.[28] It is located near the Danube, the longest river in Europe, and it is one of sixty-five landfills in Serbia which are sited in the direct vicinity of rivers.[29] As I have already noted, there are several estimates of how much of Serbia's waste is recycled, but the highest is only 5 percent. One can conclude that the greatest proportion of Belgrade's plastic waste ends up in Vinča or in one of the illegal landfills. The Vinča site is not properly taken care of, as evidenced by its propensity for combustion, nor is the waste treated in any way. The most problematic and most widely reported combustion was in 2017, when the fire burned for more than two months. A story from news portal *Blic* gives an account of the fire at Vinča, along with testimonies from residents of the village of the same name, near the landfill. People complained about breathing issues, about red, watery eyes, and about fruit trees ruined by ash falling from the sky.[30] A report titled "Thick smoke over Belgrade: Landfill in Vinča has been burning for 20 days" explained that the fire was hard to extinguish because, as the deputy mayor said, "it is located at a depth of several tens of meters, in the part of the landfill with landslides ... There is a danger of air pockets and that people and machines could fall in, so it is not possible to extinguish the fire with the usual method, that is with water, but with putting earth over it."[31] It is also noted that particular attention was paid to the municipalities of Karaburma, Višnjica, Višnjička Banja, Lešće, Rospi Ćuprija, parts of Krnjača, Kotež, Slanci, Mirijevo, and Zvezdara, as the smoke reached these most often. In another story, titled "The fire in Vinča is visible from space!," a screenshot from Google Maps shows a huge cloud of smoke passing over the river, and it is noted that it is not only plastic garbage burning, but organic waste the size of five hectares, and that it has been burning with varying intensity for years.[32]

The same story includes a statement from the then-mayor of Belgrade, who said that the only systemic solution for the landfill problems in Vinča would be to find a private partner who would close the existing landfill and build a waste-to-energy plant, turning the garbage into heating and electric power. He adds that the estimated cost of the project would be 300 million euros and that "the largest waste management companies responded. It is a new process: they bring know-how and new technologies, and it is good for us because we will learn something and implement the knowledge later."[33] However, as the Ne da(vi)mo Beograd (Don't Let Belgrade D(r)own) citizens' initiative claims, the incinerator project would not only be harmful in itself, it would raise the cost of communal waste disposal and the cost of energy production, and it would decrease employment in the recycling industry. Finally, incineration of plastic and other kinds of waste found in the Vinča landfill would increase greenhouse gas emissions, as well as other harmful substances, if not safely carried out.[34] Warnings were not heeded, and the contract for the incinerator was signed in September 2019 with French waste and water management giant Suez. According to a Ne da(vi)mo Beograd report from March 2020, Suez is contracted to invest several hundred million euros in Vinča, while the city government should pay Suez 1.6 billion euros by 2043. As an activist from the initiative says: "We, the citizens, will pay. Taxes for the collection and management of waste will rise. In Belgrade, these public-private partnerships that put the interest of private, foreign companies above the public general interest have become the rule, and they keep increasing the cost of living for citizens."[35] Despite all this, they further note, international monetary institutions like the EBRD not only keep defending projects like these, but actively invest in them.

This brings me to the second event—the announcement of the opening of chapter 27 of EU accession negotiations, addressing the environment and ecology. In the middle of the events in the urban olfactoscape, local and national news were filled with information about the chapter and pundits explaining the details. Importantly, it was said that Serbia is obligated to invest up to 1.5 billion euros in strategies, laws, and facilities for water purification and waste recycling, and other environmental protection measures. Many of the pundits also mentioned that Serbia does not have the necessary funds, and that it would need private-public partnerships, as well as EU funding, to fulfill the obligations even partially, as is happening with the landfill in Vinča. What these events show, I argue, is that we have two discourses: one of the "ruined" environment, particularly the urban environment, as a consequence of a long history of infrastructure neglect, due to lack of funds and misapplied strategies, and the other with a promise of a "clean" environment. This promise of a clean environment hinges on the idea that the urban environment will improve after investing as much money as possible; that it will improve as long as projects are financed either from public or, even better, private sources, and where the profit from these investments is guaranteed. Environment, in this second discourse, is imagined as inextricably connected to the flows of capital, and as pristine as the sum of the investment. This financialization of the environment is seen as a part of the transition from the ruined (post)socialist society to the contemporary European properly capitalist society of the free market. Vacillating between the two environments, one actual and ruined, the other still virtual and clean, there appears a question for the city inhabitants of how to lead a livable life when the actual, the present, is projected as a part of the past while the future is as uncertain as the ebb and flow of the market.

Such changes are a feature of the age of postsocialist realism, the period of social and economic restructuring following the breakdown of SFR Yugoslavia and, after the year 2000, the reorientation toward a free-market economy and attendant changes such as the precarization of the workforce and the privatization of common property. I coined the term postsocialist realism after "socialist realism," which is defined as a stylistic formation offering optimal projection or utopian vision of a future socialist society, especially as used in Soviet-era art. Postsocialist realism is a condition of possibility for imagining and living (and dying) in the spacetime of perpetual transition toward what is said to be contemporary Western European society. In this sense, there are at least three aspects of postsocialist realism, differentiated with the following terms: *investo-* for investment capitalism, which is increasingly, in the form of investment urbanism, shaping the urban landscape and (non)human life, shrinking urban green areas and urban riverside habitats (this can be seen in the Vinča landfill project, in the urban sprawl of Novi Beograd, and the Belgrade Waterfront development on the banks of the river Sava);[36] *extracto-* for extractive capitalism, which is, among other things, exploiting river ecosystems via the construction of small hydroelectric power plants and destroying the environment in rural areas (the River Rakita at Stara Planina being the most infamous example);[37] and *extincto-* marking the real possibility of extinction after the investing and extracting.

Plastic is an appropriate object for postsocialist realism and its *investo-extracto-extincto* aspects. As analyzed above, the Vinča landfill's plastic becomes part of investment capitalism through the public-private partnership with Suez. A clean environment is promised, but only via a thorough financialization of both "clean" and "polluted" nature. Plastic is one element that enables financialization, as it is perceived as polluting and harmful, especially when it burns. Extractive capitalism is inextricably interlinked with its investment counterpart, and it works to produce the largest possible surplus of value given the "raw" material it is working with and on. The incinerator is going to produce heating and electric power, much more expensively than by the usual means, and via recycling, plastic waste enters another round of surplus value production in the consumer society. Both investment and extractive capitalism are enabled by the laws and policies of both state and city, and these are continuously being adjusted to make the flow of capital easier; in Serbia this is ostensibly for the purpose of inching ever closer to full membership in the European Union. Finally, the extinction aspect of postsocialist realism points to the possibility of the disappearance of various forms of life. While the landfill is usually seen as a purely waste-filled site, it is a complex ecosystem. As is visible in many photographs of the Vinča landfill, there are huge flocks of gulls (*Larus argentatus*) finding their feeding and breeding grounds there. One can only imagine the numbers of other animals, such as dogs, rodents, and various species of snakes and insects. Moreover, it is not only animal species that could disappear but human forms of life too. There are around twenty Roma families who have been living on the landfill site for generations, since the Vinča landfill was opened in 1977. They make their living from collecting the plastic and other recyclable kinds of waste. They are to be moved because of the incinerator, without any other source of livelihood provided.[38]

It is over seventy years since plastic products were first manufactured in Yugoslavia (Jugovinil was founded in 1947). During that time, and especially in recent years, as shown by the waste-to-energy project at Vinča, plastic has become a highly complex object both to understand and to manage. Socialist Yugoslavia experienced plastic as a means of modernization and of enabling mass consumption in the present moment, with one eye on a future classless socialist society with the potential for unbounded consumption. Postsocialist Serbia has to face the after-effects of this experience, both conceptual and practical: the after-effects of plastic. Citizens of postsocialist Serbia face a different plastic object, one that both pollutes and offers potential for profit, depending on the form of materialization the plastic takes. With all these aspects in mind, plastic cannot but become a *hyperplastic object* in the age of postsocialist realism and in the epoch of the Anthropocene.

Hyper-Jugoplastika: ontopolitics of the hyperplastic object

What kind of ontopolitics is most appropriate for approaching the weird agency of plastic under postsocialist realism and the epoch of the Anthropocene? Now, plastic products from the socialist era have become nostalgic collectibles and museum exhibits, yet, oftentimes, examples of these same products are not only among the strata of waste in Vinča but also a source of thick and acrid smoke from landfill combustion, which people breathe in and complain about, and which has become a feature of the olfactoscape of Belgrade. A possible answer to this question lies in the broader context of the Anthropocene.[39] Many theorists argue that, due to human impact on the Earth to an extent previously characteristic of geological processes, we need a new ontology to comprehend the radically changed nature of both nature and the human.[40] It is argued that it is not only the human that possesses the power of agency, but that other, nonhuman beings, both organic and inorganic, as well as material and immaterial systems, are active in ways previously unimaginable. This agential re-evaluation leads to a reconceptualization of the nonhuman as much more complex and more active than was thought in earlier theoretical frameworks. So much so that the complexity of the nonhuman points toward the multimateriality of nature, in which variously materialized beings relationally participate in mutually conditioned becomings. This also entails a different conceptualization of time or, rather, temporalization, where differently and differentially materialized beings become along different timelines of speed and slowness which do not necessarily relate to each other, or if they do, they relate in hypercomplex ways without linear cause and effect. The Anthropocene epoch turns out to be a rather disorienting spatial and temporal geological marker, one which offers little solid ground and exact time to orient oneself in thinking, feeling, and acting. It is, however, only disorienting if we approach the Anthropocene and the newly discovered agency of the nonhuman through essentializing and common-sense framework. Hence the need for new ontology.

The most general way to describe Anthropocene ontology and, consequently, postsocialist realist ontology, as a subdivision of Anthropocene ontology due to its more local spacetimemattering, is to say that nonhumans are pluritemporal and multimaterial. Plastics show this changed nature of nonhuman beings. Plastics, introduced and formed through the practices of socialist consumerism, produced and shaped various human activities that impacted the environment to greater or lesser degrees. These plastics then survived socialist consumerism by several decades, changing the ways they materialized in the environment and human affairs. Namely, the plastics became, and are still becoming, microplastics and microparticles of various sizes measured in nanometers, through the spontaneous combustion of the Vinča landfill and decomposition in the environment. This changing materialization needs to be considered as one and the same phenomenon, but without substantializing, while the constitutive parts of the phenomenon radically change their materiality and temporality. Plastics need to be at one and the same time relatively stable phenomena but without essence, and to be thought of as incessantly changing while retaining the continuum of plasticity. It also needs to be added that plastics are not exhausted by their social use through signification and other practices of the production and use of signs.[41] Rather, the level of signification is only one stratum of multiplicity of multimaterial strata that compose both actual and virtual sides of plastics.[42] Plastics, then, extend through space in a manner different from substances with essences, and the duration is of an altogether different nature from the linear time of simple cause and effect. Plastics are metastable hypercomplex relational intra-acting stratified spacetimematterings.

Hypercomplexity is to be found in the multiplicity of scales, and the concept of intra-acting points to the level of virtuality and potentiality. Karen Barad defines the concept of intra-acting as "the mutual constitution of entangled agencies,"[43] in contradistinction to the concept of interaction, which assumes separate beings and their powers of agency. Intra-action, however, points to the fact that "agencies are only distinct in relation to their mutual entanglement; they don't exist as individual elements."[44] Barad further defines a phenomenon as the "ontological inseparability/entanglement of intra-acting 'agencies,'"[45] which becomes through "specific agential intra-actions,"[46] that is, it does not preexist the intra-acting relationality of the agencies. It is through the intra-acting, the agential cut, that subject and object are set up as relatively stable positions, but such that they do not exist before the entanglement of intra-acting agencies. In a word, "phenomena are differential patterns of mattering,"[47] "dynamic topological reconfigurings/entanglements/relationalities/(re)articulation of the world."[48] Phenomena are spacetimematterings, activities of the world's intra-acting and becoming. Intra-acting, then, is the becoming of the world before the separation of subject and objects, and it is only through these reconfigurings of relationalities that something we can recognize as a metastable phenomenon becomes. This intra-acting potentiality of entanglement is the same for plastic microparticles as it is for the hyperobject that is the plastic. The scale starts weirdly to fold, infold, and unfold across the spacetimematterings.

The weird folding of the intra-acting phenomenon that is the plastic brings me to another attribute of the hyperplastic object: hyperobjectivity. It was Timothy Morton who first theorized hyperobjects, defining them as "things that are massively distributed in time and space relative to humans."[49] For Morton, a hyperobject can be an oil field, a swamp, the biosphere, the solar system, but also the "sum total of all the nuclear materials on Earth; or just the plutonium, or the uranium."[50] A hyperobject may also be "the very long-lasting product of direct human manufacture, such as Styrofoam or plastic bags."[51] Hyperobjects possess certain characteristics such as: 1) viscosity, 2) nonlocality, 3) temporal undulation, 4) high-dimensional phasing, 5) interobjectivity. High-dimensional phasing tells us that the unity of the hyperobject lies in an interference pattern which cannot be seen in everyday experience. Or as Morton writes, "hyperobjects are *phased*: they occupy a high-dimensional *phase space* that makes them impossible to see as a whole on a regular three-dimensional human-scale basis."[52] That means that we see only parts and pieces of a hyperobject at any one time, while the whole is always beyond our experience as well as comprehension. In this sense, the hyperplastic object includes all instances of plastic objects of all shapes and forms, sizes and colors, but goes beyond them, not in some kind of Platonic universe of eternal Ideas but in being "massively distributed in a phase space that is higher dimensional than the equipment used to detect it."[53] It is this massiveness of hyperplastic that surpasses our experience.

This massiveness of the hyperobject is the reason why we talk about nonlocality and what enables viscosity. Morton writes that "hyperobjects seem to inhabit a Humean causal system in which association, correlation, and probability are the only things we have to go on."[54] It points to the entanglement of beings and their utter interconnectedness, which means that "there is no such thing, at a deep level, as the local. Locality is an abstraction."[55] This directly ties in with the distribution of the hyperplastic object in its multimaterial becomings from a "solid" object for everyday use to a microparticle to be breathed in, created by the spontaneous combustion at the landfill. The hyperplastic object is constantly moving, or differently put, it is so massively distributed that it is impossible to fix it at an exact point in space and time, hence the temporal undulation. Temporal undulation speaks of inhumanly long durations of hyperobjects, their massive distribution in time, which is a "time that is beyond predictability, timing, or any ethical or political calculation."[56] Viscosity is related to nonlocality, temporal undulation, and the high-dimensional phasing. Namely, viscosity points to the fact that a hyperobject envelops, surrounds, and penetrates other (hyper)objects. According to Morton, hyperobjects "seriously undermine the notion of 'away.' Out of sight is no longer out of mind."[57] The viscosity of the hyperobject indicates that it sticks to bodies, so much so that it erases the difference of outside and inside, interiority and exteriority of a body. Coupled with the nonlocality and the massiveness of the hyperobject, this has boundary-shattering consequences for the conceptualization of individual beings. It is impossible to disentangle oneself from the hyperplastic object as there is no outside to run away to. Or, for that matter, an inside to run away from.

The aspect of interobjectivity of a hyperobject reveals to us a particular relationality between hyperobjects. Hyperobjects relate to each other without the human mediation, and they relate in hypercomplex ways given that they viscously envelop human beings and thusly exceed the reach of human agency. In this sense, the interrelationality of hyperobjects that is called interobjectivity can tell us a much more interesting story about the ways in which the Anthropocene, the postsocialist condition, and plastics as hyperobjects interobjectively interrelate. If understood hyperobjectively, these three become much more complex objects which require a different conceptualization of agency. That is, humans become encompassed by hyperobjects which last an incomprehensibly long time and for which we still do not have an adequate thinking-feeling framework. What is more, we do not have means of insight into the interobjective relationality and so have to assume nonhuman or even inhuman assemblage that defies human understanding. All of this means that simple policies stemming from the liberal conceptual framework, which take humans as key actors, are far from adequate. Neither would the simple free-market economic approach of financialization suffice, since we cannot predict the "behavior" of the interobjectively related Anthropocene, postsocialist condition, and plastics. In other words, interobjective relationality escapes the instrumentalization of late liberal capitalism, as well as the objectification of the humanistically and rationally conceptualized human being.

Conclusion

The development of a new ontopolitics of hyperplastic leads to several insights. Firstly and most theoretically, the ontopolitics of hyperplastic shows that the ontology is historically contingent. In former Yugoslavia, hyperplastic has emerged as a product of socialist self-governing forms of manufacturing and consumerism, processed by the period after the dissolution of Yugoslavia—the age of postsocialist realism. The age of postsocialist realism would not be possible without the (perpetual) transition from self-governing socialism to what is said to be modern Western society, and particularly the modern free-market economy with its attendant consumerism and production of waste. In relation to this, the new ontopolitics of hyperplastic offers a depiction of the ways plastic acts and is being acted upon, potentially enabling us to map the entanglements of various human and nonhuman bodies and systems. Given the planetary context of the Anthropocene, the ontopolitics of hyperplastic shows the knotted entanglements of the Anthropocene, postsocialist realism, and plastic in such a way that allows us to imagine, and perhaps create, different forms of life, which would lead to re-evaluation of bodily boundaries as well as boundaries between waste and usefulness. We need to rethink what it means to be human when we are intra-actively entangled in the phenomenon that is the hyperplastic object.

Finally, the ontopolitics of the hyperplastic object offers critical points about the financialization of the environment, particularly in the context of postsocialist realism, where the knots of finance, environment, and plastic are tied especially tightly. As neoliberal capitalism slowly encroaches upon the commons in a formerly socialist self-governing country, the ontopolitics of the hyperplastic object shows that we need to go beyond the neoliberal discourse of financialization and marketization of the environment, and find wholly different approaches that would view both the environment and the plastic as active agents instead of just mere passive resources to be exploited and discarded as we wish.

Notes

1. Boban Cvetanović et al., "Analiza stanja i potencijali reciklaže u Republici Srbiji," in *Četvrti naučno-stručni skup Politehnika 2017: Zbornik*

radova, ed. Marina Stamenović (Beograd: Beogradska politehnika, 2017), 19–20.

2. See Marija Dimitrijević, "Problem reciklaže u Srbiji," http://www.cqm.rs/2011/2/pdf/09.pdf; Svetlana Jovanović, "Upravljanje otpadom u Srbiji—problemi, izazovi i moguća rešenja," last modified February 19, 2019, https://balkangreenenergynews.com/rs/upravljanje-otpadom-u-srbiji-problemi-izazovi-i-moguca-resenja/.

3. "EBRD obezbedio kredit od 72,25 miliona evra za deponiju Vinča," *Balkan Green Energy News*, October 2, 2019, https://balkangreenenergynews.com/rs/ebrd-obezbedio-kredit-od-7225-miliona-evra-za-deponiju-vinca/.

4. See Jurica Pavičić, "Slava Jugoplastici," last modified December 24, 2018, https://www.xxzmagazin.com/slava-jugoplastici.

5. See ibid.

6. Branislav Dimitrijević, *Potrošeni socijalizam: Kultura, konzumerizam i društvena imaginacija u Jugoslaviji (1950–1974)* (Beograd: Fabrika knjiga, 2016), 30.

7. Pavičić, "Slava Jugoplastici."

8. Dimitrijević, *Potrošeni socijalizam*, 36.

9. Ibid. 76.

10. See Igor Duda, *U potrazi za blagostanjem: O povijesti dokolice i potrošačkog društva u Hrvatskoj 1950-ih i 1960-ih* (Zagreb: Srednja Europa, 2004), 65–6.

11. See Ibid. 66.

12. See Radina Vučetić, *Koka-kola socijalizam* (Beograd: Službeni glasnik, 2012), 366.

13. See Ibid. 369.

14. For an image of a shopping basket, see Muzej Jugoslavije, "Potrošačka korpa." Retrieved from https://www.muzej-jugoslavije.org/art/potrosacka-korpa/.

15. Darko Suvin, *Samo jednom se ljubi: Radiografija SFR Jugoslavije* (Beograd: Rosa Luxemburg Stiftung, 2014), 333.

16. Dimitrijević, *Potrošeni socijalizam*, 32.

17. Ibid. 86.

18. Ibid. 49.

19. Pavičić, "Slava Jugoplastici."

20. Ibid.

21. Marina Vidas, "Poslanica modernoj ženi," in *Balkanska šaputanja*, ed. Panče Hadži-Andonov (Brisbane: PantaOz Publishing, 2009), 177.

22. Stela Đujić, *Stare hrvatske igre i igračke* (Pula: Fakultet za odgojne i obrazovne znanosti, 2017), 6.

23. Ibid. 8–9.

24. See Sonja Ćirić, "Igračke iz muzeja: Čime smo se igrali," last modified December 27, 2007, https://www.vreme.com/cms/view.php?id=554518.

25. Silvija Šesto, "I igračke ubijaju, zar ne?" last modified June 5, 2013, https://zg-magazin.com.hr/i-igracke-ubijaju-zar-ne/.

26. Quoted in Dimitrijević, *Potrošeni socijalizam*, 47.

27. See Ibid. 48.

28. Lav Mrenović, "Trostruki probem deponije u Vinči," last modified February 8, 2018, https://www.masina.rs/?p=5984.

29. Saša Petrović, "Problem plastične ambalaže u Srbiji—od zagađenja do mogućnosti," last modified May 29, 2018, https://www.masina.rs/?p= 6778.

30. Jovana Aleksić, "'Oči mi suze, ljudi povraćaju': Požar u Vinči ima sve dramatičnije posledice," last modified June 13, 2017, https://www.blic. rs/vesti/beograd/oci-mi-suze-ljudi-povracaju-pozar-u-vinci-ima-sve-dra maticnije-posledice/pezb9gj.

31. "Gust dim nad Beogradom: Deponija u Vinči gori već 20 dana," Vestionline, June 9, 2017, https://arhiva.vesti-online.com/Vesti/Drustvo /654388/Gust-dim-nad-Beogradom-Deponija-u-Vinci-gori-vec-20-da na.

32. "Požar u Vinči se vidi iz svemira!" *Dan u Beogradu*, July 4, 2017, https:// www.danubeogradu.rs/2017/07/pozar-u-vinci-se-vidi-iz-svemira/.

33. Ibid.

34. Nataša Stojanović, "Ne da(vi)mo Beograd: Koncesijom deponije u Vinči mnogi građani će ostati bez posla, a grad bez resursa i profita," last modified August 24, 2017, https://novaekonomija.rs/vesti-iz-zemlje/ne-d avimo-beograd-koncesijom-deponije-u-vinC48Di-mnogi-graC491ani-C 487e-ostati-bez-posla-a-grad-bez-resursa-i-profita.

35. Ne da(vi)mo Beograd, "Deponija i spalionica u Vinči: Otpad javni, profit privatni," last modified March 20, 2020, https://nedavimobeograd.rs/ot pad-javni-profit-privatni/.

36. Belgrade Waterfront "represents a combination of commercial and residential luxury space. It is a foreign investment project for which the state gave land and offered clear support. If realized on the scale planned, it will significantly transform that part of the city through a process of profitable gentrification. The space will be dedicated to members of the elite and foreign citizens." Vera Backović, *Džentrifikacija kao socio-prostorni fenomen savremenog grada: Sociološka analiza koncepta* (Beograd: Filozofski fakultet, Univerzitet u Beogradu, 2015), 183. The project has been controversial from the very beginning for several reasons, including an influx of capital from large foreign investors that altered the neighborhood's landscape and decades-old local ways of life; see Aleks Eror, "Belgrade's 'top-down' Gentrification Is Far Worse than Any Cereal Café," last modified December 10, 2015, https://www.theguardian.com /cities/2015/dec/10/belgrade-top-down-gentrification-worse-than-cereal -cafe; Herbert Wright, "Belgrade Waterfront: An Unlikely Place for Gulf Petrodollars to Settle," last modified December 10, 2015, https://www .theguardian.com/cities/2015/dec/10/belgrade-waterfront-gulf-petrodol lars-exclusive-waterside-development; Filip Rudić, Maja Živanović, and Ivana Jeremić, "Serbians Protest as Controvesial Demolitions Remain Unexplained," last modified April 24, 2019, https://balkaninsight.com

/2019/04/24/serbians-protest-as-controversial-demolitions-remain-unex plained/.

37. See the official site of the Odbranimo reke Stare planine (Let's Defend Old Mountain's Rivers) movement: https://novastaraplanina.com/.

38. Maja Nikolić, "Život na deponiji—adresa bez broja, struje, vode, puta . . ." last modified June 2, 2018, http://rs.n1info.com/Vesti/a393157/Ro msko-naselje-na-deponiji-Vinca.html.

39. See Jan Zalasiewicz et al., *The Anthropocene as a Geological Time Unit: A Guide to the Scientific Evidence and Current Debate* (Cambridge and New York: Cambridge University Press, 2019).

40. See Stephanie Wakefield, *Anthropocene Back Loop: Experimentation in Unsafe Operating Space* (London: Open Humanities Press, 2020); Andreas Weber, *Enlivenment: Toward a Poetics for the Anthropocene* (Cambridge and London: MIT Press, 2019); Elizabeth DeLoughrey, *Allegories of the Anthropocene* (Durham, NC and London: Duke University Press, 2019); Dipesh Chakrabarty, "The Human Condition in the Anthropocene," https://tannerlectures.utah.edu/_resources/docu ments/a-to-z/c/Chakrabarty%20manuscript.pdf.

41. See Andrija Filipović, "How to Do Things in the Plasticene: Ontopolitics of Plastics in Arendt, Barthes, and Massumi," *AM: Journal of Art and Media Studies* 23 (2020): 91–101.

42. For the concept of stratification, see Gilles Deleuze and Félix Guattari, "10,000 BC: The Geology of Morals (Who Does the Earth Think It is?)," in *A Thousand Plateaus* (London: Continuum, 2004).

43. Karen Barad, *Meeting the Universe Halfway: Quantum Physics and the Entanglement of Matter and Meaning* (Durham, NC and London: Duke University Press, 2007), 33.

44. Ibid. 33.

45. Ibid. 139.

46. Ibid. 139.

47. Ibid. 140.

48. Ibid. 141.

49. Timothy Morton, *Hyperobjects: Philosophy and Ecology after the End of the World* (Minneapolis: University of Minnesota Press, 2013), 1.

50. Ibid. 1.

51. Ibid. 71.

52. Ibid. 70; emphasis in original.

53. Ibid. 77.

54. Ibid. 39.

55. Ibid. 47.

56. Ibid. 67.

57. Ibid. 36.

7

Failed Infrastructures, My Little Ponies, and Wadden Plastics: The Eco-Intimacies of the *MSC Zoe* Container Disaster

Renée Hoogland

During the night of January 1 to 2, 2019, the *MSC Zoe*, one of the world's largest cargo ships, lost a staggering 342 shipping containers in the sea. Many of these containers, while en route from the south of Portugal to northern Germany, washed ashore on the Dutch islands of the Wadden Sea—an intertidal zone in the southeastern part of the North Sea—already hours after the spill. As a result, inhabitants of the Wadden Islands started witnessing the debris from some of the lost containers gradually washing up on the shore early the next morning: flat screens, Ikea furniture, all sorts of plastics, and many children's toys were found by the hundreds of people who immediately volunteered to clean the Northern beaches. As the *MSC Zoe* container disaster deeply affected the symbiotic relationship between islanders and the Wadden landscape, it forged new relations of intimacy between material (plastic) waste, humans, and their surroundings.

By engaging with the *MSC Zoe* container disaster, this chapter explores the emerging intimacies in a landscape where humans exist in co-constitutive relationships with their ecologically damaged surroundings. Human responses to such newly damaged surroundings that are produced by projects of modernity, in this case that of intermodal infrastructure, open up new reflections on the intimate effects and affects that these projects have on human lives and the ecologies in which they find themselves intertwined. Such ecological intimacies are what the anthropologist Kath Weston tracks in her book *Animate Planet*, where she details new types of intimacies between humans and the animate environment. These "eco-intimacies" are emerging at a time in which the "high-tech ecologically damaged world"[1] that humans have made is simultaneously remaking them. Building on Weston's interpretative frame for reading the human entanglements with animate materiality in a high-tech world, I ask: how did the people of the Wadden Islands make visceral sense of the container spill that essentially led to their ecologically damaged surroundings? What happened, moreover, when the functional invisibility that is so typical to cargo shipping infrastructures suddenly turned highly perceptive and entered human life?

My purpose here is not so much to focus on the political embeddedness of the *MSC Zoe* accident, but rather in taking the contemporary fascination with ecologically infused intimacies worthy to study in its own right. My intention with this, moreover, is not to enter scholarly debates on new materialism, posthumanism, or what anthropologists have termed the "ontological turn," as interpretative frameworks that are all concerned in their own way with decentering the human. Instead, I employ a close reading of the *MSC Zoe* case as a heuristic method to explore the ecological entanglements of logistical infrastructures, plastic waste, and human life.

After all, the intimacies that I discuss here are, as Weston spells out, "not about separate-but-equal. Neither are they the products of relations between entities."[2] Instead, I understand them as highly compositional and historically located. This is why the chapter starts with figuratively tracing the shipping container before it actually entered the surroundings of the Wadden Islands as a washed-up "lost" object. After accentuating how the container disaster marks a dialectic of visibility and invisibility, I explore the social, artistic, and technological bids to make sense by people of the, mostly plastic, content of these spilled steel boxes. In this way, I aim not only to reflect upon the ways that the materiality of plastic has entered our lives but also to contribute to the growing conviction that "creatures co-constitute other creatures, infiltrating one another's very substance, materially and otherwise."[3] Indeed, such "infiltrating creatures" are conceived in the case of the *MSC Zoe* container disaster as the washed-up plastic products of failed logistical infrastructures, promptly producing altered environments and new identical forces (see fig. 7.1).

The shipping container

The cargo vessel that lost its containers in the North Sea on the first of January 2019 ranks among the largest container ships within the world's shipping industry. The *MSC Zoe* stretches a length of almost 1,300 feet, and its cargo is made to carry around 19,000 shipping containers at the same time. Not coincidentally, the vessel is property of the second largest container transporter in the world: the Mediterranean Shipping Company (MSC), founded in Italy and headquartered in Switzerland. For the population of the Wadden Islands, however, the shipping container is anything but out of the ordinary. When standing on the beach overlooking the North Sea, one would

Figure 7.1 *Visuals of the* MSC Zoe *container spill that were taken by the Netherlands Coast Guard (Kustwacht Nederland) on 3 January 2019: IMG_0064. Published online by the Assignor at the following link: https://www.kustwacht.nl/ dossiers/msczoe. Netherlands Coast Guard, 2019.*

notice the dozens of cargo vessels that sail in front of the islands every day.[4] In fact, the Wadden Islands lie in the midst of an extraordinarily busy sailing route since, despite its relatively low waters, the cheap close-to-coast route enables vessels to avoid taking more expensive detours that maintain more distance from the Dutch shoreline.

In other words, the amphibious shipping container is, although never quite noteworthy, always there surrounding the islands. The way, then, that the ubiquitous shipping container can be considered as an almost non-recognized object for the people of the Wadden Islands is perhaps mostly due to its ability to be moved from A to B in an almost seamless fashion: simultaneously the primary reason for its global rise over the last five decades. As Craig Martin spells out, "[a]bove all the container's primary innovation is the ability to deliver goods door to door without needing to unload them every time they are moved from ship, to trains or trucks."[5] It is precisely this feature of the shipping container that is central to the logistical transport innovation known as intermodalism, by which "at least two different modes of transport" are used in an "integrated manner in a door-to-door transport chain."[6]

Yet there is much more to the shipping container that accounts for this corrugated steel-made box's omnipresence over the globe. They trundle through railroad terminals, highways, and parking lots on the back of wagons and trucks; they serve as cheap student accommodation, but also as expensive lodgings in the desert; they lie in rural farmlands, urban gardens, and industrial parks, in which they function as storage lodgings or perhaps as an artistic workshop space; they rest on the bottom of the sea, and they feature as subject in some post-punk lines by the English band The Fall.[7] The shipping container, in other words, is globally ubiquitous and yet very much taken for granted. How can it be, then, that this object is so ubiquitous and yet habitually unrecognized by people on the Wadden Islands?

Following Tim Cresswell's discussion on the architecture of shipping containers, I similarly propose that it is its design of invisibility that makes the box seamlessly pass by human imagination and attention. After all, as Cresswell writes, the "particular mobilities of the container and the ships that carry them rely on their blankness, their invisibility. This invisibility enables the smooth operation of capitalism. It is at the center of our everyday lives without us even noticing."[8] And indeed, while systematic containerization arguably started as early as the 1830s, it was in the year of 1967 that the shipping container as we know it now was fully designed. Before this, containers were designed to regularize and homogenize their content, albeit the containers *themselves* were never uniform in physical appearance. The 1967 McKinsey & Co. report expressed a particular discontent against this container diversity that was maintained by the shipping companies that still used different container designs, all depending on the transported goods. The containers' diverse appearance, in other words, was believed to be logistically slowing down the entire shipping process. From then on, Marc Levinson argues, the standardization of the container would create a globally recognized design that is characterized by the corrugated metal side panels with a standard corner fitting that is applied to all shipping containers.[9] It is, then, the abstract and featureless materiality of the standardized container box that essentially sets in motion a seamless global economy of commodity, embracing a sense of anonymity and general state of not-knowing.

Many scholars as well as artists have explored this generalization of cargo shipping as a broader invisible industry. Documentaries such as Allan Sekula's *The Forgotten Space* and books with titles such as Rose George's *Ninety Percent of Everything: Inside Shipping, the Invisible Industry* all seem to point to an industry that embodies a concealed world that is nevertheless crucial to our everyday existence. However, I am more interested in the "container box" itself, which is central to a politics of infrastructural invisibility. After all, it is the shipping container itself that is moving through abstract space: spaces of precisely mapped locations and estimated times of departure and arrival. The standardized container can, in other words, be described as a type of liminal space itself, floating in between the "footloose, drifting, and purposelessness"[10] of mobility and capitalist efficiency and goal orientation. The container box, then, literally embodies a space of not-knowing, since it is inherent to an infrastructure of mobility that simultaneously conceals its very purpose: the shipping containers' ultimate function is that of containment, but we have no idea what they contain.

In the process of meaning-making when it comes to cargo infrastructures, formulations and "common metaphors present logistical infrastructures as a substrate, something upon which something else 'runs' or 'operates' underneath the surface."[11] Despite its ubiquity, such systems of infrastructure tend to weave into the fabric of everyday life while still functioning beneath the conscious awareness of their users. The sociologist Susan Leigh Star famously articulated this by noting that at the core of logistical infrastructure is its embeddedness into other structures and that its enduring invisibility only "becomes visible upon breakdown."[12] For Leigh Star, the ubiquitous container nevertheless achieves a state of invisibility precisely because people experience logistical infrastructure as being behind the scenes, as the background of processes that necessarily go unnoticed in order to function smoothly. I would extend this argument by suggesting that it is the 'natural' environment that additionally becomes part of the material infrastructure through which cargo is transported. The ubiquitous shipping container, in other words, blends in with the natural landscape of the Wadden Sea so that this landscape becomes an infrastructural component of cargo transport, in the sense that "it is matter that enables the mobility of matter."[13]

What happens, then, when the supposedly invisible shipping container promptly turns highly visible? For many inhabitants of the Wadden Islands, this question turned into reality when dozens of containers, including their contents that had been previously concealed, washed onto dry land in the morning of January 2, 2019. While people initially suspected the debris to be leftovers from the recent New Year's celebrations, general concerns increased when news reached the islands about a container spill of a yet unspecified number of containers. The first news items that covered the accident, however, were still limited and imprecise: whereas it was first communicated that the vessel had lost 30 containers, the coast guard eventually agreed that 270 spilled boxes had to be located somewhere in the water. It ultimately concerned 342 lost shipping containers.

Soon after the spill, people on the Wadden Islands became actively involved with the dropped containers, which had to be rendered visible in order to grasp the dimensions of the spill. The Dutch graphic designer Frédérik Ruys created a reconstruction of the accident by visualizing the locations and timeframe in which the MSC Zoe sailed along the Wadden coast.[14] This visualization starts with intense color lines that represent the extremely crowded navigational routes for cargo vessels along the coast. It then imaginatively follows the MSC Zoe to estimate, by means of prepublished logistical information, at what point roughly the container spill must have occurred. Interestingly, Ruys's data visualization also indicates that the time between the spill occurring and the vessel crew taking notice of the accident was an estimated four hours. This suggests that while the standardized container is perhaps mapped and located to the minute during mobility, once it disappears in the sea the shipping container is instantaneously off the grid. The container spill, then, signals toward a threshold of visibility: while it marks the logistics of global visuality from above, the spill—as a disruption into the normalization of logistical infrastructures—similarly marks the invisibility of animate materiality from below.

In the days following the accident, other visualization videos of the spill circulated widely over the internet as well as local news. It was only after official film footage by the Dutch coast guard was publicized that people realized that the spill involved far-reaching impacts. In the footage, published three days after the accident, one can see the unstable containers balancing on the vessel before disappearing into the water.[15] Such visualization of the spill, however, also enabled concerned citizens to viscerally grasp what it entails when cargo infrastructures fail to maintain their invisible efficiency. Such attempts to make sense of the MSC Zoe container disaster remind us of the patterns of invisibility and visibility to which infrastructures of mobility give rise. It is only when the shipping container materially turns invisible that it concurrently becomes visible and enters human imagination. And with this dialectic of invisibility and visibility came new engagement practices too for the Wadden population: that of working with the *content* of the spilled containers, once figuratively so far away, and now gradually washing ashore as physical objects.

My Little Pony

In the early morning following the night of the container spill, people on the Wadden Islands witnessed the debris that the lost containers no longer contained gradually washing onto the shore. Thousands of objects, mostly plastic, were found by the hundreds of volunteers who immediately started to clean the northern beaches. At this point people also began to imagine what sort of content the dazzling array of steel boxes actually concealed: at the moment that the seamless infrastructure of the shipping container was interrupted, the question of what was inside the containers became a pressing concern. Both the Dutch national newspapers and the island communities actively engaged with the mysterious contents of the containers that now exposed themselves to the people, animals, and natural environment of the Dutch Northern Islands.

As Allan Sekula notes of the shipping container in his semi-photographic book *Fish Story*, "the contents [are] anonymous: electrical components, the worldly belongings of military dependence, cocaine, scrap paper (who could know?) hidden behind the corrugated steel walls emblazoned with the logos of global shipping corporations . . ."[16] Sekula here brings to our attention the miscellaneous imaginations that are generated by the normally enigmatic inside of the gigantic shipping boxes. Thus, while the standardized container and its cargo are characterized by functional invisibility, this state of not-knowing is simultaneously widely captured by the cultural, political, and social imagination; from popular culture such as the television show *The Wire*, which features the Port of Baltimore,[17] to political debates on international smuggling suspicions,[18] to alarming messages from the human trafficking circuit—perhaps one of the darkest aspects of containerization. In other words, the containers' content, which naturally bypasses any clear sense of visual perception, is a source of endless surprising and often terrifying imaginations. It is, as Sarah Hirsch notes, the "not knowing what is inside the containers [that] is intriguing but also potentially alarming precisely because what awaits you is unknown."[19]

It is perhaps due to the generally darker imaginations invoked by the mysterious content of the shipping container that many inhabitants of the Wadden Islands were rather surprised when they noticed thousands of My Little Pony toys washing ashore. This glittery invasion of an impressive amount of plastic Applejacks, Fluttershys, and Pinkie Pies did not initially meet people's expectations of a symbolic materialization of the spill's damaging results.[20] While it might seem rather exploitative to focalize on children's toys for the purpose of this chapter, I argue that in fact the drifted pony dolls effectively bring the topics of globalized infrastructures, ecological damage, and "fantasy worlds" together. Similar to the invisible "other" world of container shipping, the plastic My Little Pony toys embodied a non-reality that suddenly turned very real to the Wadden population. In this way, the fantasy element of these shiny toys that floated in the North Sea and ultimately washed up on the beaches as ecologically damaging objects precisely captures people's attempts to grapple with the aftermath of the container accident. In these efforts, both the mysterious other world of the plastic My Little Ponies and the cargo shipping industry visibly expose themselves to the Wadden population and thread their ways into their environment.

The plastic My Little Pony toys, however, proved to be challenging to people's sense-making of the container spill. Soon after various clean-ups of the surrounding beaches were organized, the local governments of the islands as well as the Netherlands coast guard allowed the participating volunteers to take home as much as possible of the found bulk of trash and, quite paradoxically, dispose of it privately. Whereas the washed-up content of the spilled boxes were initially the property of someone else (i.e. foreign companies), the objects had now lost their value alongside any claim of ownership due to contact with the water.[21] Yet while people generally were not so concerned with disposing of most of the trash they found on the beaches, the plastic My Little Pony toys were engaged with in an alternative way.

Perhaps most remarkably, on the Wadden island of Terschelling an actual My Little Pony emergency shelter was opened where people abundantly brought their found plastic toys. The bulk of glossy dolls was accordingly exposed to visitors of the shelter, who wanted to visually experience the consequences of the container spill. Such opening of a viewing space not only took the form of an ecologically concerned initiative but equally gestured toward a means of recognizing the newly emerging intimate relationship between the Wadden population and the legacies of the cargo accident. As such, an awareness circulated that even though the My Little Pony toys were deemed plastic trash, they also had intimate and profoundly human (or planetary) stories to tell. The toys were indeed the silent witnesses to the invisibility of logistical infrastructures, and by exposing the pony dolls these same infrastructures turned just a little more visible and graspable to the human mind. The engagement with these toys, then, not only produced alternative affects of plastic trash, but also caused a recognition of the intimacy between the washed-up plastics and the people finding them: as planetary waste that was essentially created by themselves, and would not simply disappear when disposed of.

As an artistic response to this symbolic materialization of the spill, the Dutch artist Maria Koijck created a new statue, introduced in April 2019. The horse sculpture, composed entirely of plastic materials and with the conforming name *Zoe de Ponie*, is not only an artwork against plastic violence; it also seeks to visualize the shipped trash that was at that moment still floating in the North Sea. After all, the thousands of My Little Pony toys most likely originated from merely one of the 342 lost container boxes. The plastic-filled horse, almost 17 feet high, is an artistic consideration of the excesses of Western capitalist production. Its body is covered with white plastic shrink film and a large letter P (Pony? Plastic?) directs the attention of the viewer. The sculpture's long manes of colored cleaning cloths, all collected by the artist on the Wadden coast after the accident, makes up a rather impressive pendulum of kitchen materials that invites an ethics of consumption in response to environmental crisis.

Similar to the shipping container itself, *Zoe de Ponie* impresses the viewer with its magnitude, which precisely situates the human viewer as a subject rather than an agent in relation to its surroundings. In fact, the large sculpture calls for a likeness in appearance with a Trojan horse, in Greek legend the ultimate wooden weapon that hid its warriors inside until the latter jumped out of the massive wooden construction to open the city gates from within enemy territory. Koijck's updated plastic Trojan horse embodies this same function of concealment: it gestures toward the logistical invisibility of the shipping container, signifying an outbreak from the inside just like the spilled boxes lost their content. As the Greek warriors make place for plastic cargo, Koijck's plastic Trojan horse is a visual consideration of what counts as waste. The sculpture, after all, marks the very thin line between what were considered transported commodities *before* the spill and how these commodities turned into environmental waste *after* the accident happened. The argument that I attempt to make here is that this thin line precisely captures how at the very moment that the containers disappeared from the radar of logistical infrastructures, they entered into a newly emerging intimate and co-constitutive relationship between people and their (now environmentally damaged) surroundings. Koijck's plastic artwork, then, articulates forms of connection between plastic waste, the (failed) infrastructural systems that contribute to this waste, and, therefore, the intimate co-extensiveness of people and their surroundings.

It is important to note, however, that situated modes of ecological intimacy in the case of the *MSC Zoe* container spill do not automatically lead to forms of acceptance or identification.[22] Thus, rather than utilizing the concept of *ecological intimacy* as "an ontological claim"[23] or some sort of universal descriptor that dwells in the realms of friendship and kinship,[24] I instead use the category of intimacy in this chapter as a heuristic tool,[25] exploring the ways in which people of the Wadden Islands attempted to make sense of their intimate entanglements with the ecological situation of which they were inherently part. As such, linking intimacy to forms of plastic pollution leads to complex cases of interconnectedness that are not necessarily desired.

This is precisely what Kath Weston captures with the idea of "unwanted intimacies" presented in her compelling case study on the Fukushima Daiichi nuclear disaster; she understands people's fear that escaped radioactive pollutants would become part of the bodies exposed to the toxins as an example of an unwanted intimacy.[26] Such unwanted intimacies lead people to realize their sometimes menacing intimate entanglements with a disrupted environment, and Weston essentially asks how people accordingly "throw their bodies into the mix by *viscerally* engaging with a socially manufactured, recursively constituted 'environment' that is also, crucially, them."[27] While the toxicity that Weston covers in her study might be of different intensity, scale, and bodily effects than the plastic pollution at the Wadden Islands, in the latter ecological intimacies similarly thread their way through people's attempt to grapple with the container disaster. Imagining the polluted water and beaches ultimately infiltrating their bodies, the social and creative responses to the *MSC Zoe* disaster all prompted alternative visions of the Wadden landscape.

And indeed, Koijck's artwork *Zoe de Ponie* visions a seascape that spits out plastic waste. In using this waste as a resource for engagement, the social and creative responses that emerged after the accident precisely illustrate how people on the Wadden Islands felt something deeper than merely "exposed to" the plastic waste caused by the container spill. Their understanding of a symbiotic relationship with their surroundings made it that simply cleaning up and disposing of the spilled container content that they found was not enough. These responses, in contrast, were concerned with rendering visible the waste in order to confront the invisibility of logistical infrastructures.

Wadden Plastics

A week after the container spill, the Dutch Ministry of Infrastructure and Water Management and the Wadden Academy introduced a collaborative research project to determine the anticipated presence of petrochemical substances in the Wadden Sea, and the overall ecological condition that the accident had precipitated. A few days later it was revealed that one of the spilled shipping containers carried thirty bags of more than 22,000 kilos of polystyrene micro pellets, a synthetic hydrocarbon in the form of plastic that is used for the production of a wide variety of consumer products. The public knowledge of these micro pellets caused a next phase of concern for many people, including fishers and local environmental organizations, especially after the Wageningen Marine Research Centre republished its previously conducted research on the effects of a possible leak of microplastics on the food chain, animal health, and the North Sea ecosystem.[28] Even though these washed-up plastic pellets were partially visible in the Wadden landscape, in the sense that people encountered bulks of pellets on the beach, the nurdles were nevertheless even more problematic to keep track of. As devoted groups of independent volunteers, at this point joined by deployed military personnel who provided assistance with the clean-up, were actively involved in undoing the beaches from the accumulated spilled waste, the plastic pellets multiplied into smaller fragments and easily traveled from one location to the other beneath people's feet or by tidal movements.

In response to these still officially unlocated but clearly ubiquitous plastic pellets, many concerned volunteers involved themselves in a new project, *Wadden Plastic*. The goal of this project was to make relatively fast estimations on the amount and locations of the plastic pellets that had occupied the Wadden landscape. While a team of ecologists collaborated and launched an accessible web application,[29] the many citizen volunteers who were already involved in cleaning the beaches took the lead in counting and localizing the plastics. One of the volunteers created a simple instruction video for the process which users could watch within the *Wadden Plastic* web application. As demonstrated in the video, users of the app could specify their precise location and date for research by means of the GPS tracker that was part of the application. After this, the volunteers roughly counted the number of plastic pellets that they found within a 15 x 15 inch area and selected the fitting category within the app. Additionally, the *Wadden Plastic* app encouraged users to submit some observational remarks on whether larger types of plastic material were visibly present in the immediate surroundings.

Hundreds of volunteers used the app to locate the plastics and sent their observations to the research team, which devoted itself to data collection. Ultimately, by means of the accumulated citizens' data, a spatial image was created that marked 300 "crisis points" of plastic waste along the northern Dutch coast. Based on the results from *Wadden Plastic*, estimations were made that the container spill involved a total of 24 million plastic pellets, which have spread over the Wadden landscape.

The *Wadden Plastic* app and project, which I understand as another bid to make sense of the *MSC Zoe* container spill, embodied a form of *technostruggle*—a term I borrow from Kath Weston to describe how "ordinary people avail themselves of technology to produce knowledge about their visceral engagements with potentially lethal

derivatives of the 'resources' upon which they rely."[30] While Weston
defines technostruggle in a different context (that of Fukushima's radiation), I use the term to understand how users of *Wadden Plastics*
seized upon technology in order to grapple with their ecologically
damaged surroundings.

Technostruggle, in this sense, fostered a citizen-based initiative by
which people attempted to seize the means of perception and scale
in the aftermath of the container spill. The technostruggle of the
project *Wadden Plastic*, moreover, generated new forms of participation that enabled households on the Wadden Islands to make more
informed decisions, based on the now localized presence of plastic
pellets, on where to locate themselves, which places to avoid letting
their children play, or where not to let their dogs run loose. It also
emphasized the high complexity of the newly emerging eco-intimacies
I have outlined in this chapter; the technostruggle that the users of
Wadden Plastic were involved with essentially situated technology as
a required instrument to observe the plastic waste, while the waste
was essentially caused by projects of that same technology in the first
place. After all, the double-edged terms of technological innovation
and industrial growth are typically imposed as the only conceivable
horizon within intermodal transport industries. The glistening promises but wavering uncertainties that technology and automatization
give rise to in this industry seem to come to the surface only briefly
when a shipping container loses its cargo.

Nevertheless, the technostruggle that emerged at the Wadden
Islands—one that fostered a material participation with the plastic
waste resulting from the container spill—can best be understood as
novel appropriations of already existing technologies to make sense
of the complexities and often paradoxical nature of an ecologically
damaged environment that humans have made, but that is simultaneously remaking forms of human life in these environments. As such,
the technological response that the citizen-based project *Wadden
Plastic* prompted to the *MSC Zoe* container accident embodied a critical intervention that offered the people involved with the aftermath
of the spill a cross-section not only of the material excesses of global
plastic consumerism, but also of the present ecological condition itself
in their local surroundings.

It is important to note that, while this chapter explores the various
ways in which the Wadden residents attempted to make sense of the
emerging eco-intimacies that the *MSC Zoe* container spill generated,
my aim here is not to provide a commentary on whether or not these
bids to make sense can be considered successful. In fact, the asymmetri-
cal relations between failed logistical infrastructures, altered ecological

conditions, and humans grappling with this precisely compose new identical forces that resist immediate comprehensibility. This I consider an ineluctable part of the attempts to make sense of such new forces, an experience of a present-day planet that is both familiar and uncanny, accessible and concurrently off the map. An experience, as the *MSC Zoe* container disaster illustrates, that is additionally embedded in an enduring politics of visibility and invisibility.

This threshold of visibility still dominates the aftermath of the container spill. Weeks after the accident, when a significant number of spilled boxes were mapped and located, concerns were raised about the remaining containers that were still off the grid, floating somewhere in the alien ocean. Instead of children's toys or plastic pellets, at least two of these missing boxes contained the hazardous substance organic peroxide, a carbon-based compound that is relatively unstable and can easily break down into free chemicals. Three months after the spill, on March 29, 2019, one of these containers carrying the toxic substance was located along the German coast. The other one, together with dozens of containers that have partly, or not at all, been located, remain missing. With the location and condition of these boxes still invisible and unknown, search operations and clean-ups are still active today. It is perhaps these still missing shipping containers that, as carrying modes of uncertainty, precisely capture the complexity of eco-intimacies when things are not mappable anymore, and locations go beyond imagination. They ultimately situate humans, as Gayatri Spivak writes, as "planetary subjects" instead of "global agents"[31] within these newly emerging intimacies that briskly arise in contemporary damaged environments.

Conclusion

In the fifth episode of *Leviathan*, a ten-part film cycle directed by the artist Shezad Dawood, a more futuristic envisaging of the shipping container is projected upon us. In a scene set in a post-apocalyptic future in which half of the world's population is wiped out and the other half finds itself in extreme survival mode, two characters stand on a dock platform looking out on an enormous MSC cargo ship that is just leaving the port, stored with the familiar dazzling amount of colorful shipping boxes. The characters exchange distressed looks as one of them notes that "there is just something off with these ships."[32] Remarkably, while in *Leviathan*'s post-disaster society consumerism, globalization, and transnational infrastructure are no longer a given, this episode ambiguously signals toward at least one of their legacies: we see the same "invisible" shipping containers as we know them now still being transported. The scene leaves the question of what the steel boxes now conceal to the viewer's imagination: endangered species, both human and nonhuman? The fundamentals of a new world order (who could ever know?), or the last ever delivery of children's plastic toys?

More than this, Dawood's scene marks the importance not only of historicizing the shipping box and the content it conceals, but also of imagining what its particular future entails. The emerging eco-intimacies I have discussed in this chapter, then, do not confine themselves to set activities or fixed categories. Future entanglements of logistical infrastructures, ecological conditions, and plastic consumerism can best be imagined by understanding the present-day emerging intimacies and sense-making projects by humans of the damaged environments that projects of modernity have created.

At the Wadden Islands, these social, artistic, and technological bids to make sense simultaneously emerged once the new eco-intimacies that the container spill precipitated turned more perceptible. The sea, while filled with an alienating complexity of more-than-human ecologies, is also one of the most powerful public highways there is due to its embeddedness in globalized plastic consumerism, and it is this that many people along the Wadden coast were forced to intimately come to terms with. The shipping containers of the *MSC Zoe* that all reached the wrong address by washing ashore not only displayed an inventory of artifacts of our plastic culture, but likewise emphasized the intimate understanding of the uncomfortable yet inevitable entanglements between humans and their damaged surroundings in a warming world.

Notes

1. Kath Weston, *Animate Planet: Making Visceral Sense of Living in a High-Tech Ecologically Damaged World* (Durham, NC: Duke University Press, 2017), 2.
2. Ibid. 33.
3. Ibid. 33.
4. Marine Traffic, "Global Ship Tracking Intelligence," *AIS Marine*, https://www.marinetraffic.com/en/ais/home/centerx:4.2/centery:53.3/zoom:9.
5. Craig Martin, *Shipping Container*, Object Lessons (New York: Bloomsbury Publishing, 2016), 3.
6. Organisation for Economic Co-Operation and Development, *Intermodal Freight Transport: Institutional Aspects* (Paris: OECD Publications Service, 2001), 7.
7. Martin, *Shipping Container*, 4.
8. Tim Cresswell, "Ergin Çavuşoğlu and the Art of Betweenness," in

Spatialities: The Geographies of Art and Architecture, ed. Judith Rugg and Craig Martin (Chicago: Chicago University Press, 2012), 78.

9. Mark Levinson, "Setting the Standard," in *The Box: How the Shipping Container Made the World Smaller and the World Economy Bigger* (Princeton: Princeton University Press, 2016), 127.

10. Cresswell, "Ergin Çavuşoğlu and the Art of Betweenness," 72.

11. Shaylin Muehlmann, "Clandestine Infrastructures: Illicit Connectivities in the US–Mexico Borderlands," in *Infrastructure, Environment, and Life in the Anthropocene*, ed. Kregg Hetherington (Durham, NC: Duke University Press, 2019), 47.

12. Susan Leigh Star, "The Ethnography of Infrastructure," *American Behavioral Scientists* 43, no. 3 (1999): 386.

13. Brian Larkin, "The Politics and Poetics of Infrastructure," *Annual Review of Anthropology* 42, no. 1 (2013): 329.

14. Frédérik Ruys, "Cargo Vessel *MSC Zoe* Loses up to 270 Containers," *Vizualism*, January 9, 2019, http://www.vizualism.nl/containers-msc-zoe/.

15. Marcus Hand, "Video: Damage to Containers on the *MSC Zoe*," *Seatrade Maritime News*, January 4, 2019, https://www.seatrade-maritime.com/europe/video-damage-containers-msc-zoe.

16. Allan Sekula, *Fish Story* (Dusseldorf: Richter Verlag, 2002), 12.

17. *The Wire,* season 2, directed/written by David Simon, Ed Burns, George Pelecanos, and David Mills, aired June 1, 2003, on HBO, https://www.hbo.com/the-wire.

18. For example, in February 2019 the Dutch authorities believed 90,000 found vodka bottles in a container to be for the leader of North Korea, Kim Jong-Un. See Daniel Boffey, "Dutch Customs Seize 90,000 Bottles of Vodka Believed to Be for Kim Jong-Un," *Guardian*, 26 February, 2019, https://www.theguardian.com/world/2019/feb/26/dutch-customs-seize-90000-bottles-of-vodka-believed-to-be-for-kim-jong-un.

19. Sarah Hirsch, "Inhabiting the Icon: Shipping Containers and the New Imagination of Western Space," *Western American Literature* 48, no. 1 (2013): 30.

20. Nevertheless, the My Little Pony toys do bring back memories of the nearly 29,000 rubber ducks that were spilled in the Pacific Ocean in 1992 after a cargo vessel lost one of its containers; see Donovan Hohn, *Moby-Duck: The True Story of 28,800 Bath Toys Lost at Sea and of the Beachcombers, Oceanographers, Environmentalists, and Fools, Including the Author, Who Went in Search of Them* (New York: Viking Books, 2011).

21. "Mensen mogen aangespoelde spullen meenemen" (People are allowed to take home washed-up things), *Leeuwarder Courant*, January 2, 2019, https://www.lc.nl/binnenland/Mensen-mogen-aangespoelde-spullen-meenemen-24014584.html.

22. Weston, *Animate Planet*, 7.

23. Ibid. 7.

24. Ibid. 7.
25. I take my inspiration for this approach from Weston, *Animate Planet*.
26. Weston, *Animate Planet*, 73–81.
27. Ibid, 8.
28. For the dossier of published research, see "Microplastics and Nanoplastics," *Wageningen University & Research*, https://www.wur.nl/en/Dossiers/file/Microplastics-and-Nanoplastics.htm.
29. This team of ecologists is affiliated with the University of Groningen (RUG), the Netherlands Institute for Sea Research (NIOZ), and Wageningen Marine Research (WUR). For more information on the project, see "Help Map Plastic in the Wadden Sea," *University of Groningen*, 11 January, 2019, https://www.rug.nl/news/2019/01/help-mee-met-in-kaart-brengen-van-plastic-vervuiling-in-en-om-de-waddenzee?lang=en.
30. Weston, *Animate Planet*, 21.
31. Gayatri Chakravorty Spivak, *Death of a Discipline* (New York: Columbia University Press, 2013), 292.
32. Shezad Dawood, *Leviathan Cycle*, "Ismael" (2018).

Part III

Plastics in Art

8

The Pioneers of Plasticraft: When Artists Found Plastics in the United States

Danielle O'Steen

In the 1960s and 1970s, many artists working in the United States found plastics. "The new pioneer of plasticraft has invaded a domain," declared art critic Palmer French in an exhibition review from 1967.[1] He wrote that "[the artist] is actually experimenting and, in some cases, innovating techniques and potentially useful modifications not hitherto explored in commercial applications."[2] From 1965 to 1972, more than a dozen museums and galleries across the US hosted exhibitions focused on plastics in art.[3] Several guidebooks for artists were published in this period, outlining methods for working with plastics in an effort to close the knowledge gap between plastics and more traditional artistic media.[4] Artists were tapped by universities to teach plastics to students.[5] Even the plastics industry saw the value in plastic arts. In a 1968 editorial in the trade publication *Modern Plastics*, Joel Frados claimed that artists were finding new directions for plastics, making artwork that was "fast becoming the most exciting expression of our times."[6] Frados continued: "Where's the plastics industry in this revolution? Why hasn't it used plastics to similarly blast through the limitations of shape and form that have too long restricted consumer and industrial product design?"[7] Frados argued that it was the art community that would write the next chapter in America's history of plastics. The rise of plastics as artistic materials created a significant, if short-lived, "plastics moment" in the 1960s and 1970s.[8] Yet the collision of plastics and fine arts began much earlier, in an often overlooked chapter of American history in the 1930s, when artists *first* found plastics at the beginning of the modern plastics industry.[9]

In the 1930s, newly formed plastics companies were fighting to become relevant. Early on, these corporations saw the value in engaging with the arts and design to reach consumer and grow their emerging industry. By relying on visual displays, in advertising and public events, they found new opportunities to teach, to wow, and to engage customers. Their work with artists and industrial designers even drove innovation and generated new applications for these materials. This article will consider the early relationship between the arts and the plastics industry through the lens of one particular event in 1939: a competition for plastic sculpture, organized by plastics producer Röhm & Haas Company and the Museum of Modern Art (MoMA). This partnership shows the important, and often complicated, roles of artists in the burgeoning field of synthetics, anticipating collaborations between the arts and plastics for the decades to come.

When plastics found artists

In 1939, the Museum of Modern Art hosted a competition for sculpture in plastic. The contest was organized on the occasion of the 1939 World's Fair in Flushing, New York, as a collaboration between the museum and Röhm & Haas Company, a chemical corporation based in Philadelphia and Darmstadt, Germany. Röhm & Haas was launching their new product—Plexiglas—at the fair, and had decided to use the chosen sculpture as the centerpiece of their exhibit. The competition was the brainchild of industrial designer Gilbert Rohde, who had been hired by Röhm & Haas as the technical advisor and creative mind behind the fair exhibit. The winner of the competition, American artist Alexander Calder, was chosen by a jury from 250 entries. Calder was awarded $800 and his plastic arrangement in red, purple, and clear Plexiglas was installed at Röhm & Haas's display in the Hall of Industrial Science (see fig. 8.1). Lit from below, exhibiting the illuminating qualities of Plexiglas, Calder's sculpture was surrounded by interactive and didactic displays lauding the abilities of the plastic. The abstract artwork was the centerpiece of a visual agenda to sell this product, boasting the aesthetic as well as practical virtues of Plexiglas.

This early collaboration between the arts and a plastics company occurred at a crucial juncture for the nascent industry. Companies such as Röhm & Haas were continually exploring creative ways to sell their products to the public, as plastics were still new to many consumers and not yet a part of everyday life. These corporations were pressed to cultivate new audiences and find new ways of emphasizing the innovation and creative power of plastics. The enthusiasm they cultivated for plastics in the 1930s is unrecognizable today. That early—even naïve—sense of wonder around plastics has since been lost, as twenty-first-century life is filled with the epidemic of plastics, from the floating island of synthetic matter in the Pacific Ocean to the microplastics that wreak havoc on our ecosystem. Yet when modern plastics first emerged in the United States in the 1930s, the

Figure 8.1 *Alexander Calder's winning sculpture, Röhm & Haas exhibit, 1939, New York World's Fair, detail, Gilbert Rohde Collection.© 2022 Calder Foundation, New York/Artists Rights Society (ARS), New York. Photo: Matt Flynn @ Smithsonian Institution. Photo credit: Cooper Hewitt, Smithsonian Design Museum/Art Resource, NY.*

conversations surrounding the materials emphasized man's ability to create matter, pitching plastic as an abstract future rather than a particular present.

An article in the March 1936 issue of *Fortune* magazine announced that plastics had displaced "the divine monopoly on primary substances," and continued: ". . . He had brought forth the first plastic. And only man can make a plastic."[10] In 1940, *Fortune* magazine described the early years of the plastics industry, up until the 1930s, as "something like the world on the first day of creation."[11] Proclaimed the future of US industry, plastics were secular substances imbued with spiritual significance.[12] The early excitement of plastics' potential drove the vast expansion of synthetics, with fervor likened to a material religion. In the 1940 article, the *Fortune* author described the wide reach of the industry, where no one company was capable of cornering the market: "The plastics industry, indeed, is something special in the way of industrial chaos."[13] Citing the 240 trademarks in circulation, representing approximately fourteen types of plastics, the article casts the field of plastics—even in 1940—as a runaway train of exponential growth, and fiercely competitive for US companies like Monsanto, DuPont, General Electric, and Union Carbide.[14] And no wonder: the *Fortune* reporter writes that from 1935 to 1939, plastics production grew from 127 million to 247 million pounds of matter, with the finished objects valued at around $500 million.[15]

Even with such early growth, the American public struggled to understand modern plastics as a new set of industrial materials with no easily identifiable sources. Plastics could not be attached to one particular natural resource or production process. Rather, synthetics were developed through closed-door, laboratory-based innovations—an engineered, new resource for a modern era. Plastics were—and are in some cases, still—considered an all-encompassing entity, where "plastic" could stand in for a range of substances, such as polymethyl methacrylate (acrylic or Plexiglas), polystyrene (Styrofoam), and polyvinyl chloride (PVC). In the 1940s, the use of "plastic" to describe a range of materials continued to create a challenge for the industry. In a report on the Annual Fall Conference of the Society of the Plastics Industry, held in New York in November 1944, Colonel Willard Chevalier, publisher of *Business Week*, wrote that the industry had to educate consumers, as the "biggest difficulties in marketing plastics is the very word 'plastics'" as a generic term.[16] Two years later, a report on the Society of the Plastics Industry's National Exposition in New York in April 1946 stated: "Unfortunately the word *plastics* means only one material to too many people. They have not yet learned that cellulose acetate, polystyrene, phenolics, ureas and other materials are as different from each other as iron, steel, copper or lead."[17] Even while the industry was growing, corporations struggled to teach their consumers about the specificity of their products. To reach the public, plastics producers turned to visual corporate strategies that relied on exhibitions, displays, performances, design, and artworks, often drawing on the aesthetic vision of industrial designers at events such as world's fairs.

Plastics at the fair

Plastics had its most public debut at the 1939–40 New York World's Fair, held on over 1,200 acres in Flushing, New York.[18] Fair organizers played an important role in elevating plastics. They capitalized on the emerging industry by making it a major attraction. "Plastics have been called Nature's most formidable competitor," announced the first line of a press release sent out in anticipation of the fair.[19] The declaration was part of a ten-page document circulated by the Department of Feature Publicity on the presence of plastics at the fair, which would "offer America its first opportunity to see plastics used on a large scale—in art, in design and in practical applications."[20] Letters from fair organizers to industry heads, in their pitch to cultivate exhibitors, similarly showed their enthusiasm for plastics. Walter M. Langsdorf from the Department of Exhibits and Concessions for the Chemical and Plastics Building—later renamed the Hall of Industrial Science—wrote to a company in Michigan that the "fifty million expected visitors" to the fair would see the future potential for plastics and would "get a new appreciation of the possibilities in years to come."[21] In another letter from Langsdorf, he writes of the decision to situate the Chemical and Plastics Building in a prominent location at the fair, since "there are no two industries which will affect the everyday lives of our 50 million visitors more than Chemical and Plastics."[22]

The relationship between American industrial producers and the public was a major catalyst for the content at this world's fair. Organizers and vendors operated under the fair's theme, "Building the World of Tomorrow," which promoted progress, economic growth, and community strength, all to elevate the visitor as a responsible consumer.[23] The strength of the consumer was framed as a necessary condition to fight the still-present shadow of the Great Depression.[24] Run and supported on private funds, the fair gave corporations the power to build their own narratives. This event was driven by the visual agendas of American industry and its industrial designers. Figures such as Walter Dorwin Teague, Henry Dreyfuss, Norman Bel Geddes,

Raymond Loewy, and Gilbert Rohde drove the majority of the visual programing. They favored a streamlined aesthetic that maximized the use of new materials.[25] These designers came of age during a great expansion of America's material culture. Their businesses were fueled by the sudden need to explain newly introduced substances to the public, and their displays imagined a future in which every aspect of daily life would abound with the innovations of materials like plastics.

These industrial designers shone at the World's Fair. The exhibits they designed for their clients, however, were not wholly virtuous. These displays, in their celebrations of science at the fair, revealed corporate appropriation of scientific discoveries in dramatic fashion, often compounding innovation with magic and spectacle. This approach overran the efforts of the scientific community to promote a progressive narrative, a message of science aimed at bettering the livelihood of the American citizen rather than constructing the consumer.[26] Privileging spectacle, the chemical company DuPont took a cue from alchemy, a practice aimed at turning base matter into gold. Their exhibit included a demonstration transforming an everyday teaspoon into "gold" through electroplating. DuPont's stunts at the fair were also part of a push to improve public relations after the US Senate had called members of the DuPont family to testify in 1934 about war profiteering.[27] Other displays relied on theatrical props to sell their products, such as the transparent car at General Motors. For their exhibit designed by Bel Geddes, GM built the car on the foundation of a 1939 Pontiac Deluxe Six, using Plexiglas sheets specially tailored by Röhm & Haas.[28] Exhibitors like GM hoped to sell the public on plastics, making synthetics accessible and exciting. In this context, the strategy used by Röhm & Haas, to incorporate fine art into their exhibit, seems radical, as it privileged the role of artists in their corporate agenda.

The New York World's Fair declared how plastics in particular would be introduced to the American public, which included the acrylic sheet product Plexiglas. The plastic was first launched in 1936 as a cast acrylic sheet product developed by Röhm & Haas. It quickly found wide appeal for industrial use, outselling its rival Lucite, a product by DuPont.[29] The material, generically known by its chemical name polymethyl methacrylate (PMMA) but more commonly called acrylic, quickly garnered interest for industrial and military use as a lighter and more durable stand-in for glass.[30] Plexiglas is made by injecting the plastic in a liquid state into a mold of glass sheets on either end, and then immersing it in water for polymerization. The resulting hard plastic possesses a glossy, reflective surface like glass, as the material is a direct, positive impression of the glass mold, but with the durability of a synthetic substance. In short, Plexiglas is the progeny of glass, born from its surface with similar reflective and visual properties. To make tinted sheets, color is added to the initial liquid substance, making the hue inherent to the plastic. Initially developed as Plexiglas by Röhm & Haas, by the 1960s acrylic sheets were available from a variety of companies under trade names such as Perspex (from Imperial Chemical Industries) and Acrylite (from the American Cyanamid Company).

Cast acrylic sheets are lightweight, durable and well suited to use in arts and crafts. Comparable to glass but lighter, the material possesses great transparency and clarity and comes in a range of readymade colors. Plexiglas also has the unique ability to conduct light from one edge to the other, passing through the lens-like, slick surface. A 1945 corporate text from Röhm & Haas described this effect as "edge-lighting" or "light-piping," where light transmitted from edge to edge and emanated from "any surface that is interrupted by sanding, machining, carving, engraving, or even painting."[31] Plexiglas was also developed with a sister product, Crystallite, a powdered acrylic for molding. To sell Plexiglas and Crystallite to the public at the New York World's Fair, Röhm & Haas presented a multilayered exhibit that was interactive, educational, and material-driven, lauding the abilities of its products. At the center: the winning sculpture in Plexiglas by Alexander Calder.

As the facilitator and technical advisor for the Röhm & Haas exhibit, industrial designer Gilbert Rohde led the plastics contest and subsequent collaboration with Calder, working with the Museum of Modern Art (MoMA) as a co-sponsor. The competition garnered significant interest from the public. One reviewer claimed that around 600 people had inquired and 250 proposals were received from artists as well as teachers, engineers, lawyers, chemists, and medical students.[32] This response signaled a wide interest among people of all professions in engaging with plastics creatively at this time. In asking for entries, Röhm & Haas was attracting the attention of a new audience of consumers seeking to use the material in arts and crafts. Applicants were asked to mail in sketches, and five finalists were selected by the jury to make sculptures using Plexiglas provided by Röhm & Haas: Calder, along with photographer and graphic designer Herbert Matter, painter Werner Drewes, designer C. K. Castaing, and Bauhaus artist Xanti Schawinsky. A requirement stipulated that the sculpture had to be made by hand, without using molds, and had to fit within a 42-inch cube, limiting its size.[33] The judges—arts patron Katherine Dreier, sculptor Robert Laurent, and MoMA curator James Johnson Sweeney—declared Calder the winner in April 1939.[34] The artist was awarded the $800 prize and Röhm & Haas installed his Plexiglas sculpture at the center of their corporate display.

Calder at Röhm & Haas

Exhibited without partition, Calder's sculpture offered visitors the closest encounter with Plexiglas in the display, as the most direct link fairgoers had with the physical presence of the material (see fig. 8.1). Rohde placed the sculpture at the heart of the exhibit, surrounded by educational displays on Plexiglas. To the right, a large window display celebrated the industrial applications of Plexiglas in airplanes, boats, and trains, including models of an airplane with a plastic cockpit and a motorboat with a plastic shield. This display was oddly static and flat, as the objects were not functional, but rather prophetic props, as if part of a future theatrical production. The models anticipated but did not accurately represent the major civic and military uses of Plexiglas that were yet to come. It is perhaps telling that Rohde originally had another plan for this space. His design had included a "large Plexiglas molecule" for the center of the exhibit that would be raised and lowered—a plan that was left unrealized.[35]

The display to the left was more interactive, with shadow boxes highlighting the strengths of plastics, activated by touch buttons. Running in a horizontal line along the left wall were seven circular Plexiglas windows with assemblages inside. Each view presented the properties of Plexiglas and Crystallite with interactive features that could be operated by push buttons. Placed at eye level, the windows acted as portholes. This aspect of the exhibit was a draw for visitors, as evidenced by a photograph of busy fairgoers around the displays (see fig. 8.2). For instance, one window demonstrated illumination by allowing fairgoers to shine a light through the tips of Plexiglas rods at the push of a button. This display showed the material's ability to carry direct light through square bars, round rods, rectangular and hexagonal shapes, as if a manmade geode. Deemed "edge illumination," the transmission of light exhibited in this display paralleled the use of light in Calder's sculpture nearby.[36]

Calder's work was larger than life, towering over its visitors. It embodied the artist's experimentation with this new material, built from bisecting planes and circumventing rods in three colors of Plexiglas: clear, purple, and red, colors lost in the only remaining black and white photographs of the sculpture (see fig. 8.3). The sculpture represented several types of production—cutting, bending, smoothing, and attaching—using two different forms of the material:

Figure 8.2 Gilbert Rohde (1894–1944), Focal exhibit at the 1939 New York World's Fair, designed by Gilbert Rohde, detail, Gilbert Rohde Archive, Gift of Lee Rohde, P7521-0093. Photo credit: Cooper Hewitt, Smithsonian Design Museum/Art Resource, NY.

flat sheets cut into shapes, and round rods of varying diameters. The central figure was composed of flat sheets interlocked at their centers, one clear and the other tinted purple. Three different types of round rods, in varying thicknesses, orbited around the central figure. A thick vertical rod stood alongside, culminating in a red disk. A series of silhouettes and lines, the work gave the impression of volume while capitalizing on Plexiglas's transparency. Light acted as an ancillary medium, as the circular base was illuminated from below by an unseen source and transmitted light out through the round rods circling the sculpture. The entire work emitted enough light to bathe the onlooker in an ethereal glow. Calder's use of colored Plexiglas, as previously described, had also pushed expectations; the artist had intended

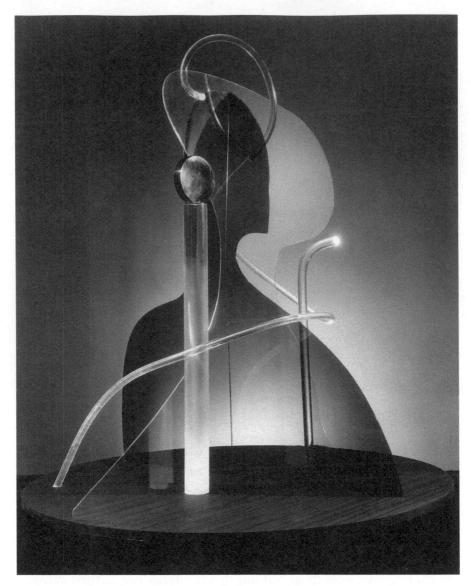

Figure 8.3 *Alexander Calder's entry which received first prize among the 250 entries, May 1939, photograph, 25.5 × 20.6 cm, Katherine S. Dreier papers/Société Anonyme Archive, Yale Collection of American Literature, Beinecke Rare Book and Manuscript Library. © 2022 Calder Foundation, New York/Artists Rights Society (ARS), New York.*

to use even more color and colored lights, as he claimed in a later autobiography, writing that the company had deemed his plan "too complicated."[37]

In hosting this competition, Röhm & Haas and MoMA were charging artists to become quick experts on a new material. For instance, they included very specific instructions in the call for entries, which was written by Rohde. This four-page document instructed artists how to bend, cut, polish, carve, and cement the material. Based on Rohde's own methods, this how-to guide reveals some of the physical challenges of the material that were likely new to the artists in submitting their entries, with some unusual instructions. For two-dimensional curves, Rohde suggested bending warmed Plexiglas over household items like "pipes, washing machines, waste baskets." For three-dimensional curves, he advised the artists to place the softened plastic between two pillows or press the plastic up against the body with a pillow or pad.[38]

Such specific, informal instructions are surprising for the context of this industrial exhibit. In sharing his trade secrets, Rohde hoped that "a new sculptural technique may be developed which will express the unique properties of this plastic material," as he writes in the guide.[39] He also addressed the difficulties artists would face in tackling the new material: "Plexiglas is a transparent, colorless plastic having an exceptionally high light transmission. Its sculptural properties are entirely different from the natural materials that have heretofore been available to the artist."[40] With this document, Rohde invited artists to become his collaborators. Plexiglas was presented as both a great and new future for artistic materials.

In May 1939, a month after the contest was over, Röhm & Haas and Gilbert Rohde circulated a document that announced Calder as the winner of the contest, in an effort to drum up media attention. The document was attached to a studio photograph of Calder's sculpture taken by commercial photographer Louis Werner in New York, showing the work in a darkened space illuminated by a mysterious light source in the background. The accompanying text summarized comments from the judges on why Calder's work had been chosen. First, the judges noted that the sculpture was strong when "viewed from all sides," as "an important factor in its selection."[41] This criteria is certainly related to the planned position of the sculpture in the center of the World's Fair exhibit, a quality relating more to the spatial logic of the corporate display than to the artwork. Furthermore, the document notes the strong use of light from a concealed source in the base, maximizing the material's ability to carry light and exhibiting "edge-lighting," stating: "The manner in which this sculpture exploits these properties makes light an organic part of the design."[42] For the judges, Calder moved beyond the static properties of Plexiglas in his design, using light as a secondary medium and opening up possibilities for artists working with this material in the future.[43] The text concludes by celebrating Calder's "contribution to the technique of sculpture" in the use of Plexiglas rods as well as sheets "to give motion and sweep to the whole design."[44] For the judges, as well as Röhm & Haas and Gilbert Rohde, Calder succeeded in creating a sculpture that represented innovation for this new plastic, in the form, composition, and material choices.

Despite his involvement, Calder's own account of the competition suggests a somewhat fraught collaboration. In his 1966 autobiography, he writes: "About this time [in 1939], a plexiglass manufacturer wanted to get some publicity, cheaply. So he put up a modest prize and published a lot of rules about what you could do with plexiglass ... I entered this competition, but I did not like their suggestions on how to work it. I just used a hacksaw and a file. Anyway, they finally took my object and reproduced it—they thought—nice and smooth."[45] Even though Rohde's document specifically states that Röhm & Haas would not manufacture objects for the artists, Calder tells a different story, recounted decades later. The resulting work emerges from this tension between an artist's exploration of a new material and corporate expectations of how Plexiglas should be presented.

After the competition, Röhm & Haas retained ownership of Calder's sculpture, though its whereabouts today are unknown. As part of the call for entries, MoMA also stated that it would exhibit the winning designs after the fair, but this intended plan never materialized.[46] In 1940, Röhm & Haas shuttered their display at the fair, citing lack of attendance and frustration with the prominent placement of their competitors in a letter to fair organizers.[47] Calder's sculpture was subsequently included in the *Plastics* exhibition at the Institute of Modern Art, Boston, in 1940.[48] Exhibited among everyday objects made in plastic, the sculpture introduced the experimental possibilities of Plexiglas to a new audience, continuing the work done by Rohde and anticipating the exhibitions decades later.

The next pioneers of plasticraft

Exhibits at the World's Fair celebrated the novelty and optimism of plastics. Röhm & Haas's exhibit and the competition organized with MoMA embodied the same utopian ambitions, looking to artists to share ideas, innovate, and push the boundaries of their materials. In this early period, plastics companies were willing to be experimental and explore new territories, which contributed to the industry's ability to grow and adapt for the years to come. As one example, Röhm & Haas, as well as Rohde, imagined a future of great partnerships between the arts and industry. Unfortunately, these ambitions would not materialize until decades later, in the "plastics moment" of the 1960s and 1970s. However, these early collaborations would spark a diversification of the use of plastics, which would take off in the encroaching wartime era.

In World War II, plastics took on increasingly important roles, as scientists used synthetics to develop more efficient materials for everything from the largest vehicles to the smallest detail on a soldier's uniform.[49] The armed forces of the modern age were redesigned to be the carriers of a range of synthetic objects. The wartime effort also fanned the flames of the plastic industry's growth, creating high demand for innovation. In the postwar era, plastics became at once the most accessible technology for Americans to understand, live with, and invest in, while at the same time the most threatening. By the 1940s, plastics were growing in prominence; one 1942 article claimed the substances were "emerging as *materials in their own right* . . . [and] may one day become a dominant material, just as steel did in the immediate past."[50] It was with these grand ambitions that plastics entered the national landscape, anticipating the daily presence of synthetics that is now part of twenty-first-century life. This shift came with the end of the war, when plastics producers were pressed to find a new, national need for synthetics, so as to keep their many factories around the US—built for the war effort—active and growing. A report on a conference for the Society of the Plastics Industry in 1944 read that "the peacetime possibilities of plastics are virtually untouched," opening up a "broad field of new markets" to redefine and even solve the problems of the postwar era, at least in the eyes of the industry.[51] To stay afloat, synthetics corporations sold plastics as the future of the American home, manifested in products like Tupperware.[52]

With great comfort in plastics came unbridled abundance for inexpensive, convenient, and disposable objects, transforming the materials of domestic life in the United States. At the same time, artists were finding the value in using synthetics to create their objects. The November 1944 issue of *Modern Plastics*, for instance, included an article on plastics and sculpture, claiming that synthetics could "liberate" artists from traditional materials and that the substances "combine the colorful clarity of glass, the massiveness of stone, the permanence of marble, the mobility of clay, the warm surface of wood, the yielding rigidity of ivory."[53] In this context, plastics were presented as composed of the best qualities of other materials to support their potential for sculptural innovation.

By the 1960s, artists had taken up the call and further honed their craft with these synthetic materials. As critic John Perreault noted in 1969: "When an artist tells you he works in plastic—and many do—the next question—if you want any information at all, has to be what kind of plastic. Uvex, acrylic, plexiglas, fiberglas, vinyl, lucite? And then you have to find out how he uses it."[54] This specialized attention to plastics allowed artists to expand the potential of their materials, creating objects that strayed far from the manufactured consumer products of plastics past, bringing a new awareness to an art-going public. Perreault remarked on this phenomenon: "We have finally begun to adapt to plastic . . . and some of this has to do merely with the presence of plastic all around us, not to mention the fact that artists recently have done a lot to wake us up to it."[55]

The many exhibitions on plastics and art in this period, particularly those held across the US from 1965 to 1972, presented the American public with a different side of plastics through immersive environments, absurdist architectural forms, and unlikely monuments constructed by artists.[56] *A Plastic Presence* offered the largest survey of sculpture in plastics by forty-nine artists working in the US and Canada, organized by the Milwaukee Art Center with stops at the Jewish Museum in New York and the San Francisco Museum of Art from 1969 to 1970. Milwaukee Art Center director Tracy Atkinson, who organized *A Plastic Presence* with the institution's Director of Exhibition, John Lloyd Taylor, stated what was at stake with their show: "Plastics have earned a bad name from all the flimsy toys, all those endless and deadly countertops and lighting fixtures, those boxes and appliance housings which split and shattered under normal use, not to mention the early, less-than-satisfactory 'synthetic' fabrics and leather."[57] They position the artists in the exhibition as essential to correcting this situation.

In many of these exhibitions, organizers took on the role of educators, even providing didactic material to educate the public on plastics. *Plastics West Coast* at Hansen-Fuller Gallery in San Francisco in 1967 offered visitors a glossary of terms, defining resins, acrylics, vinyl, and other types of plastics as well as laminating, fabricating, and vacuum forming as methods in plastics production.[58] *Plastic as Plastic* at the Museum of Contemporary Crafts (now Museum of Arts and Design) in New York in 1968 published a catalogue that included the scientific compositions and characteristics of synthetics as well as a historical survey adapted from a document by the Society of the Plastics Industry.[59] In a foreword for the exhibition *Made of Plastics* at Michigan's Flint Institute of Arts in 1968, director G. Stuart Hodge cited a select list of inventors and corporations as a short history of plastics.[60]

Even when plastic artworks appeared in art museums or gallery spaces, the presence of the industry was always looming. Several exhibitions worked directly with industrial companies. *Plastic as Plastic* was sponsored by Hooker Chemical Corporation, a subsidiary of Occidental Petroleum Corporation, later exposed in the late 1970s as responsible for the Love Canal disaster.[61] *A Plastic Presence* paired up with Philip Morris and their plastics subsidiary Milprint for financial support, to the dismay of critics. Grace Glueck of the *New York Times* lambasted the exhibition's corporate arrangement with Philip Morris, which led the artist to be compelled into "image-building for the plastics industry."[62] Finally, the aptly titled *The Last Plastics Show* at the California Institute of Arts in Valencia, California in 1972—which presented itself as the finale of these exhibitions—stemmed from a collaboration between the university, Hastings Plastics in Santa Monica, California, and artists De Wain Valentine, Judy Chicago, and Doug Edge, who organized the show of twenty-four artists. With these sponsorships, the industrial companies did not require artists to be complicit in corporate activities, as many of the works privileged experimentation over political statement. However, the very practice of working with plastics often came with an implied affiliation to the American industry, calling to mind the earlier collaboration between Calder and Röhm & Haas.

These 1960s and 1970s exhibitions introduced the public to a new generation of plastic sculpture made through a range of novel methods and techniques. Critics struggled to make sense of the artworks and often scoffed at the very presence of plastics in a museum setting. In a review of *Plastic as Plastic* for *The New York Times*, Hilton Kramer asserted that artists working in plastics were "primitives" rather than innovators, in comparison to the innovations of industrial designers.[63] He wrote: "Instead of a grand synthesis, they have achieved only a public display of divided loyalties."[64] On *A Plastic Presence*, *Chicago Tribune* critic Thomas Willis wrote: "If plasticult belongs to all of us, these works seem to say, let's do the best we can to find its inner poetry and logic."[65] The critics reflected an anxiety with this "plasticult," even as artists continually found their own forms of innovation among plastic materials.

This "plastics moment"—and the rise of the "pioneer of plasticraft"—fulfilled the earlier ambitions of Röhm & Haas and Rohde from the 1930s: to link the plastic industry with artists and creative fields, and to encourage the growth of the industry through aesthetic as well as practice means. In the 1960s and 1970s, corporations were primarily exhibition sponsors, not organizers in their own rights. Plastics companies were no longer the driving force behind artistic innovation, as had been the case at the New York World's Fair. That role later fell to artists, particularly as synthetic materials became increasingly available to find and buy in the latter part of the twentieth century. Yet the partnerships of the 1930s had laid the groundwork for these intersections between the arts and plastics, showing how essential artists and exhibitions were to the foundation of the modern plastic industry.

Notes

1. Palmer French, "Plastics West Coast," *Artforum* 6, no. 5 (January 1968): 48. French was reviewing the exhibition *Plastics West Coast* at Hansen-Fuller Gallery in San Francisco in 1967.
2. Ibid. 48.
3. The exhibitions included: *Plastics*, John Daniels Gallery, 1965; *Plastics West Coast*, Hansen-Fuller Gallery, San Francisco, 1967; *Plastic as Plastic*, organized Museum of Contemporary Crafts (now Museum of Arts and Design), New York, 1968, and shown at the Arts and Industries Building, Smithsonian Institution, Washington, DC and Oberlin College in Oberlin, OH, between 1969 and 1970; *Plastics: Los Angeles*, California State College at Los Angeles, 1968; *Made of Plastic*, Flint Institute of Arts, MI, 1968; *Plastics and New Art*, organized by the Institute of Contemporary Art at the University of Pennsylvania, Philadelphia and shown at the Marion Koogler McNay Art Institute in San Antonio, TX, 1969; *Hard, Soft and Plastic*, 1969, Wilcox Gallery, Swarthmore College, Swarthmore, PA; *A Plastic Presence*, organized by the Milwaukee Art Center (now Milwaukee Art Museum) and traveled to the Jewish Museum in New York and the San Francisco Museum of Art (now San Francisco Museum of Modern Art), between 1969 and 1970; *Editions in Plastic*, 1970, University of Maryland Art Gallery, College Park, MD; *Works in Plastic*, 1970, Trinity College Art Gallery, Hartford, CT; *Tony Delap/Frank Gallo/Eva Hesse: Trio*, Owens-Corning Fiberglas Center in New York, 1970; and *The Last Plastics Show*, California Institute of Arts in Valencia, CA, 1972.
4. Thelma Newman, *Plastics as an Art Form* (Philadelphia: Chilton Book Company, 1964); Nicholas Roukes, *Sculpture in Plastics* (New York: Watson-Guptill Publications, 1968); Clarence Bunch, *Acrylic for Sculpture and Design* (New York: Van Nostrand Reinhold Company, 1972); and Newman, *Plastics as Design Form* (Philadelphia: Chilton Book Company, 1972).
5. As one example, artist De Wain Valentine recounts being hired by UCLA in 1965 to teach students how to work in plastics, a position he held for two years. Tom Learner, Rachel Rivenc, and Emma Richardson, *From*

Start to Finish: De Wain Valentine's Gray Column (Los Angeles: Getty Conservation Institute, 2011), 7.

6. Joel Frados, "Plastic for Art's Sake," *Modern Plastics* 45, no. 12 (August 1968): 45. The editorial cites *Plastic as Plastic*, the 1968 exhibition at the Museum of Contemporary Craft in New York, as proof of artists' innovations in plastics.

7. Ibid.

8. On this topic, see: Danielle O'Steen, "Plastic Fantastic: American Sculpture in the Age of Synthetics" (PhD diss., University of Maryland, 2018).

9. I use the term "modern plastics" to distinguish these from earlier iterations of the synthetic materials that appeared before the 1930s, which were more unstable products that are no longer popular today, such as celluloid developed in the mid nineteenth century.

10. "What Man Has Joined Together ...", *Fortune* (March 1936): 69, 149–50.

11. The same article illustrates the growing field of plastics as a fictional continent called "Synthetica," which includes "Rayon Island" and a country of Acrylic, with the capital of Plexiglas. Other synthetic nations—Petrolia, Alkyd, and Petrolia among them—include a smattering of "cities" representing industrial brands that are no longer extant. "Plastics in 1940," *Fortune* 22 (October 1940): 90, 92–3.

12. Here I borrow from David Nye and his discussion of sublimity and an almost-religious veneration of technology and infrastructure in the United States. He writes: "The technological sublime is an integral part of contemporary consciousness ... In a physical world that is increasingly desacralized, the sublime represents a way to reinvest the landscape and the works of men with transcendent significance." David E. Nye, *American Technological Sublime* (Cambridge, MA: MIT Press, 1994), xiii.

13. "Plastics in 1940," 94.

14. Ibid. 94.

15. This figure does not include plastics in liquid form, which would bring the 1939 figure to more than 300 million pounds of material. Ibid. 89.

16. Willard Chevalier, "When Plastics Comes Home from the Wars," *Modern Plastics* 22, no. 4 (December 1944): 129.

17. "The Plastics Industry Has Come of Age," *Modern Plastics* 23, no. 8 (April 1946): 132.

18. Richard Wurts, *The New York World's Fair 1939/1940 in 155 Photographs* (New York: Dover Publications, 1977), xiii. Many plastics exhibits were placed in the Chemical and Plastics Building, later renamed the Hall of Industrial Science, one of the most central structures.

19. "Plastics at the New York World's Fair," Department of Feature Publicity, New York World's Fair 1939 and 1940 Incorporated records, 1935–1945, New York Public Library, New York, NY.

20. Ibid.

21. Letter from Walter M. Langsdorf, Department of Exhibits and Concessions, Chemical and Plastics Building, to Roy Austin, VP, Reynolds Molded Plastics Division, Reynolds Spring Company, MI, New York World's Fair 1939 and 1940 Incorporated records, 1935–1945, New York Public Library, New York, NY.

22. Letter from Walter M. Langsdorf, Department of Exhibits and Concessions, Chemical and Plastics Building, to J. V. Stauf, Solvay Sales Corporation, New York, NY, New York World's Fair 1939 and 1940 Incorporated records, 1935–1945, New York Public Library, New York, NY.

23. Joseph P. Cusker, "The World of Tomorrow: Science, Culture, and Community at the New York World's Fair," in *Dawn of a New Day: The New York World's Fair, 1939/40*, ed. Helen A. Harrison (Flushing, NY: The Queens Museum, 1980), 5.

24. Robert W. Rydell, John E. Findling, and Kimberly D. Pelle, *Fair America: World's Fairs in the United States* (Washington, DC: Smithsonian Institution Press, 2000), 11.

25. A. Joan Saab, *For the Millions: American Art and Culture Between the Wars* (Philadelphia: University of Pennsylvania Press, 2004), 135.

26. Peter J. Kuznick, "Losing the World of Tomorrow: The Battle over the Presentation of Science at the 1939 New York World's Fair," *American Quarterly* 46, no. 3 (September 1994): 341–73.

27. Ibid. 355, 361.

28. The car, which was sold at auction at Sotheby's in 2011 for $308,000, was only one of two produced. Costing a then-astronomical $25,000 to construct (approximately $460,000 by today's standards), it was hardly a practical vehicle of the future, and only had 86 miles on the odometer at the time of its sale in 2011. Katie Scott, "1939 Clear Car Showcases Miracle of Plexiglas," *Wired* (June 15, 2011).

29. Sheldon Hochheiser, *Röhm and Haas: History of a Chemical Company* (Philadelphia: University of Pennsylvania Press, 1986), 59.

30. Ibid. 59.

31. *Chemicals for Industry* (Philadelphia: Röhm & Haas Company, 1945), 118.

32. Elizabeth Clark, "Plastics Sculpture Competition," *Modern Plastics* (June 1939): 29, 68.

33. Competition for Sculpture in Plexiglas, Röhm & Haas Company, New York World's Fair 1939 and 1940 Incorporated records, 1935–1945, New York Public Library, New York, NY.

34. "Plexiglass Sculpture Prizes are Awarded," *Pencil Points* 20, no. 6 (June 1939): 56–7. Regina Lee Blaszczyk, *Röhm and Haas: A Century of Innovation* (Philadelphia: Röhm & Haas Company, 2009), 62, 64.

35. Memorandum from Dana Stuart Cole for Gilbert Rohde to Langsdorf, July 13, 1938, New York World's Fair 1939 and 1940 Incorporated records, 1935–1945, New York Public Library, New York, NY. The plan was to exhibit a "large Plexiglas molecule" in the exhibit, but due to

the restrictions in the architecture Rohde decided to use "full size models of a section of a dining car, of an aeroplane, of a motorboat, and of an automobile" to be shown in "large shadow boxes." An early application describes the Plexiglas molecule as "spheres of different sizes and colors, representing different elements," which would have had a motor to raise and lower the structure and three neon tubes for illumination. Outline Specifications, Plexiglas and Crystallite Exhibit, June 6, 1938, New York World's Fair 1939 and 1940 Incorporated records, 1935–1945, New York Public Library, New York, NY.

36. An image of this display in Rohde's archive uses the phrase "edge illumination." Gilbert Rohde Collection, 1930–1944, Cooper Hewitt, Smithsonian Design Museum, New York, NY.

37. Alexander Calder, *Calder: An Autobiography with Pictures* (New York: Pantheon Books, 1966), 175.

38. Competition for Sculpture in Plexiglas, Röhm & Haas Company, New York World's Fair 1939 and 1940 Incorporated records.

39. Ibid.

40. Ibid.

41. Competition for Sculpture in Plexiglas, May 1939, Katherine S. Dreier Papers/Société Anonyme Archive, Yale Collection of American Literature, Beinecke Rare Book and Manuscript Library, Yale University, New Haven, CT.

42. Ibid.

43. On artist Donald Judd and his work in Plexiglas: O'Steen, "Plastic Fantastic," 29–63.

44. Competition for Sculpture in Plexiglas, May 1939, Katherine S. Dreier Papers/Société Anonyme Archive.

45. Calder, *Calder: An Autobiography*, 175.

46. Richard Meyer notes that MoMA never exhibited the winning sculptures in its galleries. Richard Meyer, *What Was Contemporary Art* (Cambridge, MA: MIT Press, 2013), 248.

47. Letter from D. S. Frederick of Röhm & Haas to Walter M. Langsdorf of Department of Exhibits and Concessions, June 9, 1939, New York World's Fair 1939 and 1940 Incorporated records, 1935–1945, New York Public Library, New York, NY.

48. For a brief account of this exhibition, see Meyer, *What Was Contemporary Art*, 241–6.

49. For instance, a report on the US Navy and plastics from 1945 claims that a battleship includes "around 50,000 plastic parts." "Plastics at war—Navy," *Modern Plastics* 22, no. 5 (January 1945): 100.

50. Joseph L. Nicholson and George R. Leighton, "Plastics Come of Age," *Harper's Magazine* 185 (August 1942): 301.

51. Chevalier, "When Plastics Comes Home," 129.

52. Jeffrey Meikle has argued that the rise of Tupperware in the home in the 1950s marked the beginning of seeing "plastic as plastic," which continued through the 1960s. Jeffrey L. Meikle, *American Plastic: A Cultural*

History (New Brunswick, NJ: Rutgers University Press, 1995), 181. An article in the May 1950 issue of *Fortune* also claimed that the "greatest concentration of plastics around the average American is in the kitchen." "A 1950 Guide to Plastics," *Fortune* 41 (May 1950): 110. For further reading: Alison J. Clarke, *Tupperware: The Promise of Plastic in 1950s America* (Washington, DC: Smithsonian Institution Press, 1999).

53. Jan and Ursula de Swart, "A Medium for Sculpture," *Modern Plastics* 22, no. 3 (November 1944): 103.
54. Perreault, "Plastic—Very Present," *Village Voice*, 4 December 1969: 28.
55. Ibid. 28.
56. See footnote 3.
57. *A Plastic Presence* (Milwaukee: Milwaukee Art Center, 1969), 6.
58. "Plastic Definitions," Hansen Gallery, Exhibitions Archives, San Francisco Museum of Modern Art, San Francisco, CA.
59. *Plastic as Plastic* (New York, NY: The Museum of Contemporary Crafts, 1968), 7–8.
60. *Made of Plastics* (Flint, MI: Flint Institute of Arts, 1968), 2.
61. Andrew C. Revkin, "Love Canal and Its Mixed Legacy," *New York Times*, November 25, 2013, https://www.nytimes.com/2013/11/25/boo ming/love-canal-and-its-mixed-legacy.html.
62. Grace Glueck, "Building the Plastic Image," *New York Times*, December 7, 1969: D28.
63. Hilton Kramer, "'Plastic as Plastic': Divided Loyalties, Paradoxical Ambitions," *New York Times*, December 1, 1968: D39.
64. Ibid. D39.
65. Thomas Willis, "Bringing the Art of Plastics into Focus," *Chicago Tribune*, February 8, 1970: G10.

9

Plastic Intimacy:
Chinese Art Making as Recycling Practice

Victoria Oana Lupascu

In the edited volume *Waste: Out of Sight, Out of Mind*, Christof Mauch and his collaborators theorize the apparent absence of waste from urban areas by interrogating the facile connection between waste's awayness and the amount of critical attention it receives both from academic circles and individuals around the world. Mauch and the volume's contributors join an increasingly large community of scholars, activists, and socially engaged people who describe waste as geographically located outside habitual routes of seeing, and unveil the falsity of such an absence while highlighting plastic materials' ubiquitous presence in the world—from microparticles in the water supply to clothes to food packaging to computers and daily use objects.[1] The apparent absence has been fueled by globalization and extensive trade treaties between Asian and African countries and the West regarding waste disposal and plastic waste recycling and reusage in a transcontinental fashion: plastic and paper waste from the West arrives in Asia and Africa, where is partially recycled and remade into more packaging and tools that travel back to the West in the form of finite products, with little mention of waste itself. Increased consumerism and sanitization of urban environments as well as a generalized lack of social awareness regarding plastic waste enable uninterrupted import-export cycles which function within their own industries and local economies. One of the most salient examples comes from China, which began importing plastic and paper waste in the 1980s and reached a peak in the twenty-first century, with a 56 percent share of the global market of plastic waste imports going into mostly rural areas for repurposing and recycling. In villages such as 柳絮 Liuxu, plastic permeates bodies, lands, and the atmosphere in a complete reverse of Mauch's collection title.

王久良 Wang Jiuliang refuses to accept the idea that once waste is not in sight, it should no longer be a concern and asks "Where does the waste go?" to open a transcontinental dialogue which can inform the visual politics of waste in general, and plastic waste in particular. After directing *Beijing Besieged by Waste (垃圾围城 Laji Wei Cheng)* (2011), Wang started a new project in 2014 and released it in 2016 where he analyzes and uncovers the microeconomies, the entrepreneurship, the migrant labor routes and the deterioration of agricultural practices in China that enable former farmers and people from rural areas to begin their own plastic recycling businesses.[2] The plastic is mainly of foreign origin, from Europe and the US. From the beginning of the documentary, *Plastic China (塑料王国 Su Liao Wang Guo)* (2016), the owner of such a business, Wang Kun, tells the camera: "I'm just a farmer, I have no other skills. If I don't do this [recycle plastic], how can I make money?" His employee, Peng Wenyuan, a migrant worker from Sichuan province, echoes the same words and posits that it is easier to work in the makeshift workshop than be at home in Sichuan, where he can't find work as an unskilled laborer. The amount of plastic waste present in multiple panoramic establishing shots is breathtaking and implies there will be no shortage of work. Moreover, the plastic's overabundance offers a visual materialization of many statistics on waste imports and suggests the lack of regulation and systematic handling of such waste. At the same time, we notice the treacherous and imperfect nature of the recycling process, which is far from the cleanliness rhetorically constructed in environmental discourses in the West.[3] Another implication emerging from the panoramic shots is that of the intimate connection between plastic, the atmosphere, the land, and every form of life in the area, creating, hence, a new type of ecology characterized by a problematic symbiosis between animate and inanimate entities. Thus, plastic transmutes, changes shapes to smoke and microparticles that adhere to surfaces, penetrates bodies, going out of a traditional visual register and demanding the application of critical tools that can conceptualize the already permanent intimacy between plastic and the environment of which the human body is part.

Plastic as a pollutant is not a local but a global issue, as the death of marine life in all oceans shows. Wan Yunfeng 万云峰, in a similar effort to Wang Jiuliang, designs clothes out of plastic waste in a haute couture style and poses for pictures to raise awareness about the deleterious effects such waste has on all aspects of life. For instance, in the series *Protection of the Ocean*, Wan designs a composite outfit from construction foam, plastic, and bricks and wears it while resting in a crouching position on top of a traditional fishing boat (see fig. 9.1). The outfit's weight rests in the frontal part and bends his body to suggest painful entanglement. Also, the fishing boat appears old and unused, a sign of decreased fishing possibilities for the locals as

Figure 9.1 Wan Yunfeng.

a result of diminishing fish populations due to pollution. Wan's body attempts to balance the burden on his neck, but its length and suggested weight occupy the viewer's attention. The photograph's composition is unbalanced and the color scheme emphasizes the plastic entanglements, highlighting their impact on the environment. With works similar to this, Wan Yunfeng challenges traditional practices of raising awareness and, echoing artistic endeavors from the Caribbean and Africa, joins Wang Jiuliang in highlighting plastic's influence on cultural practices around fashion, filming techniques, the ethos of documentary films, and the role of globalization in and on human life and bodies at large, and in China in particular.

Both Wang and Wan insist on integrating plastic waste into art as a way of enhancing their creative processes and aesthetics, but also to draw attention to the planet's suffocation via waste. They are part of larger artistic collectives in China and around the world and contribute their voices and practices to the formation of global civic and ecologic consciousness. They represent the self-driven, independent researchers and artists who bridge disciplines and artistic domains to show that understanding plastic demands a transcontinental and transdisciplinary collaborative hermeneutic approach.

Ecocinema

As waste studies, environmentalism, and ecocriticism have carved a space for themselves in the academy in the past few decades and have been devising new hermeneutic tools and paradigms for analyzing humans' exploitation and extraction of resources from the environment, art and literature have remained at the heart of this new approach. Ecocinema, following the environmental turn in film studies,[4] offered a defined and classifiable domain for directors to explore notions of waste, structural violence, the effects of globalization, and the changes in the relationship between the nature and human and non-human bodies. Scott MacDonald argues that "ecocinema's [job is the] retraining of perception, as a way of offering an alternative to conventional media-spectatorship, . . . as way of providing something like a garden—an 'Edenic' respite from conventional consumerism—within the machine of modern life, as modern life is embodied by the apparatus of media."[5] MacDonald opposes ecocinema to conventional media via a return to disciplined attention and multidimensional perception, specifically the perception of the different effects our actions have on the environment around us. MacDonald sees the long take and the slow pace of a film or documentary as ecocinema's defining characteristics, elements that, with a certain amount of affective force, demand the audience's attention and engagement with the images. Additionally, Sheldon Lu and Haomin Gong highlight the "participatory, moral, activist aspect[s] of ecocinema"[6] that allow for multiple points of analytical entry into the study of films "with ecological significance."[7] Documentaries such as *Plastic China*, or *The Blind Shaft* (盲井 *Mang Jing*, dir. Li Yang, 2004) and Chai Jing's *Under the Dome: Investigating China's Smogs* (穹顶之下 *Wan Ding Zhi Xia*, 2015), in their depiction of environmental degradation and evolution of the relationship between humans and nonhumans, expose such an ecological consciousness beyond doubt. However, one important yet understudied aspect of ecocinema is its eclectic technical repertoire. In the documentaries mentioned above, techniques specific to Italian and French cinéma vérité, Russian socialist realism, ethnographic film, and anthropological research films, as well as time-vetted techniques used in Chinese cinema during the Mao era, enable the long takes and tame the images' pace. Thus, the "'Edenic' respite from conventional consumerism" imagined by MacDonald is a multifaceted cinema that turns the respite into a complex work of multimodal analysis of consumerism, globalization, and their effects on the environment.

Ecocinema represents a flexible enough framework to incorporate documentaries and films with nuanced and plurivalent national affiliations. Sheldon Lu in *Chinese Ecocinema in the Age of Environmental Challenge*, published in 2009, posits that "the study of Chinese ecocinema specifically should be placed squarely within the specific intellectual and socio-historical Chinese contexts that may be different from Euro-American settings in significant ways. . . . Chinese ecocinema is a critical grid, an interpretative strategy. It offers film viewers and scholars a new perspective in the examination of Chinese film culture."[8] While Lu, in collaboration with Haomin Gong and in the light of Sinophone studies and scholarly debates about the meaning of the "Chinese" concept,[9] elaborates in the 2020 volume a larger definition of the term "Chinese," his 2009 argument brings to the fore a few critical aspects, chief among which is the juxtaposition between a self-reflexive gesture on the directors' part regarding the environment writ large, and the Chinese state's position toward ecological exploitation. For instance, Wang Jiuliang's *Beijing Besieged by Waste* has allegedly triggered a vast cleaning operation of landfills and waste management sites around Beijing, while *Plastic China* has contributed to an import ban on multiple types of plastic into China from the USA and Europe.[10] The connection between art, politics, and economy here is reminiscent of old Confucian relations between intellectuals and the emperor and paints a collaborative exchange between art practices and a growing economy. However, although valuable, the emphasis on geographical specificity and politics draws attention away from the objects, or hyperobjects, as Tim Morton defines plastic,[11] at the core of the critique while marginally taking into consideration the aesthetic transformations necessary for the existence of ecocinema as a critical grid. Lastly, ecocinema, carrying the implied understanding that man-made and natural disasters, as well as the effects of environmental erosion, function as active agents for every type of life, is instrumental in the formation of a global, collective ecological consciousness that abounds in culturally driven regimes of visibility and modes of engagement with nature from a non-anthropocentric point of view.

On a closer look, ecocinema becomes an umbrella term housing, as Kiu-Wai Chiu asserts, deep ecology film, environmentalist film, and perception training film on environmental issues.[12] These subgenres springing from multiple environmental imaginations bring into conversation disparate disciplinary formations and redefine cinematic techniques, but, I argue, one should not lose sight of the importance of the object of analysis itself, namely plastic and the subsequent waste. Balancing the anthropocentric and non-anthropocentric perspectives on plastic simultaneously reveals the state's and the individual's relationship to one another and to nature in a race for profit accrual, territorial expansion, and positive or negative excess as a precursor to waste. Thus, ecocinema, in its multiple forms, offers a capacious framework for defining plastic waste and the depth of its entanglement with every other aspect of life.

Plastic as a hyperobject

The documentary *Plastic China* starts with a long take from inside a plastic waste pile where toddlers dig in as a game of finding materials to "build a house to sleep in at night." The next set of images shows ships in Tsingtao Harbor bringing in hundreds, if not thousands, of containers to be distributed over northeastern China. The juxtaposition of the two shots implies that the containers are full of plastic waste from abroad to be recycled in China and gives a first, large-scale view of the amount of waste entering the harbor. The images amount to a crescendo: from inside one, indistinct pile to a harbor full of containers, from the children's play to international trade, from a claustrophobic digging motion punctuated by the phrase 挖不到 (*wabudao*, not being able to reach something through digging) to the crowded harbor full of ships carrying containers full of waste. All of this indicates the ubiquitous nature of plastic, as it takes multiple shapes and forms that simultaneously enable and stunt all development. Moreover, right from the start, the documentary establishes the globalized network through which plastic circulates, thus putting China and the rest of the world on equal terms regarding their responsibility to environmental protection. With this, the documentary takes a more nuanced stance in its criticism of plastic's handling and invests the images with full narrative agency in suggesting that plastic and plastic waste permeate every aspect of life and are part of every decision conducive to international collaboration.

Plastic is a hyperobject, as Morton argues, since it is "massively distributed in time and space relative to humans."[13] With a very long lifespan that incorporates, at times, multiple generations, it functions on a different temporality than other objects and humans, and is part of any type of space, although it can be invisible to the naked eye in multiple instances. Additionally, this is a non-local material, without any geographical allegiances, but recognized/recognizable everywhere. In Wang Jiuliang's eco-documentary, sound precedes the image and our ears capture the familiar ruffling of plastic scraps before seeing the pile of cling film, bags, and other wrapping materials. The eco-documentary comments on the multisensorial engagement with plastic objects that has gained quotidian nature. As a hyperobject, plastic adheres to everything it comes into contact with and becomes integral to other materials and to human skin to the point of inconspicuousness, as Wang's work demonstrates.[14] The familiarity between plastic—from grocery bags to clothes and microparticles in food and water—and humans at a biological and environmental level translates into a problematic intimacy that challenges visual regimes of representation as well as modes of health and knowledge.

Thinking about culture and the environment, Amanda Boetzkes agrees that "hyperobjects are also changing human art and experience (the aesthetic dimension)."[15] From performance art and sculpture to paintings made of or containing large amount of plastic or plastic waste, artists in the last two decades have been proposing ways of making recycled art a pivotal heuristic in the conceptualization of new aesthetic dimensions necessary for the understanding of plastic's influence over global culture. Ecodocumentaries are part of this effort, specifically in China, where independent documentaries are raising ecological awareness while negotiating their own position in relation to the state and media policies. As plastic, in *Plastic China*, encompasses material particles in the air, fuel, clothes, toys, the manufactured landscape, and raw material for more plastic, as well as metaphorical, immaterial dimensions such as its palpable enabling of hopes for building a better life, the visual registers of representations become entangled and offer multiple possibilities of analysis. Wang manipulates the light, sound, and camera angles in his anthropological study of a recycling workshop to incorporate the material and immaterial, opening up conversations about plastic's ubiquitous nature and cinematic art's capacity to evolve in parallel with its subject.

Long takes *and* shangshui *paintings*

Ecocinema, recycled art, and environmental art all share an intricate imbrication with art and non-art disciplines for a better representation of their subject. Wang Jiuliang includes in *Plastic China* different epistemological frameworks and artistic traditions to show the extent to which plastic not only defines a new industry, shapes lives and sociocultural relations, but also necessitates composite aesthetic tools for its representation as a hyperobject. In the documentary's first half, the director offers a few panoramic shots of the mountains of plastic surrounding the village where the workshop keeps its activity. The camera travels from a very cramped space where crouching workers sort plastic for recycling to a minivan and into the fields where hundreds of tons of plastic waste await processing. The landscape has changed around the newly formed mountains; the water is highly polluted judging by its black color, and the crops appear besieged by the towering plastic waste (see fig. 9.2). While some of it is contained into large bundles and covered with white or dark blue sheets, many small pieces are loose in the field or waver in the wind, ready to break away from the bundles. The camera rests on this image for a few seconds

Figure 9.2 Plastic China.

before overlaying an explanation by Kun, one of the main characters, that the recycling shops pay a lot of taxes to the government. Next, the camera moves inside the plastic waste mountain. Here, the bundles of plastic are almost collapsing into black, stagnant water, while the temporal length of their presence there cannot be accurately estimated. The angle is, as throughout the documentary, low or towards the middle in relation to the objects or people present in focus, and creates an affectively more imposing image of the amount of waste encircling the fields. The long take of the manufactured landscape[16] brings to mind traditional landscape painting techniques (山水 *shanshui*)[17] and the relationship between humans and the environment depicted through painting in Chinese traditional culture.

With a long cultural history and precise regulations and combinations of elements, *shanshui* paintings attested to the painter's talent, providing at the same time an aesthetic description of a personal philosophical stance toward nature's might.[18] There is a highly structured form that perfectly balances all elements in a *shangshui* painting and, most of the time, the human presence is minuscule in comparison with the mountains and the waterways. By contrast, in Wang Jiuliang's long takes of the surrounding nature, plastic overshadows nature, and the landscape is not natural anymore, but manufactured. Although human presence is still dwarfed by the mountains of waste, as it is by mountains of rocks and forests, the constructiveness in Wang's stills testifies to human intervention into natural pathways via exploitation, extraction, and pollution. The logic of balance of aesthetically fluid elements becomes the rule of stagnant water harboring towers of waste with no end in sight. The comparison between such long takes in *Plastic China* and traditional paintings uncovers the director's "close connection with traditional aesthetic ideas and landscape art"[19] and his argument regarding the complete transformation of nature, its "plastification," as it were, or entering into the age of the "plastiglomerate," as Patricia Corcoran calls it,[20] which can be represented by inverting traditional visual practices and their cultural connotations.

Wang Jiuliang is not the only director and artist to rethink and intermix traditional practices with ecocinema or ecological art. For Yao Lu, traditional landscape art is a symbol which no longer serves the pursuit of beauty and harmony; rather, it brings the viewer unexpected sights in order to create a more powerful impact and to promote higher-level thinking about the relationship between contemporary humanity and nature.[21] Wang's work, both in *Beijing Besieged by Waste* and *Plastic China*, takes place outside the studio and follows an anthropological method of living in areas or with the people he researches and films. The manufactured landscapes he insists on showing are not beautiful or harmonious, but, as Yao Lu claims, cultural triggers for a timely reassessment of art's and artists' agency in ecological change.[22] The mountains of waste we see are results of slow aggregation, of labor and international exchange and collaboration, of cycles of consumption and production germane to the current global neoliberal condition. Their towering presence is a testimony to slow environmental violence,[23] as well as to a rekindling of a connection between humans and nature, albeit destructive: plastic mountains in the fields are a synecdoche for the plastic deposits in the workers' blood and lungs in the documentary. Differently put, the documentary uses multiple techniques, including *shangshui* painting suggestions, as artistic strategies to unveil hidden strata of reality[24] and to underscore the importance of non-art disciplines and areas of knowledge, such as sociology, anthropology, and civic politics, that can give art plurivalent levels of complexity.[25]

In *shangshui* paintings, the panoramic view materializes an introspective representation of the artist's inner state of mind, their construction of a dialectical relationship between the self and the surrounding, immanent nature. Mountains and waterways in such paintings gain an aesthetic dimension that characterizes the artist's affective state of mind and incites the viewer to discover it in the brush's turns, in the balance between the representation of the five elements and, sometimes, in the poem accompanying the painting. *Shangshui* paintings document and classify the history of Chinese aesthetics in a visual register, as the individual transforms himself through cultivation and interactions with nature. In the age of "plastiglomerate," nature suffered irreversible changes whose full extent is unfolding and demanding of a categoric reassessment of human intervention. Wang Jiuliang's long takes focusing on mountains of plastic waste, while extending an uncomfortable invitation to keep following the camera's lens as it captures the historic formation of the "age of plastic," do not aestheticize their subject. Plastic waste appears in a direct, unmediated manner to give a clear, intentionally shocking view of material, economic, social, and political phenomena as they coalesce in one hyperobject. The prolonged view over the manufactured landscape problematizes the existence of such changes to the environment and questions the land's possible future alongside the local flora and fauna. For instance, the black, polluted water walled in by plastic piles is not a given, a natural waterway, but a hazardous entity and a consequence of local, national, and international agendas. Its relationship with the subterraneous water sources is deleterious and poses a source of risk to the surroundings; it is a complex actant in itself.[26] The mere existence of large plastic piles covering the horizon draws immediate attention to the problem, and, by contrast with the traditional *shangshui* paintings and their aesthetic tradition, Wang highlights a subtle change in the aesthetics of the long cinematic take.

The goal of ecological art is to catalyze awareness about our actions' impact on the environment and to determine changes in personal, national, and international behaviors regarding waste production and handling alongside extractive processes and extreme consumerism. Meiqin Wang describes the situation in China: "[Artists'] creative interpretations and realistic representations of waste endow this lowly material a unique role of critically making visible its invasive presence, exposing waste as a phenomenon largely ignored in the state-controlled mainstream media and cultural production until recently. ... the significance of these artworks lies in their potential to contribute to the growth of bottom-up civic consciousness and public space."[27] Alongside Yvan Schultz and others, Meiqin Wang subsumes art into politics, leaving the aesthetic challenges of such representations aside. Plastic and e-waste dumping trigger environmental crisis in need of immediate attention and sustained effort, all formulated in the discourse portraying a better future. However, Meiqin Wang does not analyze the different results creative and realistic representations have for the creation of a bottom-up environmental consciousness, nor the ways in which plastic and e-waste as artistic subjects and objects determine an evolution of artistic techniques.

The long take technique, as seen in Wang Jiuliang's film, continues, albeit in a reimagined manner, the *shangshui* tradition of painting and forcefully suggests introspection and critical engagement with the subject at hand. As an eco-documentary, *Plastic China* follows in the path of cinéma vérité and adopts an observational style which favors the long take as the primary means of expression. As Romanian director Cristi Puiu has remarked about his own films in the cinéma vérité genre, directors who favor this genre are more interested in "mov[ing] the accent from what's going to happen to how it's going to happen. [Thus] the audience stops wondering where the story is leading and is forced instead to face the fact of what's *on* screen."[28] Facing what is on the screen becomes an intense emotional investment, a deep reflection over the impact plastic has on the environment and humans as well. In these moments, plastic, its existence and influence, is brought back into the audiences' sight and mind for extended viewing time and the only aspect of pollution left to wonder about is its potential minimization. The long take manifests as an image, what Kristin Thompson has called "excess."[29] The excess becomes one of the elements that differentiates Wang's long takes from the *shangshui* techniques and suggests the complete loss of harmony and balance between nature and humans via plastic. Moreover, excess comes from the repetition of long takes with similar visual composition that center plastic mountains in nature or in the home, showing nature's and humans' coexistence with plastic. Wang, thus, emphasizes ad nauseam the need for active, worldwide action toward recycling and civic consciousness, all in the present tense.

What is happening on the screen and its demands for the audience's attention comprises a certain temporality at stake with the eco-documentary's own engagement with time. In *Plastic China*, temporal progression comes from Wang Kun's accumulation of enough resources to buy a new car, which suggests that enough plastic was recycled and sold, but not from a change in the plastic mountains around his house or in the fields outside the village. By contrast, Peng Wenyuan and his family's lives do not change at all and the string of days filled with plastic sorting remains unchanged. Although Wang Kun succeeds in buying a new car, his social status does not appear to have changed dramatically in the village. The eco-documentary's epistemological progression is diegetically marked at the end and underlines the contrast between Wang Kun, the entrepreneur, and Peng Wenyuan, the migrant worker. The long takes showing the latter or his family sorting or prying suspend time's passing and describe the stagnant, precarious condition he finds his household in. As a migrant family, Peng, his wife, and their children are stuck in an exploitative circle which revolves around ad-hoc recycling stations. Screen time

Figure 9.3 Plastic China.

does not favor one family over another, but the contrast implied through long takes and their montage indicates a salient social, political, and economic problem. Lastly, the plastic's innate temporality further complicates the relationship between the two families, the environment, the state, and the international community, as plastic's material persistence challenges the effectiveness of recycling itself and the human body's durability in performing recycling (see fig. 9.3). Thus, Wang's penchant for the long take as his main cinematic aesthetic tool highlights these contrasts and tensions while focusing the attention on the quotidian, repetitive activities he became very familiar with during the pre-filmic research period.

The difference in temporal progression as a sign of social advancement is part and parcel of what Meiqin Wang sees as a problem in the development of civic consciousness. The documentary suggests that plastic waste recycling and "wasted" lives are deeply connected. Arguing for civic consciousness without a holistic plan of addressing waste production, without devising equal opportunities for all and doing away with structural violence and neo-liberal upward (im) mobility, borders false consciousness. Some of the issues presented in *Plastic China*, namely migrant workers' search for jobs, the decline in agricultural endeavors and agriculture as a profession in a traditional sense, and the exploitation embedded in entrepreneurship, problematize the concept of civic consciousness and unveil the class aspect hidden behind the ecological activism. Differently put, when living off waste and when plastic stands for work, food, and dreams, the discourse of ecology uncovers its need for nuanced expansion. Wang Jiuliang, by subtly referencing the *shangshui* tradition and by equally dividing screen time between the two families, makes the claim that neither humans nor the environment should be wasted, in all senses of the word. Civic consciousness as a mindful practice beyond anthropocentric, class-driven positions, and as proposed in the eco-documentary, is intersectional and its urgency emerges from the repetitive long takes that focus on poverty, environmental pollution, and human bodies transformed by plastic microparticles and contorted in crouching positions for sorting the plastic better.

Recycling the private and the public

The documentary film has established itself in the Chinese tradition of the twentieth century through its didactic potentialities and capacity for information delivery. Zhang Zhen distinguishes between two types of documentary films, the public (公共 *gonggong*) and the private (私人纪录片 *siren jilupian*) documentary, depending on the level of generalization possibility, topic, and location.[30] By these definitions, a large part of eco-documentaries is public-facing, since they document spaces outside the home which are being exploited or polluted. However, in Wang Jiuliang's case, *Plastic China* recycles the definition and bridges the public and private categories in its depiction of plastic as a constitutive part of environmental and human life, inextricable from neither of the two spheres.

Luke Robinson theorizes the concept of 现场 *xianchang* as a complex nexus of livelihood, cinematic techniques, aesthetic of location, and contingency that defines documentaries in China roughly in the new millennium.[31] *Xianchang*, while sharing traits with observational cinema, cinéma vérité, and neorealist film, is a concept specific to Chinese cinema and its conceptualization of documentary films in the post-Mao era. One important aspect emerging from *xianchang* is its capability of cinematically describing both private and public spheres using handheld cameras, long takes, no extradiegetic sound, and little artificial light. In Wang Jiuliang's case, plastic is everywhere and holds one signification for both levels: the means for advancement in society through small-scale recycling, or, on the international economic stage, via imports from the West into China. At the beginning of the eco-documentary, one establishing shot offers a view over the Tsingtao Harbor; then the image cuts to a panoramic view of a truck transporting a container full of, presumably, plastic, after which the camera lowers its angle and rests for a few seconds to show us a girl's face. The girl is Peng Wenyuan's child, as we learn in the next scene. The harbor and the individual container signify China's international trade and are a public matter. However, the continuous camera movement in one long take between the moving truck, its passing by a small roadside market, and the girl waiting by the side indicates from the very beginning the intimate interconnection between the public and the private temporalities and spheres driven by plastic's presence.

In *Plastic China*, the intimacy between humans, nature, and plastic is telling. Earlier in this essay we have seen how manufactured landscapes represent a new type of geography, but the aestheticization of nature and waste as it appears in works by other artists, such as Liu Xintao, is almost absent here. The aesthetic choices rest on direct, observational style which recycles the boundaries between nature, waste, private and public and represents a continuous space of altered life. Wang Jiuliang, when filming in private areas, especially in the sorting, melting, and cutting zones, favors very low angles or high angles, medium-distanced or close-ups. Despite showing the very claustrophobic work and living conditions, lacking proper heating, insulation, and living amenities, the camera angle provides a close reading of facial expressions during work hours and makes visible microparticles flying everywhere (see fig. 9.4 and fig. 9.5). These plastic particles

Figure 9.4 Plastic China.

Figure 9.5 Plastic China.

are harmful to the human body, but in the recycling workshop they are a given. They form a thick layer on the skin, are inhaled, pollute drinking water and food, as well as the environment. All the workers live and work in these conditions, but only Wang Kun brings up the potential health issues such an intimacy with plastic might have. The director films from a low angle, allowing sunlight to make visible the plastic microparticles and the smoke in the workshop, a cloud released into the air and further into the village. The particles become visible only in these instances and expose, hence, yet another, otherwise invisible, relation between private and public spheres alongside intimacy between the skin, the particles, human bodies, and the environment.

The recycling of the private-public dichotomy in independent cinema in China functions very well in this instance and furthers the visible field of knowledge about plastic. The long take is communal as it moves from one worker to another, from one crouched body to the children playing in the plastic heaps, and insists on showing an intimacy between human bodies and plastic that can no longer become undone. The visual negotiations between the private and the public, between microparticles and air, between plastic as waste and plastic as a means for subsistence in the workshop, coalesce in this film and inscribe themselves into an aesthetic performance of *xianchang*. This becomes one of Wang's artistic contributions to the independent documentary and ecocinema in China, an artistic position that aims to create large-scale visibility to what has been a ubiquitous undercurrent, namely plastic's penetration of both private and public life.

Photography and plastic haute couture

Ecocinema's efforts are matched by performance art and photography in China and all over the world. From the Caribbean to Africa to Asia, artists repurpose plastic and e-waste by integrating it into their installations and artworks to draw attention to conspicuous material consumption, faulty and polluting waste management plans, dumping and invasive practices conducted on land and water. Artists such as Ma Li, Wan Yunfeng, Xu Bing, and others[32] have recuperated and reused construction materials and plastic objects and presented them across the world to raise civic consciousness and, implicitly, to highlight the world's interconnectedness when it comes to the environment. More specifically, criticizing China and its recycling and dumping practices is not the most productive endeavor, as dealing with waste in a physical form and conceptualizing it in an intellectual manner constitute shared transcontinental and globalized responsibilities by the entire world.

Wan Yunfeng 万云峰, a Chinese artist from 辽宁 Liaoning province, designs unwearable clothes out of plastic waste to draw attention to the deleterious effects these materials have on the human body and on the environment. His works are unwearable by anyone but himself and are staged to blend the earthy or oceanic background with his understanding of high fashion. Inaccessible for most people, haute couture represents the embodiment of time and expensive materials coming together under the hands of the most experienced sewers. This practice and way of life, provided to private clients by highly exclusive fashion houses with at least one workshop in Paris, epitomize excess and conspicuous consumption, implying the spending of large amounts of resources for outfits that, more often than not, are worn only a handful of times. Despite impeccable technique, the highest material quality, and the most qualified sewers, haute couture collections are also known to propose unwearable clothes for quotidian activities—too daring, too abstract, too excessive in their statements. The human body, in this instance, becomes a mere prop, while the clothes require all the space to shine and reveal their intricacies. This conceptualization of the body and resources represents the artistic entry point for Wan Yunfeng.

In a series entitled *Protect the Ocean*, Wan collects plastic nets and objects found in the ocean as a result of dumping and unethical fishing practices (see fig. 9.6). The nets are very deleterious to marine life,

Figure 9.6 *Nets*, Protect the Ocean.

while all the other objects are pollutants that kill marine fauna. Their existence under the water triggers the unfolding of silent and invisible deaths, suggesting the similar trajectory of landlocked waste and its disposal outside of urban dwellers' sight, and out of their minds. Wan collects, sews, and assembles the random materials he finds to resemble haute couture creations—in other words, highly intricate pieces with multiple hidden details, visually impactful but with questionable practical application. He wraps himself in his own creations and poses in front of a camera, not in a studio setting but outside, in the very environment that provided the plastic materials. He uses heavy, dramatic make-up and poses that resemble those of professional models during photography sessions for high-fashion publications. Each series contains multiple takes with the purpose of capturing different affects, especially representations of pain, sadness, or confusion. In fig. 9.6, for instance, we see Wan's creation comprising small wheels, medicine bottles, a water gun, a plastic bird, plastic bottle caps, forks, bags, and nets. The small objects are in the center, a position suggestive of them being captured by the nets. The human body, with only a portion of the head visible, represents, via metaphorical association, the marine environment suffocated by plastic. Wan's knitted brows and closed eyes imply a painful relationship between any living creature and plastic, a suffocating excess that threatens life itself. Moreover, his tilted position and the aquatic background highlight the inescapable bind in which plastic pollutes and interrupts life of all kinds, from human lives to those of waterways and all living creatures.

Wan mounts a multilevel critique in this piece directed at fashion, specifically high fashion, and its ephemeral trends, large expense of resources, and wasteful practices. He focuses against society's inability to recycle plastic better and stop its spread in the oceans. Haute couture, due to its highly creative aesthetic design tendencies, allows for unusual associations of materials and forms, although it has not yet subscribed wholeheartedly to notions of sustainability. Thus, Wan's creation of unwearable pieces from discarded plastic waste subverts haute couture's intellectual and abstract position: it puts plastic under the limelight to redefine fashion's choice of raw materials, its handling of said materials, and its relationship with the environment. Also, the unusual thematic association in Wan's artworks centers the body in its loss of control over plastic waste, over its own garments and activities, reaching a point of almost disappearance in its production and consumption of plastic as a hyperobject. The threat is underlying, as the body tilts toward the water.

Wan's art can be categorized as recycled art in the sense of converting waste into reusable material for his creations. Most of the time he makes use of nets, discarded fabric, rocks, metal grids, and discarded plastic objects to conceive his outfits. The painful positions he puts his body in when posing for photographs, an uncomfortable intimacy, is suggestive and raises the alarm about ethical recycling practices related to waste, but also to the production of waste and its impact on the human and nonhuman body. Some of the poses remind us of the crouched positions in which we see Peng's family in *Plastic China*. These efforts hope to bring plastic waste back into public attention and contribute to the manufacturing of civic consciousness. Wan's innovation for Chinese art consists of recentering the human body inside plastic structures, as opposed to presenting the outfits themselves as individual works of art separated from their creator and audiences. However, the artistic practice of tying one outfit to one particular client, specific to haute couture, creates a hidden tension between the art's message and its own recycling endeavors and existence. While salvaging, sewing, assembling, and wearing are part of the recycling effort, after the photography session the outfit's life continues in the art gallery, in the artist's home, or in the garbage can. This raises questions about art's ability to raise awareness or increase recycling efforts through its own engagement with plastic waste. It also suggests, in an oblique manner, that complete recycling and transformation of all extant waste is impossible, thus making plastic into an object that has already begun to influence Earth's geological progression. Wan shows the limitations of both art and well-intended recycling initiatives and silently agrees with Wang Jiuliang's meditation on pollution's irreversibility.

Wan's works borrow the love of intricate detail from haute couture. He spends his time assembling each work and wearing it in front of the photo camera, and encourages us to discover the outfit and its significance through attentive analysis. He manipulates time—viewing time and performative time—so as to draw attention to the myriad plastic objects ending up in the ocean. The photographs are diverse, from portraits to landscape frames, and the body's centrality under the nets causes a shock and invites even more detailed engagement with the art. As we've seen, in *shangshui*-style paintings, the viewer's appreciation comes with an understanding of the harmony between nature and humans. Here, on the contrary, the message rests on the lack of balance, on the silent destruction triggered by plastic waste. Personal time and geological evolution are co-constitutive elements in these photographs, with the former coming to a possible sudden end and the latter transforming toward the "plastiglomerate" era.

Conclusion

Artworks such as *Nets* or *Protection of the Ocean*, as well as *Plastic China*, find counterparts across genres and continents. Artists such as Ma Li, Wan Yunfeng, and Xu Bing in China are in constant dialogue with artists such as Wang Jiuliang or Wang Bing, or designers and photographers such as Doulsy from Senegal, Nils-Udo or Caitlin Esterby and Simon Pascoe from Africa and Mongolia, to create a plurivalent network of multi-genre works that raise ecological and civic awareness. Wang's insistence on the long take as a commentary on plastic's deleterious effects on the body and the environment underlines the temporal dimensions germane to recycling and repurposing bodies, tools, knowledge, and attention in order to overcome pressing social, economic, and political issues. Wan's innovation in presenting plastic waste stems from his centering of the human body in landscape-type photographs, as well as from his association of different knowledge domains (fashion, photography, ecology, and art) to bring plastic back into the spotlight, into public consciousness, from its "out of sight, out of mind" framework. Doulsy stages much more dramatic installations in his collaboration with NGOs to draw attention to the extensive destruction of the environment in Africa brought about by consumption, neoliberalism, and postcolonial exploitation. Opening up such dialogues has the potentiality to build transcontinental solidarity for common action and modes of seeing plastic beyond its utility that can inflict change and delay massive ecological catastrophes. Here, plastic is both the object and the subject of critique, the material that makes the critique possible, the support in the tools used for art, and the poisonous element in nature. It creates and enables social and economic hierarchies as it travels across oceans and land and underlines the global interconnectedness at the basis of our daily lives. Plastic demands complex cultural approaches and its ubiquitous nature imposes an integrated collaboration between isolated cultural fields of knowledge, all for the hope of a future in which water and food will be cleaner, plastic nets will not kill marine fauna and flora, and animals will regain their natural habitats.

Notes

1. Christof Mauch, Introduction to *Out of Sight, Out of Mind: The Politics and Culture of Waste*, ed. Christof Mauch (RCC Perspectives: Rachel Carson Center, 2016), 5–13.
2. *Beijing Besieged by Waste* is Wang's first eco-documentary, in which he tracks all the dumping sites around Beijing and the scavenging prac-

tices in these areas. For more information, see Kiki Zhao, "China's Environmental Woes, in Films That Go Viral, Then Vanish," *New York Times*, April 28, 2017.

3. Zygmunt Bauman, *Wasted Lives: Modernity and Its Outcasts* (Cambridge: Polity, 2017).

4. Sheldon H. Lu and Haomin Gong, *Ecology and Chinese-Language Cinema: Reimagining a Field* (Abingdon: Routledge, 2020).

5. Scott MacDonald, "The EcoCinema Experience," in *Ecocinema Theory and Practice*, ed. Stephen Rust, Salma Monani, and Sean Cubitt (New York: Routledge, 2013), 109.

6. Lu and Gong, *Ecology and Chinese-Language Cinema*, 4.

7. Ibid. 14.

8. Ibid. 2.

9. For more information on the *Sinophone* concept, see Shumei Shi, *Visuality and Identity: Sinophone Articulations across the Pacific* (Berkeley: University of California Press, 2007).

10. S. L. Wong et al., "Current Influence of China's Ban on Plastic Waste Imports," in *Waste Disposal & Sustainable Energy* 1 (2019): 67–78.

11. Timothy Morton, *Hyperobjects: Philosophy and Ecology after the End of the World* (Minneapolis: University of Minnesota Press, 2017), 2.

12. Kiu-Wai Chiu, "Screening Environmental Challenges in China: Three Modes of Ecocinema," *Journal of Chinese Governance* 2, no. 4 (2017): 437–59.

13. Morton, *Hyperobjects*, 1.

14. "Plastiglomerates," at *Oceansplasticleanup.com*, http://www.oceansp lasticleanup.com/History_Plastics_Anthropocene/Plastiglomerates_Sto nes_Plastics_Rocks.htm.

15. Amanda Boetzkes, "Plastic, Oil Culture, and the Ethics of Waste" in *Out of Sight, Out of Mind: The Politics and Culture of Waste*, ed. Christof Mauch (RCC Perspectives: Rachel Carson Center, 2016), 51–9.

16. The phrase originated from the film *Manufactured Landscapes*, directed by Jennifer Baichwal with photography by Edward Burtynsky (USA, 2007).

17. Robert J. Maeda et al., *Two Twelfth-Century Texts on Chinese Painting* (University of Michigan: Center for Chinese Studies, 1970).

18. Ibid. 16.

19. Jing Yang, "Rising Ecological Awareness in Chinese Contemporary Art: An Analysis of the Cultural Environment," *Tahiti* 6, no. 1 (2016): n.p., http://tahiti.fi/01-2016/tieteelliset-artikkelit/rising-ecological-awareness -in-chinese-contemporary-art-an-analysis-of-the-cultural-environment/.

20. Ben Valentine, "Plastiglomerate, the Anthropocene's New Stone," *Hyperallergic*, November 30, 2015, https://hyperallergic.com/249396/pl astiglomerate-the-anthropocenes-new-stone/.

21. Yang, "Rising Ecological Awareness."

22. Ibid.

23. Rob Nixon, *Slow Violence and the Environmentalism of the Poor* (Cambridge, MA: Harvard University Press, 2011).
24. Ban Wang, *Illuminations from the Past: Trauma, Memory, and History in Modern China* (Stanford: Stanford University Press, 2005), 45.
25. Linda Weintraub, *To Life! Eco Art in Pursuit of a Sustainable Planet* (Berkeley and Los Angeles: University of California Press, 2012), 5.
26. Jane Bennett, *Vibrant Matter: A Political Ecology of Things* (Durham, NC: Duke University Press, 2010).
27. Meiqin Wang, "The Socially Engaged Practices of Artists in Contemporary China," *Journal of Visual Art Practice* 16, no. 1 (2017): 33.
28. Ryan Gilbey, "Chasing the Ambulance: The Death of Mr Lazarrescu," *Sight & Sound* (August 2006): 28–30; cited in Dominique Nasta, *Contemporary Romanian Cinema: The History of an Unexpected Miracle* (London: Wallflower Press, 2013), 162, emphasis added.
29. Kristin Thompson, "The Concept of Cinematic Excess," *Film Theory and Criticism: Introductory Readings*, ed. Leo Braudy and Marshall Cohen, 5th ed. (New York: Oxford University Press, 1999), 513–24.
30. Zhang Zhen, "Introduction: Bearing Witness: Chinese Urban Cinema in the Era of Transformation (*zhuanxing*)," in *The Urban Generation: Chinese Cinema and Society at the Turn of the Twenty-First Century*, ed. Zhang Zhen (Durham, NC and London: Duke University Press, 2007), 1–45.
31. Luke Robinson, *Independent Chinese Documentary: From the Studio to the Street* (London: Palgrave Macmillan, 2013).
32. For a longer and more comprehensive list of artists engaging with environmental issues, see http://www.chinaphotoeducation.com/Carol_China/2._Artists__Urbanisation_and_Environmental_Degradation.html.

10

Plastic Poetics: Challenging the Epistemologies of Plastic Waste in the Artwork of Maria Roelofsen

Nathan Beck and Jeff Diamanti

Maria Roelofsen is an artist based on the island of Texel in the Netherlands. She constructs sculptural works in the form of animals using materials that wash up on the strand, predominantly wood and plastic (see fig. 10.1 and fig. 10.2), and exhibits and sells these sculptures in the RAT Gallery (RAT standing for Recomposed Art Texel) in Den Burg, the main town on the island. Roelofsen's works are whimsical, engaging, and accessible, and yet they perform a vital function in surfacing the circuitous nature of global waste and pointing to the universal human complicity in ineffective plastic disposal.

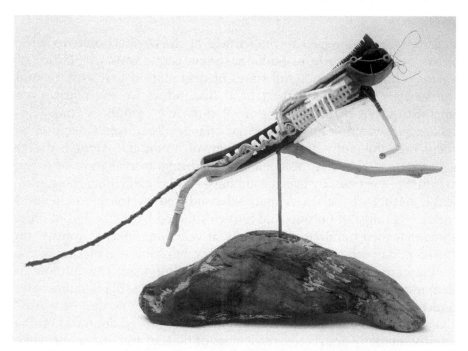

Figure 10.1 *Maria Roelofsen,* Lizard. *Washed-up plastic and wood,* *70 × 46 × 26 cm. Photograph by Maria Roelofsen.*

Figure 10.2 *Maria Roelofsen,* Blue Pelican. *Washed-up plastic and iron, 53 × 53 × 33 cm. Photograph by Maria Roelofsen.*

Plastic, literally dragged to the surface of the ocean, points to what exists below the surface, both the unknowable tons of plastics of various shapes and sizes and states of degradation that drift around the world's oceans, and the ethical discomfort we must feel, even unconsciously, in the recognition of our own complicity and help-lessness in effecting any substantial change. Roelofsen's decision to focus predominantly on interpretations of nonhuman animals draws a correlation between the dynamism and migratory nature of living creatures—particularly oceanic animals—and the uncontainable, mal-leable nature of plastic. A material conduced by the petrochemical industry's range of outputs and in turn sourced by the bodies of crea-tures that died hundreds of millions of years ago, plastic appears on the scene as both our most contemporary substance—a marker of the unique capacities of modern industry—and immortal. The ontological demarcation of plastic as waste or toxic or disposable is temporary, and rewritten in Roelofsen's art by introducing a "poetics of waste" that engages in aesthetic play to displace the obvious discourse on the environmentally destructive tendencies of plastic, and instead provoke a sense of beauty and renewed purpose in plastic in its ontological recomposition with other, natural elements. The object form of plastic itself will, in this essay, put pressure on the classic arrangement of the aesthetic encounter as it is enumerated in the Kantian tradition, even as its resurfacing as art object in turn verifies that arrangement. If oil, as Imre Szeman terms it, is the "ur-commodity" of modernity, then plastic will *surface* the concrete abstraction of petrocultural periodization in our analysis.[1] We advance this position by analyzing the form and context of the sculptural objects before moving on to the ways in which Roelofsen challenges our epistemology of waste itself, and finally suggest how the poetics she introduces can allow us to live with the ecological reticulation of the hydrocarbon surface in a future plastic world.

The post-natural world of plastic play

Plastic's unique material affordances, malleability, and adaptability—the veritable *plasticity* of plastic—come from its elemental bondability or its *polymerizationability* to the periodic table. Polymers bond, attract, and repel. As a bonded chain of smaller molecules, plastic names (most typically) the nearly unbreakable chain formed through the cracking of complex compounds of hydrogen and carbon and their subsequent reintegration through adhesive polymerization. This process of bonding continues long after the substance leaves the lab as polished object, though. Pellets, nodules, toothbrushes, grocery bags, and so many seals bond through, and *as*, environments in perpetuity. Microscopic attraction is just as agential as larger, cartographically discernible objects like the floating landmasses of plastic accumulating in all the globe's oceans. As larger plastiglomerates form out of "molten plastic," their shape and relational character look not so different from geological objects such as rocks. Here the chemical, thermal, and temporal intensity of their bonding in the industrial process meets the geological; the sublime forces of human industry emulate the forces that shape the world: heat, pressure, time, movement, juxtaposition. Synthetic or not, plastics surface the *plasticity* of chemical elements when reformatted by the elemental force of industry.

Thought of as a kind of mimetic and monstrous orientation toward ecological form, plastic is paradoxically built in the image of bonded matter and a hostile interruption of it. Pierre Baranger of the École Polytechnique in Paris—an early whistleblower on the intense industrial pollution baked into the production of plastic—worried openly in an oft-quoted article that the new material verified "the deep and unconscious tendency of man [*sic*] to do without nature and to adore himself through his own creations in the center of an ersatz pantheon."[2] In Baranger's worry is a recognition, too, of the latent world-building affordances laced through the periodic table: once the polymer becomes a motif through which industrialists can begin to reimagine the materiality and scale of production, the habitable world becomes *plastic*: an infinitely malleable container for human creativity. Both the object world and the elemental ingredients of production become resolutely *post*-natural. The environment as such becomes, in the age of polymers, a giant tapestry of aesthetic play. Amanda Boetzkes, who to our minds has the strongest account of the intimacy of plastic to contemporary aesthetic form, terms this a kind of mimetic recursion where "even waste is recuperated for its potential resourcefulness, so that the concept of a "resource" subscribes to a paradigm and ideal of "zero-waste."[3] If plastic troubles the eco-logic of waste by both necessitating and making possible a closed loop of resourcability, then our open question here regarding plastic's surfacability at the interface of the beautiful and the abject is bound up with the political ecology that its ubiquity brings into focus: mimetic of bonded matter and an unboundable knot in matter's distribution. Plastic, in short, remakes the material (and hence immaterial) ontology of capital's environment through an aesthetic of plasticity.

Moving beyond a discourse of conservation and sustainability helps explain why the cute and curious come coupled with plastic's play on beauty. Roelofsen's Picassoesque animals and light objects invite the viewer to participate in an interpretive game, requiring us to twist our necks, stroll around the sculpture, look up close and from a distance, and confer with another person in order to fit the pieces of the gestalt puzzle into place. With each new animal, the viewer asks "now what is this?" It is, of course, an assemblage of driftwood and plastic fragments in its immediate materiality—but it is also a game in which we identify and point out the patterns, or features that draw the eye to the patterns, in order to render something coherent from something abstract. It is only through our culturally imbued semiotics that we understand these configurations to represent a shorebird or a mouse or a camel. When satisfied that our assumption is correct, we move to the next interpretive level, "what *was* this?" Brush bristles, fence panels, tool handles, twine, headphones, jerry cans, bottles, beach spades, bike seats, golf tees, and so on. We angle ourselves to try to read the print on a soap bottle to decipher which land it has traveled from, or whether we recognize the brand. And it is at this point we move from interpretation to the construction of false histories. We wonder, perhaps, how a bike seat ended up in the ocean. Was it thrown in? Did it separate itself from a decomposing bike in a river years ago? Who did it belong to? This is the imaginative game Tim Cresswell plays in his poem "Plastiglomerate": "clams and cowries/

orange-lipped lava flows/the toothbrush Esme shared before she spat//
foam into the breakers/bleached/pummeled coral/thin plastic forks//
from when you barbequed."[4] There is joy to be found in the play of
world-building as opposed to discovering the truth, the uncomfortable reality. In this way these objects alert us to their past whilst also
addressing their absence of decomposition; that we do not know how
old they are suggests we cannot know how long they will continue to
persist, thus scaling the indefinite temporal span of plastic life to the
immediate moment.

It is for this reason that Giuliana Bruno prefers to speak of surfaces
rather than images, "to experience how the visual manifests itself materially on the surface of things, where time becomes material space."[5]
Distinct from the rather more antimaterialist impulses of late 2000s
surface reading, Bruno's interpretive framework helps bridge the gap
between the surface and chemical play of plastic expression. The verb
"to surface" in Bruno's account names an activity by which time,
phrased *by* materials, gets laid out for a spatial encounter. Surface *is*
depth but rendered into a counter to form. A tree's outer bark, for
instance, is a formal index of its seasonal embeddedness in soil and
wind and sun, *surfacing* its temporal entanglement with various elements. With the tree in mind, a surface presents itself as a text to read
time, but not through a linear or mechanical set of integers. With
plastic, however, surface becomes an industrial dynamic divorced pro-
visionally from seasonal and hence elemental flow. To read the surface
of the Barbie doll washed up on shore as an index of solar, subsoil,
and atmospheric time would be to make a category error. But the
Barbie doll is no less ecological for the logic of its surface. Its material ontology emerges *through* the latent affordance of materials when
pressed through an industrial apparatus—no less earthbound for its
radical alterity from organic compounds whose surfaces read time for
form. A new mythos is surfaced in plastic, one roughly as old (or new)
as fossil-fueled industrialization.

Campness and cutesy: subverting plastic waste discourse

Satisfied with the on-the-spot myths we write around the component
object histories, we may return to the initial level of interpretation
and look again at the sculptural figure from a distance to discover the
humorous aspects the artist has framed. Roelofsen's sculptures do not
simply represent animals, but animals made ridiculous by their poses,
the distorted scales of their respective bodily features, and the cartoon-
ish nature of their construction. The affective dimension of humor
settles like a veneer upon our initial interpretation of the object and

the material surface rewritten as the textures in front of us—smooth and coarse and brittle and shiny—shift from signifiers to the signified. A plastic appendage that no longer speaks to us from the ontological perspective of what it was originally made or used for may signify an eye or a beak, and once our gestalt interpretation allows for this, the thing itself becomes an eye or a beak. Like when we are shown how an optical illusion can produce a different image when "looked at" from another perspective—such as the famous rabbit-duck illusion—we then struggle to return to the previous interpretation. This would require diving back under the surface of the object and in doing so erasing the humor. Why would we want to do that?

There is something of a campness about these works. They are both whimsical and at the same time very serious. The animal forms they depict stand in like avatars for the animals themselves whose existence is threatened by climate change and various anthropogenic disasters, scaling the global reach of the consequences of the Anthropocene into purchasable aesthetic objects. Their construction is not of materials made fresh, but that wear upon their surface the residue of the damage such materials can cause, particularly in their mass-produced form. For example, the particular bottle top used in a sculpture may be innocent, so to speak, but it represents all bottle tops, many of which end up in the stomachs of seabirds. Plastic is both particular and, always, a synecdoche of itself. These materials and the sculptures they compose are phantoms of the ongoing sixth mass extinction. And yet they are fun, they are playful, they evoke a sense of joy. Susan Sontag, in her *Notes on "Camp"*, states that "camp taste turns its back on the good-bad axis of ordinary aesthetic judgement. Camp doesn't reverse things. It doesn't argue that the good is bad, or the bad is good. What it does is to offer for art (and life) a different—a supplementary—set of standards."[6] When one works through the perceptive game of identifying a sculpture, reading its materials, and constructing an on-the-spot history of its various origins, the focus is shifted from meaning and message to form. They are not ideas and words that demand a dialogue and deep engagement, but something immediately tactile that invites one to come and play, indiscriminately and without judgment. Instead of worrying about what the artist is trying to tell us, or what we are supposed to feel, we just *feel*, relishing the aesthetic composition and the artist's skill in selecting the right materials to generate the illusion. Roelofsen performs a sleight of hand, and we are left marveling at the way a vibrant yellow toothbrush forms a bird's beak, or fishing rope hangs off a panel of fiberglass like hair off a bison's back, instead of musing on the ethical dilemma of plastic usage. Camp, Sontag continues, "incarnates a victory of 'style' over 'content', 'aesthetics' over 'morality', of irony over tragedy."[7] It is in this way that Roelofsen plays with the poetics of waste, toying with the viewer's expectations of ecocritical metalanguage to extract a more positive, even hopeful, affective response.

At no point in the playing of the interpretive game is it deemed necessary to ask what these objects are saying to us. We become intimate with the materials, investing our attention in form over message, aware all along that the sculptures foreground the uncomfortable truth of plastic's environmental toxicity without necessarily demanding recognition or a change in behavior. The surface is just that—surface. Plastic turns the immediacy of surface into a smooth map of time, exhibiting an interpretive tension that is punctured with a smile or a laugh. Is this sculpture a commentary on the impact of global supply chains and inadequate recycling programs? No. It is a heron, clearly. Roelofsen chooses to shun the overtly didactic route that one may expect to experience in larger-scale artworks in museums and public gallery spaces, understanding that people do not need to be told once more what is happening "out there." The Rat Gallery is instead a celebration of the potential for beauty in plastic's materiality. The biography on the gallery website explains that Roelofsen "has the talent to recognise beauty in al [sic] sorts of materials that have been washed up on the beach. She brings together wood and plastic from the sea into 'assemblage art' full of colours, shapes and stories."[8] What is immediate about this aesthetic is the vibrancy of petrochemical hues, surfaces that refract and reflect light, juxtaposed against the desaturated, lifeless tones of the driftwood that seems to absorb and give little back. It appears to be plastic's own plasticity that lends it the ability to broadcast itself metaphorically in order to evoke an emotive response. It is a strange irony that in these assemblages the plastic components exhibit vitality with their luminous colors that strike the eye like birds of paradise, whereas the wood, rigid and unforgiving, often serves merely as structure and scaffolding. Plastics do the heavy lifting of bringing the animals to life—they are malleable, able to mimic organic elements like whiskers and hair, wings and eyes. And in this immediate context each plastic piece, once mass-produced in facsimile, becomes particular.

This camp aesthetic is essential to Roelofsen's poetics of waste. In contrast, the work of Chilean artist Cecilia Vicuña, whose 'Precarios' sculptures—"created with objects found on the beach that were meant to be erased by the ocean"—embody a more serious approach, addressing "pressing concerns of the modern world, including ecological destruction, human rights, and cultural homogenization," according to the biography on her website.[9] Looking at their biographies, we see how the ways in which these artists identify with their work differs—one leaning on aesthetics, the other on activism. Between the two appears an artificial distinction between aesthetic autonomy and political commitment coded as commentary. Vicuña's art foregrounds critical dimensions that may engage the viewer, provoke an affective response, and even teach something. But because it is often presented in a gallery or viewed through the detached image of a photograph, deprived of tactile materiality, its impact is diminished; it becomes harder to return to. Roelofsen's objects are always open to the public. The RAT Gallery is a shop, and the objects therein are replaced when sold. There is no window in which to catch a temporary exhibition. And their status as commodities encourages people to purchase the objects and take them home, thus extending their discursive lifespan. We remove an object from the gallery and exhibit it in the environ of our home, inviting others to play the gestalt game, returning waste from the outside to the inside, to the domestic sphere, where we have space and time to contemplate the overt and yet complex ramifications the art-object presents. Worthless plastic, a form of dead capital, is imbued with value by merit of the artist's intention and technique, its placement within a gallery context, and our ability to recontextualize it, if we wish.

Environments of art

It is worth considering also the locational context in which Roelofsen exhibits her sculptures. The RAT gallery is in close proximity to the beaches from which she combs her materials. The island of Texel is popular with tourists, who likely account for a large proportion of the gallery's visitors. Texel is a popular destination for wildlife lovers, boasting birdwatching and seal tours, and is of course renowned for the thirty kilometers of sandy beaches that fringe its western coast. As can be imagined, there is a craft scene on the island that targets tourists with shops selling woolen products (byproducts of the abundance of sheep farms), sea life-themed trinkets, and local craft beers. These contextual aspects act as paratextual signifiers in Roelofsen's work, the island playing a role in its construction of meaning. If one were to try to craft such sculptures from trash found in the city or countryside, the composite elements would be different—in different stages of decomposition, probably from less distant origins, and less materially diverse. These signifiers could add to Roelofsen's sculptures a sense of exoticism, or conversely diminish their status from art to tourist trinkets; mementos of a holiday on the island. The sort of thing that sooner or later finds its way back into the bin. It is for this reason that she has established a price point that rescues the objects from a crass interpretation as souvenirs, and as something demonstrably of economic and artistic value. Without getting into a discussion on the semantics of art and artistic merit, it must be appreciated that although her work shares characteristics of handicraft, the fact that each object is of an intentionally unique design (rather than produced en masse with the unavoidable signature of an artisanal hand) lends to their status as art objects. The plastic's surfacing on *these* beaches is neither spontaneous nor independent of the trade routes that circulate it amidst particular ocean currents. Its surfacing was, if not intentional, certainly not without meaning. As art objects, then, Roelfosen's works also get shaped through a particular interface between shoreline, trade route, and current.

How does the fact that these sculptures are not exhibited in a more formal setting, such as a museum or urban gallery, impact our reading of them? What if they were not available for purchase? Is Roelfosen, by commodifying her sculptures, diminishing their aesthetic work? Plastiglomerates, as referred to in the Tim Cresswell poem above, are geological-plastic hybrid objects that wash up on beaches, notably in the Hawaiian Islands. Boetzkes notes that the plastiglomerate "has come to stand as both a scientific measure of the Anthropocene and a cultural signifier of its impact."[10] The parallels between plastiglomerates and Roelofsen's sculptures are immediately obvious. The artist Kelly Jazvac places plastiglomerates in a gallery context and, like Roelofsen, draws attention to their formal qualities, inviting viewers "to train their eyes along the variegated surfaces that yield competing senses of porous basalt rock, rough shell fragments, and the suffocating smoothness of plastinated rope, net pellets, or otherwise undefined melted plastic."[11] In both instances, these art objects demand to be worked through, investigated, instead of being immediately dismissed as waste, and they both surface the scale of the global consequences of international trade and consumption and condense it into a digestible, tangible object of value. Nonetheless, whereas Jazvac's work establishes its message through contextualization, Roelofsen's work is differentiated by the artist's focus on composition.

Plastiglomerates narrate their own story. They may be pretty and interesting to look at, but they are the result of autopoietic processes that literally entangle the object with the moral complexities of plastic, following Boetzkes' suggestion that "plastics can be seen in the convergence with the economic, energetic, and value-producing assemblage by which they are intertwined, restricted, prescribed and destined into the geological future."[12] In the gallery context, plastiglomerates are systematized and problematized. They say to the viewer "look what you have done." Do not touch, you have already done enough. There is, in the bricolage arrangement of Roelofsen's work, an embedded suggestion that one could dismantle the sculptures and refashion them again, over and over, as if they were made of Lego (an artifact that must surely have found its way into her work at some point). They are imbued with a potential futurity as well as a mysterious past, and the act of reading them is more participatory. Our argument here is that, far from cheapening her work by commodifying it, Roelofsen turns the purchase into an iteration of camp ethics. Through her artistic practice she stalls the circuit of waste by surfacing form and in doing so also amplifies the persistence of plastic's mythological hostility to waste and an easy ethics of sustainability. In counter to the disposable morals of mainstream recycling, which in effect make *my* problem *someone else's* problem—the government, recycling plants, scientists, other countries to which recycling operations are outsourced—Roelofsen pulls from the planet's greatest rubbish dump, the ocean, and in her hands pivots waste into an asset.

Plastic animals: odysseys and environing

Waste exists in the past tense. It is an afterimage of purpose, something that persists after we have been exposed to it and discarded it, averting our gaze. *Refuse* derives from the Old French *refus*, meaning "denial, rejection." In its nominal, adjectival, and verbal forms, waste is weighed with negative connotation. Plastics serve their purpose in a short span of their lifetime, before their social status switches from miracle material to disposable waste. So what is the effect of Roelofsen pulling plastic waste from the past into the present and stripping it of negative association? How is something detestable and valueless made desirable and valuable by the act of recomposition and recontextualization? Roelofsen's poetics renews purpose in outcast materials, and in doing so points to ways we can learn to live with oceanic plastic.

The ocean is a conduit for global waste. It cannot always be known if the plastic washing up on the beaches of Texel found its way into the ocean from the Netherlands in the first place. Oceanic plastic is like nuclear fallout, in that it "indicates dispersion, scatter, being strewn from a source."[13] It could be from the UK or France, from fishing vessels, ferries, or freighters. Its ultimate origin is of course the petroleum extracted from beneath the earth's surface, refined into plastics shaped into infinite configurations and transported around the world in the form of products and packaging—what Stacey Alaimo calls "the banal but persistent detritus of consumerism."[14] But because waste management and recycling regulations differ from country to country, countries that pride themselves on being "greener" than others still suffer the waste of other nations, borne on air and oceanic currents. Like the radiation particles blown across Europe in the wake of the Chernobyl disaster, plastics travel, disperse, and embed themselves in the environment, eroding the viability of national borders and demarcations of sovereign governance and regulation, eluding and refusing control. As Boetzkes explains, "there is no outside to which plastic can be relegated, only a 'recycling' within a closed system."[15] Considered in this way, plastic has its own agency. Stacy Alaimo draws attention to this, looking at artist Jonas Benarroch's video "The Ballad of the Plastic Bag," which follows a bag[16] floating over various landscapes with a cheery musical accompaniment reminiscent of a drifter's ballad.[17] As Alaimo goes on to explain:

> The clever conceit of the plastic bag as ramblin' man dramatizes the agency and "freedom" of this supposedly inanimate object, stressing that these flimsy things have gotten away from us, escaping human control. Rather than demonizing the object, the video invites the viewer to take vicarious pleasure in the bag's free-roaming, aesthetically pleasing travels.[18]

We extend this notion of vicarious pleasure to Roelofsen's work. Her decision to focus on animals as the models for her sculptures not only speaks to their ontological precarity in the face of ecosystemic disaster, but points to the ways in which plastic is imbued with agency, movement, and a migratory nature. Indeed, plastic's very plasticity. Fish that travel the ocean, land animals that traverse the savannah in herds, birds that migrate halfway around the globe. Animals are as dynamic and uncontainable as plastic. It is not only in the choice of animals as subjects that Roelofsen draws attention to this dynamism, but in the choice of anatomical parts she figures with plastic pieces: the eye of a tiger; the bristled wings of a bird; beaks, fins, tails and mouths. The parts of bodies that most signify vitality, movement, and digestion.

Although Roelofsen's large collection of sculptural works and light objects includes much more than animal figures, they do predominate. Nonhuman animals speak to the human viewer in a way non-animal or inorganic sculptures simply couldn't. They sneak under our skin and compel us to anthropomorphize and empathize. The empathy invoked by Roelofsen's works asks us to meet in the middle with plastic. Their composition of both human-made and natural waste points to the unavoidable reality that the impact and entanglement of plastics and the environment is inevitable and irreversible. Plastic waste refuses to be environed, by virtue of size or sheer quantity and, unhindered by organic degradation, circumnavigates the planet to end up on distant shores. The stories behind the composite objects of the sculptural assemblages are open to construction by the viewer, and in this way we are permitted space to read the sculptures in our own way. And because Roelofsen imbues them with a sense of humor, beauty, and agency, we are willing to spend more time with these objects. Images of horror dull and repel, and by reframing the conversation through a positive poetics, Roelofsen's sculptures encourage us not to turn a blind eye, but to come to terms with the fact that we must learn to live with plastic.

Regardless of what miraculous technological solutions we invent in the coming years, plastic waste is not going anywhere. It will continue to shore up on beaches and bury itself into the undergrowth, contaminating soils and suffocating animals, as pervasive and toxic as radiation. Roelofsen's animal-plastic sculptures acknowledge that our collective actions are globally intertwined, without necessarily making it explicit. Like the artworks discussed by Boetzkes in *Plastic Capitalism*, they offer a way of reading plastic waste that allows for an understanding of its role and our complicity in ecological devastation whilst appreciating its formal qualities. Boetzkes suggests that, by considering plastic in aesthetic terms, such artworks can "loosen it from the presumption that it is either a morally contemptible substance or a testament to technological prowess."[19] They may even contain a curative quality, what Michael Marder terms "the grace of art," a form "powerless to effect a change in reality, to penetrate its core and decontaminate the bodies of the earth of animals, plants and humans" and yet one which "strokes the surface of things—the superficies of the remains, including the fragments of the thing called *psyche*—consoling them, patting them, offering gentle contact, caress."[20] If so, it would be easy to decry that, when it comes to relieving the shame of being human, Roelofsen's work is a mere drop in the ocean. That may be so, but by acting as a lens that scales the sublime impact of global human waste into tangible objects of humorous beauty, it widens the discursive frame and illuminates new ways to engage in critical, aesthetic, and activist practice.

Notes

1. Imre Szeman, "Introduction to Focus: Petrofictions," *American Book Review* 33, no. 3 (2012): 3.
2. Pierre Baranger, qtd. in François Jarrige and Thomas Le Roux, *The Contamination of the Earth* (Cambridge: MIT Press, 2020), 242.
3. Amanda Boetzkes, "Resource Systems, the Paradigm of Zero-Waste, and the Desire for Sustenance," *Postmodern Culture* 26, no. 2 (2016): n.p.
4. Tim Cresswell, *Plastiglomerate* (London: Penned in the Margins, 2020), 13.

5. Giuliana Bruno, *Surface: Matters of Aesthetics, Materiality, and Media* (Chicago: University of Chicago Press, 2014), 3.
6. Susan Sontag, *Notes on "Camp"* (London: Penguin Books, 2018), 22.
7. Ibid. 24.
8. "Maria Roelofsen," Museum Galerie Rat, https://www.museumgalerie rat.nl.
9. Photograph by Cecilia Vicuña, https://www.ceciliavicuna.com/obje cts/uitl2nj66w2quxeq24k3tjo4mnjj79; Jess Araten, "'Cecilia Vicuña: About to Happen': The Art Exhibit Addressing Climate Change," *34th Street*, March 19, 2019, https://www.34st.com/article/2019/03/cecilia-vicuna-about-to-happen-institute-of-contemporary-art; Cecilia Vicuña, "Biography," *Cecilia Vicuña*, https://www.ceciliavicuna.com/biography.
10. Amanda Boetzkes, *Plastic Capitalism: Contemporary Art and the Drive to Waste* (Cambridge, MA: MIT Press, 2019), 196.
11. Boetzkes, *Plastic Capitalism*, 196.
12. Ibid. 198.
13. Michael Marder, *The Chernobyl Herbarium: Fragments of an Exploded Consciousness* (London: Open Humanities Press, 2016), 44.
14. Stacy Alaimo, *Exposed: Environmental Politics and Pleasures in Posthuman Times* (Minneapolis: University of Minnesota Press, 2016), 130.
15. Boetzkes, *Plastic Capitalism*, 182.
16. "The Ballad of the Plastic Bag," *YouTube*, https://www.youtube.com/watch?v=vQdpccDNB_A.
17. Alaimo, *Exposed*, 134.
18. Ibid. 134.
19. Boetzkes, *Plastic Capitalism*, 198.
20. Marder, *The Chernobyl Herbarium*, 36.

11

The Performance of Plasticity:
Method Acting, Prosthetics, and the
Virtuosity of Embodied Transformation

David LaRocca

[His] mind was endowed with unusual plasticity, with unusual spontaneity and liberty of movement—it was a fairyland of thoughts and fancies. He was like a young god making experiments in creation.
—George Santayana, "Emerson"[1]

Film is considered an exemplar of the plastic arts. What follows here involves a variation on the ontology of the medium—that is, with what appears "in" or "on" film, namely, the plasticity of those who are said to act, portray, and embody. Hence, this plastic art is treated in conversation with the plastic artist—call her an actor. As an agitation to further reflection on this relationship (between medium-specific art and specificities of the human figure), one notes what is revealed in the forms of praise that accompany works of performative plasticity—especially those that travel under the appellation "method acting." Such positive estimations provide markers of aesthetic achievements, that is granted, but they also often carry with them a tinge of moral judgment—as if to pun the notions of "acting well" or "good acting." Thus, the lauded performance translates almost invisibly and without resistance or contested interpretation to a sense of the laudable performer. With this seemingly unchecked transitive move, we appear to inadvertently stumble upon, or over, the virtue of virtuosity. As the labor of film criticism makes its choices known in each new season, so often we find those actors who "transform" themselves, who "become" who they play, who "inhabit" their roles, or who—as the saying goes—simply "disappear," leaving nothing but the character, are most vaunted. In these moments, we seem to discover not a respect for the character played so much as an infatuation with the virtuosity of the actor playing the character. Hence, the public record of favoring, celebrating, and standing in awe of the plastic artist—and the more plastic the better. And yet, differences should be struck between Meryl Streep's accent (*Out of Africa* [1985],

Julie & Julia [2009], *The Iron Lady* [2011]); Gary Oldman's jowls (*Darkest Hour* [2017]); weight gain/loss by Charlize Theron (*Monster* [2003] and *Mad Max: Fury Road* [2015]) and Joaquin Phoenix (*I'm Still Here* [2010], *The Master* [2012], and *Joker* [2019]); and variations of costuming—from prosthetics to make-up to hair (as we find them, for instance, in the neo-peplum genre). Even this quick sketch, to be suitably developed, shows how vocality, prosthetics, embodiment, and plastic supplements occupy different categories of performative expression, and in turn, artistic achievement.

In this chapter, while acknowledging and integrating key contributions on the subject by A. H. Olenina, Karen Hollinger, Michael Wedel, and Stanley Cavell, among others, I will critically explore this taxonomy with the aim of making a case for why these differences in plasticity matter for aesthetic appreciation and ethical assessment of cinematic acting performances. In short, we are set upon an exploration of the cognitive impact of the plastic-as-attribute and plasticity-as-idea-and-guiding-metaphor. The very materiality of plastics has given rise to certain tropes picked up and applied in conceptualization and theorizing, and yet in the broad sweep of cultural interpretation we mainly focus on *effects* (the performance itself, actor versus character, etc.) and not the reasons we find ourselves filled with such admiration. I suppose the question that hovers is whether readers see *consciousness* as part of the (natural, physical) environment. For instance, even the notion of "climate change" relies on the metaphor, or some variant, of plasticity. Likewise, the present work explores how our commitments to certain figurations—such as plasticity—lead to behaviors, actions, ideas, and valuations. At base, environmental policy is directly a result of the discourse *on* the environment. How we speak about the nature and meaning of plasticity, for instance, as we know it in the arts—in cinema, specifically—informs our capacity to make aesthetic and moral judgments. As with climate change, the stakes for the criticism of art and culture are high.

Conceptualizing plasticity

The notion of plastic is itself plastic, used in a range of diverse circumstances and to varied ends. In biochemistry, we learn of "neuroplasticity" (as an acknowledgment of cognitive flexibility and changeableness). "Plastic surgery" alerts us to the specter of the "fake" and fabricated (as with face "lifts," tummy "tucks," and breast augmentation) but also to the revelation of the genuine and true, as in sexual reassignment surgery; Paul Beatriz Preciado's experiments with prosthetics, physical manipulation, and psychopharmacological change seem apt on this last front, for example, in *Testo Junkie: Sex, Drugs, and Biopolitics in the Pharmacopornographic Era*.² Over the last century, the prevalence and predominance of plastics as a family of petrol-based materials has given rise, not surprisingly, to a range of figurative uses of "plastic"—some of them tied to or still based in the literalness of plastic materiality. In a way, akin to the presiding metaphors of mind that track over generations (the humors for medieval thinkers, the clock for Descartes, the computer for our day), so the plastic and plasticity provide salient tropes for thinking about our relationship to the world, including our own sense of self. For example, is one's identity fixed (e.g., by way of genetics) or can it—does it, must it—evolve? This now-classic nature/nurture tandem is caught up with what we might call the notion or even the dream of plasticity. Are we fixed entities living out a prescribed fate (again, not so much ordained by a divine creature as by the sheer forces of physics), or can we navigate freely—or even be knocked off course by some other, compensatory force?

Catherine Malabou's ongoing experiments with the conceptualization and experience of plasticity are touchstones here—*Plasticité* (1999)—especially her thinking about stem cells and epigenetics, but also her perception that evolutionary traits are found in politics (as in the vexing grammar of *What Should We Do with Our Brain?* [2004]). While intellectual history is full of theories of essentialism and identity (e.g., the fixedness and permanence of the self, the soul, etc.), plasticity suddenly affords a condition for speaking of pliability and thus of freedom. As the politics of identity has become *the* politics of late capitalism and the Anthropocene, it is wholly significant to acknowledge a countervailing vision, one that makes space for fate's confrontation with freedom, identity's challenge from plasticity. It thus must be counted an irony that so much of contemporary politics champions both identity and freedom, since the two are oppositional, indeed, aggressively so. In a series of probing questions from Malabou's *The Ontology of the Accident: An Essay in Destructive Plasticity* (2014), we encounter a rhetoric that lies beyond such internalized contradiction: "What ontology can [plasticity] account for, if ontology has always been attached to the essential, forever blind to the *aléa* of transformations? What history of being can the plastic power of destruction explain? What can it tell us about the explosive tendency of existence that secretly threatens each one of us?"³

Partly the feeling that nature is perpetually abundant, endlessly regenerative and thus inexhaustible relies (again, ironically) on a notion of nature as static—that is, as an entity whose scale is too vast to be affected by human action and intervention, whose distended temporal and material reality is, in fact, beyond our comprehension. Therefore, to awaken to consequential, that is, *real* environmental destruction is to awaken to nature as pliable, as perishable, as plastic. In this moment of the anti-static, we also see how plasticity invokes and ratifies the anti-essential and anti-foundational; in a word, the plastic is impure (and invites impurity). Little surprise, then, that plasticity has about it the latent charge of attack or corrosion, since purity is so often understood as encoding or confirming positive moral status (hence the rhetoric of a fixed self or an imperturbable conscience). The static affords purity, while the plastic insists on pliability and admixture. In this way, plasticity is an affront to the reality and the aspirations of purity.[4]

Indeed, it may be advantageous to claim that part of what it means to believe in the reality of the Anthropocene involves a sense of the validity of plasticity as a prevailing fact of nature and consciousness. The static is the illusory, the myth of the given (from antediluvian times); the plastic by contrast—that which has been known as the changing and changeable—is "the *aléa* of transformations." Along these lines, John Caputo has called Malabou's rendering of the concept an "explosive plasticity" since "catastrophe, breakdown, destruction without remission, repair or promise ... sculpts a new deformed form, a deviation in being as a form of being, an adieu to life while still alive, each with a phenomenology of its own."[5] Moreover, Malabou's exploration presents the very notion of "accident as a category of being."[6]

Environmental *consciousness*, thus, would seem to require at once a perception of the material we call "plastic" and an admission that *plasticity* is the fundamental mode of nature; when the tandem is pointed out, it may seem reasonable, even something a person can support politically. And yet, the long-standing impression that earthly life is ruled by forces beyond human control (God, gods, laws, fate, Nature, etc.) is an easy default, and returns one quickly enough to bromides that have shaped civilization for millennia—that "this too shall pass away," "life goes on," "time will tell"—not realizing that these pictures of extended temporal continuity are, in fact, coded messages of temporal change. The research on microplastics bears this out, since we read more than once that by 2050 there will be "more plastic than fish" in the oceans.[7]

Though most entries in this volume explore our "life with plastics" by addressing the physical manifestation of the material (not only how it gives shape to our lived environment but also how it may pollute that same environment), I turn our attention to "plastics" as a contribution to modern cognitive experience. One salient scene of instruction: plasticity as we know it in the production of cinematic works, whether in the plastics of prosthetics (make-up, hair, etc.) or the virtual plasticity of objects (in computer-generated imagery), or perhaps most notoriously, the actual transformations of human physiology in the form of the actor's body. Here actors don prosthetics as a way of transforming what is seen on screen, or have their shape modified by computational means, or simply—or not so simply—explore the plasticity of the body by means of diet, exercise, mental training, vocal modulation, and whatever else it takes to shift one's shape to dramatic effect.

The plastic actor or the actor under plastic?

There is a scene—a single shot, really—in Sofia Coppola's experimental Hollywood metacommentary *Somewhere* (2010) that haunts these proceedings, since it seems overfull with significance for thinking about the plastic and plasticity; the experience of time (in lived experience, and in movies); the nature of film and the nature of acting; the meanings of stardom and celebrity as they intersect with the everyday; and how humans occupy space on film and the way other humans watch those "occupations" in the vernacular art we lovingly call movies. The shot opens (see fig. 11.1).[8] We have just seen celebrity actor Johnny Marco (played by celebrity actor Stephen Dorff) being given a full, some may say old-fashioned plaster of Paris treatment for the creation of a mask (or, as his manager puts it, "mold of your head"); the industrial term for this process is "life-casting" (as if to stand in contradistinction with the "death mask," its natural counterpart). While gypsum cement (a type of plaster) is used for prosthetics, rubbery cousins—foam latex, gelatin, and silicone—afford sought-after refinements for rendering the nuances of skin and faces. In the selected scene from *Somewhere*, three members of a special-effects crew quickly but carefully apply ample amounts of a white, pliable paste to Johnny's face and head, tell him he must "remain rigidly upright during this entire process," then leave him in silence and isolation for about forty minutes. Here, with studied deliberation, the A-list actor (as character) is effaced while an impression of his actual face under wraps (as an A-list actor?) is made or "taken"—a fugitive act of indexicality that reminds us of certain traits of the photographic arts (and again, the death mask).[9]

As these thirty fingers eagerly apply the viscous material to Johnny's face and head, the claustrophobic among us may begin to breathe harder. His eyes are covered, then his mouth, with those two modest holes—his nostrils—coming perilously close to being occluded. The specter of suffocation looms, adding its own kind of dramatic tension

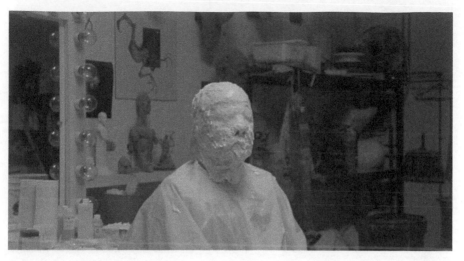

Figure 11.1 Johnny Marco (Stephen Dorff) effaced/defaced in Somewhere *(2010).*

to the otherwise inert scene. As it happens, the passageways remain open and thus function as Johnny's modest but sufficient lifelines. After the crew departs, the shot in question starts wide (see again fig. 11.1). We see the studio, once filled with active production staff, now cleared out and made silent—that is, save the sound of Johnny's labored breathing (through those tiny nasal slits). The wide shot slowly zooms in. As Johnny focuses on his breathing, so we too focus on it—and "him" somewhere beneath the pile of drying plaster (again, "plaster" as an analogue for latex, gelatin, silicone, etc.). This sound (of breathing) and this attention (ours) become the space of the shot itself; there is no music (nondiegetic or otherwise), only room tone to accompany breathing. The wordless long take lasts eighty-four seconds. Then, in the next shot, a reveal: "Let me show you how you look." Johnny spins in his chair and faces a mirror—now appearing as if decades older, maybe twice his age. "Jesus," he says and sighs as he inspects the ravages wrought by the prosthetic application; as with all such tricks of the surface, the actor's voice has not aged, but remains stubbornly in its present condition. While we watched Stephen Dorff age eighty-four seconds (on film), the character he played launched "him" into another era altogether—some forty years into the future. Dorff as Johnny becomes "resurfaced"—"refaced"—as a generic older man (see fig. 11.2).[10] Those of us watching Coppola's studied combination of metacinema and transcendental cinema will never see Johnny act in the film in which he plays this older man; it remains one among many narrative lines in the film that are studiously unfollowed, that lie "somewhere" or some when else.

Figure 11.2 Johnny Marco resurfaced/refaced in Somewhere.

And yet the long take—or this pair of film stills: Johnny under wraps (fig. 11.1) and Johnny revealed anew, which is to say, as old (fig. 11.2)—signals something of a false flag in this investigation, since the plasticity we see here (e.g., of the special effects department that deploys long-standing techniques of the prosthetic arts) is the kind we may marvel at (especially in a behind-the-scenes-how-do-they-do-that-exposé-documentary) but not usually the kind we celebrate (apart from, for example, high-profile awards for costume, make-up, and the like). In other words, we seldom find critical plaudits for acting that is accompanied by such accoutrement (Gary Oldman's Oscar-winning Churchill seems a prominent exception), but more often discover that while prostheses may aid a certain willing suspension of disbelief—and thus afford a welcome, gratifying boost to realism—the actor and her acting may be perilously subsumed, even to the point where the plastic masks the performance itself. The actor is lost under all that make-up. The face de-faced. When does the addition of such prosthetics add, and when does it detract? For all the praise heaped upon Oldman's Churchill, for example, there's not much *Oldman* there. He is, like Dorff, indiscernibly subsumed beneath a prominent and obscuring layer. This is one sense of "disappearing" into a role, yet it is not the one to settle upon; rather, it motivates thinking in another direction: namely, when instead of plastic being applied to the actor, the actor herself becomes plastic.

There are many existing researches that come to mind on what we might call the figurative sense of plasticity, or the plasticity of embod-ied performance, among them Ana Hedberg Olenina's *Psychomotor*

Aesthetics: Movement and Affect in Modern Literature and Film, where she writes about Lev Kuleshov's theory of acting: "Biomechanics, metro-rhythm, and all other acting practices that promoted the 'defamiliarized' gesture and stylized corporeal plasticity were condemned as a pernicious digression that hindered the progress of Soviet film art toward a more 'natural' psychologism."[11] Indeed, Kuleshov's approach to acting techniques were derided by Russian film scholars as "unnatural," "mechanistic," and "inhumane." In the context of our discussion, we can evaluate where these categories (or epithets) reside in relation to plasticity. For instance, could a Brechtian *Verfremdungseffekt* be understood in terms of the plasticity of acting, or does it, like the assault on Kuleshov, conjure intimations of rigidity and the mechanistic?

Yet another touchstone is Karen Hollinger's "Star Acting and the Hollywood Star-Actress," part of her study *The Actress: Hollywood Acting and the Female Star*. Drawing from remarks by Barry King, Hollinger notes: "contemporary star acting has developed in very different directions from star acting under the studio system and requires somewhat different skills." For instance,

> because of the transformation of the studio to the package-unit system of production, stars are now required to take more control of shaping their careers through role choice and image construction. King argues that they must cultivate a star image characterized by "plasticity," rather than the consistent "commercial identity" associated with stardom in the studio era. According to King, contemporary star actors must be prepared to take on the identity appropriate to their current film performance. Thus, they are not expected to have a uniform star identity, but to create a vast "wardrobe of identities" from which they can pick and choose for different occasions.[12]

Hollinger and King call to mind a sturdy descriptive of the trade, the trope of "typecasting," which itself carries hallmark traits of the fixed model who makes an impression upon a pliable, receptive surface—whether ink-on-paper or a noir anti-hero on celluloid, a type is "cast" and we are there to read and receive it. The unexpected punning of "casting"—life-casting and typecasting—doubles our frames of reference: we are left to wonder about the making and keeping of impressions. In the case of typecasting, reliability is underscored by fixity (or at least minor variation). Bold deviation from type unsettles our comprehension of the actor's identity as someone capable of playing a certain (type of) character. That said, contemporary actors of high repute can sometimes cultivate two, and perhaps more, lines of character types (in the language above, vestments in the wardrobe). Thus, Keanu Reeves has a durable identity as a hunky love interest (*Something's Gotta Give* [2003], *The Lake House* [2006],

Destination Wedding [2018]), a nitwit (the *Bill & Ted* series [1989–
]), and a formidable action hero (*Speed* [1994], *The Matrix* [1999],
the *John Wick* series [2014–]). Over time, then, we can appreciate
a pattern that slots him into these three dominant categories; there
may be more, and these may have subtypes. Could his performance
as Johnny Utah (*Point Break*, 1991) offer an apotheosis of all three
variations in a single character? This experimental typology is meant
to sustain Hollinger and King in underscoring that contemporary stars
may experience or achieve typecasting but still be allowed to have
several types to choose from (or to form special hybrids of distinctive,
seemingly proprietary types; this is the movie star as intellectual property). Similar splintering and hybridity, if by differently named categories, is easily discernible in the careers of other actors; such diagnostics
could be performed for them.

Still other points of reference should include Michael Wedel's
research on early film style and stereoscopic vision, where there
emerged a notion of a "plastic art in motion." We learn of an early
attempt, in the 1860s, by François Willème to create "photosculp-
tures," which would be what they sound like: "photographic images"
turned into "sculptured objects."[13] Not only does Willème anticipate
experiments by Étienne-Jules Marey and Eadweard Muybridge by a
decade, he seems to overshoot the twentieth century altogether, as 3D
printing would appear to have realized his uncanny vision of making
three-dimensional objects by means of photographic renderings.
Early twentieth-century innovations such as the Kinoplastikon and
Alabastra-Theater, with their attempts to achieve "a synthetic effect
of three-dimensionality," were harbingers.[14] And yet, as Wedel notes:
"While the experiments with the Kinoplastikon may not have had a
great technological influence on cinema," it was suggested that "directors should nevertheless draw on the options at their disposal, such as
mise en scène, lighting, and expressive shot composition, in order to
achieve comparable effects of relief and plasticity."[15] You would be on
the right track if experiments in German expressionism come to mind,
such as Robert Wiene's *The Cabinet of Dr. Caligari* (1920).[16] As
Wedel says, the "uncertainness concerning the relationship between
surface and depth can be understood in the context of stereoscopic
vision as a powerful influence on the stylistic paradigm of expressionism." As Noël Burch wrote in "Primitivism and the Avant-Gardes:
A Dialectical Approach," "[t]he film's famous graphic style presents
each shot as a stylized, flat rendition of deep space, with dramatic
obliqueness so avowedly plastic, so artificially 'depth-producing' that
they immediately conjure up the tactile surface of the engraver's page
somewhat in the manner of [Georges] Méliès."[17]

Another point of reference, Erwin Panofsky (that crucial theorist of acting and performance), wrote in "Style and Medium in the Moving Picture" that "[t]he character in a film ... lives and dies with the actor,"[18] whereupon Stanley Cavell, teasing out the logic in "Audience, Actor, and Star," notes in reply: "If the character lives and dies with the actor, that ought to mean that the actor lives and dies with the character."[19] Cavell conceptualizes the interaction and identifications of actor and character this way:

> For the stage, an actor works himself into a role; for the screen, a performer takes the role onto himself. The stage actor explores his potentialities and the possibilities of his role simultaneously; in performance these meet at a point in spiritual space—the better the performance, the deeper the point. [... Meanwhile, the] screen performer explores his role like an attic and takes stock of his physical and temperamental endowment; he lends his being to the role and accepts only what fits; the rest is nonexistent.[20]

Quite consequentially for our thinking here, what this means for Cavell is that "[o]n the stage there are two beings, and the being of the character assaults the being of the actor," whereas on the screen, the "performer is essentially not an actor at all; he *is* the subject of study, and a study not his own. (That is what the content of a photograph is—its subject.)"[21] Notice the ground covered in these few moves: we go from Panofsky's "[t]he character in a film ... lives and dies with the actor" to Cavell's assertion that the on-screen "performer is essentially not an actor at all."[22]

If not an actor, then what? Well, a star. "After *The Maltese Falcon*," Cavell explains, "we know a new star, only distantly a person."[23] In the wake of the 1941 film, "'Bogart' *means* 'the figure created in a given set of his films.'"[24] Thus, Humphrey Bogart's "presence in those is who he is, not merely in the sense in which a photograph of an event is that event; but [also] in the sense that if those films did not exist, Bogart would not exist, the name 'Bogart' would not mean what it does."[25] Cavell finds it surprising that "actor" is used as a term of art, since "star" seems "the more beautiful and more accurate word": "the stars are only to gaze at, after the fact, and their actions divine our projects."[26] The poetry here is, at last, philosophical, since we are not watching someone "act" so much as "be." It is for this reason that the kind of plasticity we explored above in *Somewhere*—that is, of latex, make-up, and the prosthetics of aging—felt like a false lead: it did not give us the plastic performer so much as the performer-made-of-plastic. In short, we cannot discover a star (or the characters—what Cavell calls, in yet another variation on our accumulated lexicon, "types"[27]).

The kind of plasticity we are intrigued by is of the being-sort, not the acting-sort; hence, the opening lines of the chapter announcing our subject as the ontology of medium (whether it be cinematic or humanly embodied—or both). Cavell theorizes this too when he asks and answers: "Does this mean that movies can never create individuals, only types? What it means is that this is the movies' way of creating individuals: they create *individualities*. For what makes someone a type is not his similarity with other members of that type but his striking separateness from other people."[28] Thus, on the screen we encounter "particular *ways* of inhabiting" [again, Cavell's emphasis] as opposed to ways of recognizing only "the role." Cashing this out, Cavell's notion of type (or individuality) means we understand Robert De Niro as Jake LaMotta, not De Niro as boxer (in *Raging Bull*, 1980); De Niro as Al Capone, not De Niro as mobster (in *The Untouchables*, 1987), and so on. The matter is complicated somewhat by the fact of historically situated people such as Jake LaMotta and Al Capone, so to clarify, we can swap in De Niro as Neil McCauley (*Heat*, 1995) or Moe Tilden (*Cop Land*, 1997). De Niro's ontological "separateness" is what stands out: the way he is, as it were, himself. On screen, De Niro is prominently Neil and Moe and, like Bogart before him, "only distantly a person." The demotic speech of moviegoing captures this transitivity, for example, in the way people speak of going to "see De Niro" rather than any character (historical or not) that he portrays. Indeed, Cavell appears to ratify this claim when he says in "Opera in (and as) Film": "in a film the actor is the subject of the camera and takes precedence over the character."[29] In theatre, that relationship is reversed.

The plasticity of embodied performance

With these observations by Olenina, Hollinger, King, Wedel, and Cavell (in company with Kuleshov, Burch, Panofsky, and others), let us push further into the virtues of virtuosity when it comes to the plasticity of human performance on film. For the sheer tensility of his body and the athleticism of his art, Matthew Barney might become a veritable patron saint of such an investigation—whether he is slicing through whale blubber in *Drawing Restraint 9* (2005) as "Occidental Guest" or exploring muscularity and animal hybridity in the five-part, six-hour-plus *Cremaster* cycle (2003). Despite Barney's fame, notoriety, and critical acclaim, I will continue to draw points of reference from the mainstream, indeed, to use the inheritance of the (popular) art of movies as a guide for thinking through the traits that most draw the praise of audiences, including critics.

In many cases, when we refer to the notion of "the actor as plastic"—in this phrasing and its correlates—we seem to default to variations on method acting, sometimes reverentially rendered as The Method. Thus, from De Niro to Charlize Theron, from Joaquin Phoenix and Christian Bale to Daniel Day-Lewis, here, acting-as-athleticism is expressed as plasticity. As noted, the dominant metaphor of the actor "disappearing into the role" reads differently in the wake of Cavell et al., namely, that we have it backwards: rather, it is the case that the role disappears into the actor (or better, star). In short, the role—even a very specific one—does not allow for the creation or expression of an individuality. This may be counted a first, and significant, discovery of looking closer at the interaction of praise for acting performances and our own (again, popular) sense of what is, in fact, happening on screen before us.

As a point of reference, consider that, in practice, it is the nature of a "character actor" to disappear into a role, since so often we may know the face we see but not the name of the actor—indeed, when we see the actor in another film, we are tempted to call out the name of a prior character: "There's Daisy from *Downton Abbey*." Ian Holm, despite a long and illustrious career, is exemplary on this point, as he "had a kind of magical malleability, with a range that went from the sweet-tempered to the psychotic."[30] When Holm himself commented "I am a chameleon," he called to mind a fitting trope of the character actor, but not the star. (These points can be separated from the related but distinct issues of typecasting and "playing to type.") In an industry not known for pulling its punches or sparing large, if fragile, egos, the phrasing "character actor" is, in fact, a generous way of speaking of a person who is not a star. The practice of cameos and guest (star) appearances underlines the long-standing industrial habit of suppressing character; in this way, we easily recognize Meryl Streep in *Big Little Lies* before we know anything about the nature of her role. Hence, we have a handy contrast in the chameleon and the cameo.

One senses that with character actors the "disappearance" is literal, actual, whereas for the star, it is figurative—something of an affectation meant as praise. To gloss what is happening here, recall the moment in Walker Percy's *The Moviegoer* when we are treated to an analysis of a photograph: "The elder Bollings—and Alex—are serene in their identities. Each one coincides with himself, just as the larch trees in the photograph coincide with themselves. . . ."[31] If we think of "actor" and "character" as two images, then the "character actor" finds a way to line them up—to present the impression of a unified image; actor and character coincide. By contrast, after Cavell, we see that for the star, "actor and character" do not coincide with themselves; instead, the former prevails over the latter.

When I made a distinction, just above, about De Niro playing a historical figure versus a fictional one, I was aiming to mark out how there is something strangely literal about these performative achievements, these embodied attempts at representation—e.g., in the case of historical figures. The creative urge seems to aim for matching an actor with the empirical characteristics that made the historical figure known to us—from face to gait, from the timbre of voice to the accent of speech. Along these lines, some highly praised performances include Daniel Day-Lewis as Lincoln, Philip Seymour Hoffman as Truman Capote, Meryl Streep as Margaret Thatcher, Gary Oldman and Brendan Gleeson as Churchill (though, with many fewer prosthetics and arguably more emotive force, John Lithgow offers the most achieved instance in *The Crown*), Russell Crowe as Roger Ailes (again, Lithgow provides a more charged likeness in *Bombshell*), Christian Bale as Dick Cheney, and Theron as Aileen Wuornos (in *Monster*). The repetition of Churchill—as we have already seen in myriad incarnations of John F. Kennedy—appears as a professional competition that conjures an (unnecessary) crisis of imitation. Aiming for some kind of dream of "absolute translation," the creators—directors, producers, cinematographers, actors, speech therapists, make-up and CGI technicians—go for a kind of radical isomorphism.[32] Our praise, then, seems tempered by not just our willing suspension of disbelief but the creators' capacity to maneuver acts of mimetic brilliance that trick the eye and ear. Without rancor or complaint, we can say this is the cinema of *trompe l'oeil*.

If we would, instead, keep with the *figurativeness* of acting—that actors are themselves figures in the business of figuration—then we would appreciate, instead, the heightened significance of adaptation and translation, and thus of *interpretive* translation. Not only would we appreciate an evocative (rather than a strictly imitative) performance, we would also afford some space for the "untranslatable."[33] In the previous paragraph, the examples are drawn from those acting performances that rely on known historical figures. But there is much in the sense of "known" here, namely, that the better known the actor, the harder creators tend to push on mimetic doubling—not just offering a representation of a person but a veritable re-creation, or re-instantiation. The "as" shifts to "is." No longer Daniel Day-Lewis as Lincoln (the way I phrase it above), but Day-Lewis is Lincoln.

While the rhetoric here carries significant theoretical weight, it is also meant to point up the nature of performances that are more purely imaginative, that is, not tethered to specific historical figures (and the photographs, film footage, or audio recordings that press the actor for mimetic contortion). Daniel Day-Lewis's charismatic embodiment of Daniel Plainview in *There Will Be Blood* (2007) can make us like, even love, this venal murderer—find us sketching a list of his admirable traits as a counterbalance to his repulsive ones. (Indeed, this happens repeatedly with Day-Lewis—with Cecil Vyse in *A Room with a View* [1985], with Bill Cutting in *Gangs of New York* [2002], with Reynolds Woodcock in *Phantom Thread* [2017]). Plainview, Vyse, Cutting, and Woodcock are not based on historical personages we have visual and auditory access to, so our experience of them—as Cavell has noted—arrives fully formed in the performative expression of the star, Daniel Day-Lewis. We cannot, in a word, compare and contrast the actor with the historical personage (as we could, minimally, with Lincoln). And yet, Day-Lewis's commitment to these (non-historical) roles appears to suggest, time and again, that Day-Lewis himself is evacuated from the screen, or at least, suppressed (in Cavell's parlance, "only distantly a person"). At such moments, we are forced to contend with the potential inversion of Cavell's own sense that the actor (or rather, star) gains precedence over the character. Impersonation becomes something more like personification. It may be, quite simply, that when an actor must compete with the historical figure he aims to portray, the star must win out, but when there is no competition—or it is unknown to the audience—the actor has a chance to cede a bit to character. Contrast all of these notes on Day-Lewis with the problematic issues at hand for portraying exceedingly well-known historical figures, such as Barack Obama and Donald Trump; *The Comey Rule* (2020) exemplifies the challenge with this side of performative representation.

Aside from voice talent, a special signature of the historically familiar figure (such as Kennedy, Thatcher, Obama et al.), actors rely to different degrees on make-up and prostheses: for his Capote, for instance, Hoffman used hardly more than some hair grease to achieve a distinctive combover, while Oldman goes inches deep for fleshy cheeks and a massive midsection. Both won Oscars. Even so, I want to circle back to the point, or exegetical move (viz., Day-Lewis is Lincoln), to contemplate that—despite the adulation of the Academy of Motion Picture Arts and Sciences—less make-up, fewer layers of prosthetics, in fact, gives us entrée into the way, as suggested, it is Lincoln who is Day-Lewis. The inversion may be conceptually challenging, starting with the notion being obviously *avant la lettre*. And yet, holding with the proposal—and perhaps calling Cavell's remarks above to mind—we approach the kind of achievement we want to speak of. It is *Day-Lewis's* embodiment (his gait; his high, trembling voice) that we study so closely, and only then by a kind of mental trick allow ourselves to say that somehow we are in closer proximity to a long-dead historical figure. Though we are in the habit of saying Day-Lewis resembles Lincoln, perhaps we are better off—and more accurately capture the plasticity of Day-Lewis's achievement—by saying Lincoln resembles Day-Lewis. Again, the role disappears into the actor.

What of the suggestion, borne out by trends of cultural criticism and reception, that we are perhaps as (or more) impressed by the sheer *physicality* of a given transformation than we are by the artistic merit of the offering (or, indeed, the quality of the film itself). Thus, we stand in awe of, or at, the plasticity of the human form rather than the meaning of the content it is said to create. "Don't *say* things. What you *are* stands over you the while, and thunders so that I cannot hear what you say to the contrary," wrote Ralph Waldo Emerson.[34] We may be dupes for the notion that a clinical approach—radical diet, severe discipline—yields not just a svelte performer, but a laudable performance. Thus, with some chagrin we notice that for all his fine acting performances in his *own* skin, it was Oldman's performance-under-plastic in *Darkest Hour* that won him an Oscar. That said, the male winners of acting accolades at the Academy Awards (and Golden Globes and elsewhere around the world) in the last decade tend to be those who undergo the most exaggerated body and voice transformations (most prominently Day-Lewis, Matthew McConaughey, Eddie Redmayne, Leonardo DiCaprio, and Joaquin Phoenix). Only Casey Affleck's grieving father in *Manchester by the Sea* (2016) stands out as comparatively straightforward physical performance; in this case, like Marlon Brando, his torment is internalized. By comparison, we do not see the same pattern for actresses in the last decade: since 2011, only Meryl Streep's Margaret Thatcher and Renée Zellweger's Judy Garland (both biopics) stand out for their physical and vocal plasticity—and an Oscar for both; more recently, Gillian Anderson made an award-winning claim to Thatcher in *The Crown*, and notably, without prosthetics. Yet the exception proves the rule, as these are notable attempts at isomorphism with historical figures.

A further variant is emerging with increasing prominence: the digital re-skinning of actors. Again, thinking of De Niro: as the celebrated actor has aged, he no longer needs the make-up and prosthetics demanded of his character in the double time register of *Once Upon a Time in America* (1984). Now himself a septuagenarian, De Niro must be skinned in reverse, as we have seen in *The Irishman* (2019), in which the eponymous character is played by De Niro under the smooth shimmer of a digitized epidermis (along with his aging compatriots Al Pacino and Joe Pesci, similarly veneered). The technique of, as it were, digitized prosthetics has been pioneered for CGI-laden science fiction (see below remarks on Andy Serkis's work), but here is put into commission for dramatic fare as well, thus upending our inherited sense of valuation about the embodied—one wants to add *organic*—plasticity of actors and the merits of their physiological transformations. While Serkis plays fictitious creatures, De Niro (as Frank Sheeran) would appear to be playing himself as a younger man, who is, in turn, playing Sheeran.

Thinking again of Hollinger and King, it may be worth noticing how certain contemporary female actors do not pursue plasticity as a means of expression, and yet remain deeply compelling—enough to garner critical praise and institutional accolades. Kate Winslet, Cate Blanchett, Rachel Weisz, Julia Roberts, Michelle Williams, and Marisa Tomei are award-winning and critically vaunted without much comment on their lack of commitment to the physical plasticity we find among their male counterparts. Perhaps we are discovering *emotional or emotive* plasticity as a response to the aggressive physical changes we see in male actors. (Have we stumbled, then, upon a gender asymmetry in method acting? One shouldn't be surprised, yet it is illuminating, if true.) Of Tomei, "[n]ot since Barbara Stanwyck," wrote Anthony Lane, "has an actress blended zest and pathos into such expressive chords."[35] No prosthetics necessary. Such unvarnished physical presentation puts us in mind of a dominant strain of French actresses: Anna Karina, Catherine Deneuve, Juliette Binoche, Isabelle Huppert, Marion Cotillard, Léa Seydoux, Clémence Poésy, Mélanie Laurent. Of course, the word for such figures may simply be "stars," where the name feels decidedly apt: distant "objects" of such radiance that we find our attention drawn to them, perpetually transfixed by them.

We have to admire the film studies locution "figure behavior" for capturing what it is we track from film to film—the chance, yet again, to see the behavior of this figure. Our attraction, our adoration, of the acting star, then, has to do with a certain presentment of the inhabitation of the human. In a life that is said to be singular and happen "just once," it is a marvel for the cinephile to see a single human take on form after form and live it as real-seeming as if it were a native incarnation. The star, in this sense, is not born but reborn in each successive role. As so many of us barely inhabit our own lives—are still very much "trying to be oneself," what that becoming is hoped to entail—it is a marvel to see an actor become one and then another and then yet another: one bodily existence, yet many lives. Karen Silkwood, Karen Blixen, Suzanne Vale, Susan Orlean, Julia Child, Florence Foster Jenkins, and Margaret Thatcher—for Meryl Streep;

J. G. Ballard, Patrick Bateman, Trevor Reznik, Bruce Wayne, Dieter Dengler, Dan Evans, John Connor, Dicky Eklund, Irving Rosenfeld, Michael Burry, Dick Cheney, Ken Miles—for Christian Bale. We are stunned by, in George Eliot's phrase, the "makdom and fairnesse" (form and beauty) of the creatures before us.[36] A quick scan of these performances too underscores how so few of them are indebted to prosthetics (yes, Thatcher's teeth, Cheney's hair—I will second Amy Villarejo, who has appealed for greater respect paid to hair as part of the *mise en scène*[37]), but instead to radical modes of weight loss and gain, posture and gait, and perhaps most laudable of all: achieved subtleties of voice in cadence, timbre, and accent.

The parodic, with its allowance for a degree of camp, also illuminates, by contrast, the seriousness with which we take the plastic performance (however much it is "made up"). Where Tatiana Prorokova is intrigued by parody and "laughing at the body" in neo-peplum films (as perhaps a necessary way to admit the "costuming" of masculinity, the way it is "worn"—as well as worn out), we are after the sincere—and how these works are taken seriously, however much the thespians are contorted or covered in synthetic materials (and more and more so, digital artifacts).[38] If the logic holds, however, the mood of sincerity and seriousness is, in fact, anti-mimetic. The goal of the plastic artist is not to imitate but to displace: to provide a fully formed human proxy—a star—for our dedicated attention and subsequent adulation. Thus, rather than Day-Lewis as Lincoln, or even Day-Lewis is Lincoln, we move to the highest realm of achievement: Day-Lewis instead of Lincoln. Call this—and similar cases—the apotheosis of the virtuosity of plastic performativity.

Latent moral lessons from virtuous performances

I began by acknowledging how positive estimations of acting performance provide markers of aesthetic achievements, but perhaps unwittingly or unintentionally *also* often carry with them a tinge of moral judgment—as if to pun the notions of "acting well" or "good acting." It is likely too subtle a point to argue with a Saturday matinee *cinéaste* that, in the light of Cavell, we should not think of these performances as "acting" but "being," so we can, instead, reflect on the evaluative language we reach for—principally in the application of our praise with the word "good."

Consider the connection between excellent action (a phrase laden with Aristotelian resonances: arête, ἀρετή) and the tendency, or temptation, to speak of such action as "good"—indeed, as deserving of ethical praise, as possessing moral worth. In earlier work, I have called the practice of such discernment performative inferentialism.[39] Despite the lumbering philosophical locution, the idea is quite simply that our evaluative language positions us—compels us?—to allow for slippage between aesthetic judgments and moral ones. This is, therefore, a semiotic ethics: the great aesthetic moments seem to be not merely beautiful, but also good. (An Aristotelian legacy makes itself known, admittedly, but also a Confucian one, where exemplary instances carry with them beauty *and* goodness. And not to be missed, a more contemporary equivalency as noted parenthetically by Wittgenstein: "(Ethics and aesthetics are one and the same.)"[40]) We can test this continuity, or perhaps slippage, by asking about specific acting performances. For instance, a difference in aesthetic acclaim should be drawn—should it not?—between Streep's talent for accents (and let's not forget her capable singing from *Postcards from the Edge* to *Mamma Mia!* to *Ricki and the Flash*) and Oldman's rubberized face (as Churchill)? When Streep's speech and talent for songcraft are placed beside the work of actors who depend on make-up and prosthetics, we are given new purchase on the sense that these latter actors are *actors* (not stars), that they, in fact, do play characters and do not achieve individuality. Such acting is, perhaps, closer to cosplay than the excelsior embodied performance that draws our sustained interest as proof of plasticity. A chasm widens between the cumbersome presence of the fat suit and the subtleties of vocal inflection, facial expression, or even the rhythm of breathing.

Indeed, referencing remarks above, Oldman very much feels like Oldman "under" Churchill, as if Oldman were a puppeteer and Churchill was the mask or figurine he animated from within—not unlike Craig Schwartz's (John Cusack) elated control of John Malkovich's body in Charlie Kaufman's berserk *Being John Malkovich* (1999); Schwartz is "in" Malkovich as Oldman is "under" the guise of Churchill.[41] All this to say that we may wish to praise Streep's work more highly because it reveals a greater talent for plasticity (in voice, in body, in physical sensibility), whereas Oldman's feels weighted down by (actual) plastic and related synthetics. (Similar to the effect of Oldman's appurtenances, Tilda Swinton has been so thoroughly wrapped and molded that she disappears as a star presence in *Snowpiercer* [2013], *The Grand Budapest Hotel* [2014], and *Suspiria* [2018]—in the last example undertaking not only time travel but a gender transition: she plays an old man—perhaps, for some, harkening back to her earlier gender transformation in *Orlando* [1992]). In the case of *Darkest Hour*, in fact, we may wish to praise the make-up department more highly than Oldman, since we are, it would seem, more likely in awe of the skill of talented practitioners of this art—more akin to achievements of painting and sculpture than acting—than we are the thespian who wears his accoutrements like so many vestments.

Thoughts of Oldman's fleshy exterior prostheses should be coupled with a remembrance of the now even more ubiquitous practice of applying digital "skins"—overlaid by means of motion capture. The technique of rendering photorealistic representations from computationally capable sensors and their data points is famously captured by Andy Serkis (Gollum from *The Lord of the Rings* [2001, 2002, 2003], King Kong in *King Kong* [2005], Caesar from the *Planet of the Apes* series [2011, 2014, 2017], Snoke in the *Star Wars* series [2015, 2017, 2019], and Captain Haddock in *The Adventures of Tin Tin* [2011]). While the category of digital prosthetics deserves a separate, full treatment of its own, we can find an instance—such as Leos Carax's *Holy Motors* (2012)—that provides a quick clinic on the century-plus transformation from techniques of early cinema (in-camera methods as well as costume and make-up) through to the digital dominance of the early twenty-first century.[42] In *Holy Motors*, we are given a veritable tour of prosthetics. Another moment to support Villarejo's point that hair should be acclaimed for its role in the *mise en scène*: there are not just wigs taken on and off, hair cut and shaved, or applied with glue, but hair also eaten ravenously![43] While the film begins, as it were, before cinema as we know it—with some photographic motion experiments by Étienne-Jules Marey—it moves with Denis Levant's agility to the contemporary digital video studio, where the actor is suited up with sensors and ready for his performance to be interpreted by a computer processor. The nude male figure familiar from Marey's photo-sequence—the body on uninterrupted display—is traded for the sheer erasure of the human form in the digital space of the pixel array. Computer-generated imagery (CGI) is more than a description, it is an ethos. Levant—and Serkis more famously—provide occasion for thinking what gets lost when the human performer is but a series of data points, displaced by the painterly surface of rasterized realities of the aptly named "motion capture" technologies.

Both approaches—method acting and prosthetics—have about them the "makeover," the first through feats of embodied alteration, the second through elements of application (clothes, hair, make-up, physical add-ons and adjustments, and increasingly digital manifestation). Yet, in tilting toward aesthetic as well as moral acclaim, we may be admitting our implied belief that the kind of plasticity we have been identifying here as that befitting the star (as it were unadorned, or modestly so, by prosthetics) is the one we intuitively, and thus habitually praise more highly: the one we believe is truer (physically and emotionally) and therefore better (aesthetically and ethically). Hence, it is the plasticity of the performance—of the performer—that agitates for our assessments of beauty and goodness. We admire these performers so much not because they are representative of another—some imagined or actual person, a historical figure or an imagined character—but instead, representative of themselves (and no other). We are caught up in the dramatic specificity of what Cavell calls "individuality." Even so, plasticity provides a viable anti-essential appreciation of the actor as person as well as star (e.g., as we customarily speak of the actor's "off-screen presence" in contradistinction with her "on-screen persona"; or as some metacinematic works demand: that we impose a more defiant line separating fiction from nonfiction). Any individual being, any star individuality, at last, is also becoming (the double sense duly noted). With each incandescent performance, the performer is remade. As we behold the protean artist, new expressions, novel instances are steadily realized in ever-onward modes of embodied plasticity.

Notes

1. George Santayana, "Emerson," in *Estimating Emerson: An Anthology of Criticism from Carlyle to Cavell*, ed. David LaRocca (New York: Bloomsbury, 2013), 298–9.
2. Paul Beatriz Preciado, *Testo Junkie: Sex, Drugs, and Biopolitics in the Pharmacopornographic Era* (New York: The Feminist Press at the City University of New York, 2013).
3. Text drawn from publicity for Catherine Malabou, *The Ontology of the Accident: An Essay on Destructive Plasticity*, trans. Carolyn Shread (New York: Polity, 2012).
4. For more on the history of purity tropes in philosophy, religion, and literature, see my *Emerson's English Traits and the Natural History of Metaphor* (New York: Bloomsbury, 2013).
5. Remarks derived from John D. Caputo's praise of *Ontology of the Accident*.
6. Ibid. n. 3.
7. See, for example, Liam Stack, "Ocean-Clogging Microplastics Also Pollute the Air, Study Finds," *New York Times*, April 18, 2019; Tim McKeough, "A Sea Change for Plastic," *New York Times*, November 20, 2019; Sarah Kaplan, "By 2050, There Will Be More Plastic Than Fish in the World's Oceans, Study Says," *Washington Post*, January 20, 2016.
8. *Somewhere* (2010, dir. Sofia Coppola), 00:36:29.
9. See my "The False Pretender: Deleuze, Sherman, and the Status of Simulacra," *The Journal of Aesthetics and Art Criticism* 69, no. 3 (Summer 2011): 321–9; and "Representative Qualities and Questions in

Documentary Film," in *The Philosophy of Documentary Film: Image, Sound, Fiction, Truth*, ed. David LaRocca (Lanham, MD: Lexington Books of Rowman & Littlefield, 2017), 1–54.

10. *Somewhere* (2010, dir. Sofia Coppola), 00:38:18.

11. Ana Hedberg Olenina, *Psychomotor Aesthetics: Movement and Affect in Modern Literature and Film* (Oxford: Oxford University Press, 2020), 171.

12. Karen Hollinger's "Star Acting and the Hollywood Star-Actress," is part of her study *The Actress: Hollywood Acting and the Female Star* (New York: Routledge, 2006), 53. See also Barry King, "Embodying an Elastic Self: The Parametrics of Contemporary Stardom," in *Contemporary Hollywood Stardom*, ed. Thomas Austin and Martin Barker (London: Arnold, 2003), 45, 49.

13. Michael Wedel, "Sculpting with Light: Early Film Style, Stereoscopic Vision, and the Idea of a 'Plastic Art in Motion'," in *Film 1900: Technology, Perception, Culture*, ed. Annemone Ligensa and Klaus Kreimeier (New Barnet: John Libbey Publishing, Ltd., 2015), 209.

14. Ibid. 209.

15. Ibid. 211.

16. See my "Weimar Cognitive Theory: Modernist Narrativity and the Metaphysics of Frame Stories (After Caligari and Kracauer)," in *The Fictional Minds of Modernism: Narrative Cognition from Henry James to Christopher Isherwood*, ed. Ricardo Miguel-Alfonso (New York: Bloomsbury, 2020), 179–204.

17. Wedel, "Sculpting with Light," 206. See also Noël Burch, "Primitivism and the Avant-Gardes: A Dialectical Approach," in *Narrative, Apparatus, Ideology: A Film Theory Reader*, ed. Philip Rosen (New York: Columbia University Press, 1986), 483–506.

18. Erwin Panofsky, "Style and Medium in the Moving Picture," in *Film*, ed. Daniel Talbot (New York: Simon and Schuster, 1959), 31.

19. Stanley Cavell, *The World Viewed: Reflections on the Ontology of Film* (Cambridge, MA: Harvard University Press, 1971; enlarged edition, 1979), 27.

20. Ibid. 28. For more on these claims, see Noël Carroll, "Revisiting *The World Viewed*," in *The Thought of Stanley Cavell and Cinema: Turning Anew to the Ontology of Film a Half-Century after* The World Viewed, ed. David LaRocca (New York: Bloomsbury, 2020), 46–7.

21. Cavell, *The World Viewed*, 28; italics in original.

22. Ibid. 27–8.

23. Ibid. 27–8.

24. Ibid. 27–8; italics in original.

25. Ibid. 27–8.

26. Ibid. 27–8.

27. Ibid. 29.

28. Ibid. 33; italics in original.

29. Stanley Cavell, "Opera in (and as) Film," *Cavell on Film*, ed. William

Rothman (Albany: State University of New York Press, 2005), 209. See also Cavell (1971), 34–5, and *Cities of Words: Pedagogical Letters on a Register of the Moral Life* (Cambridge, MA: The Belknap Press of Harvard University Press, 2004), 117.

30. Mel Gussow, "Ian Holm, Malleable Actor Who Played Lear and a Hobbit, Dies at 88," *New York Times*, June 19, 2020, B12.

31. Walker Percy, *The Moviegoer* (New York: Vintage International, 1960), 25.

32. William Day, "Words Fail Me. (Stanley Cavell's Life Out of Music)," in *Inheriting Stanley Cavell: Dreams, Memories, Reflections*, ed. David LaRocca (New York: Bloomsbury, 195).

33. See Barbara Cassin, ed., *Dictionary of Untranslatables: A Philosophical Lexicon* (Princeton: Princeton University Press, 2014) and *The Geschlecht Complex: Addressing Untranslatable Aspects of Gender, Genre, and Ontology*, ed. Oscar Jansson and David LaRocca (New York: Bloomsbury, 2022).

34. Ralph Waldo Emerson, "Social Aims," in *The Complete Works of Ralph Waldo Emerson* (Boston: Houghton, Mifflin and Company, 1904), vol. 8, 96.

35. Anthony Lane, "Remembrances," *The New Yorker*, June 22, 2020.

36. George Eliot, *Middlemarch: A Study of Provincial Life*, ed. Gordon S. Haight (Boston: Houghton Mifflin Company, 1956), 107.

37. Amy Villarejo, *Film Studies: The Basics*, second edition (New York: Routledge, 2013), 32, 39–40, 58, 131.

38. Tatiana Prorokova, "Laughing at the Body: The Imitation of Masculinity in Peplum Parody Films," in *The New Peplum: Essays on Sword and Sandal Films and Television Programs Since the 1990s*, ed. Nicholas Diak (Jefferson, NC: McFarland & Company, Inc., 2018), 205.

39. See, for example, my "Performative Inferentialism: A Semiotic Ethics," *Liminalities: A Journal of Performance Studies* 9, no. 1 (February 2013).

40. Ludwig Wittgenstein, *Tractatus Logico-Philosophicus*, trans. D. F. Pears and B. F. McGuinness (London: Routledge, 1961), 71, §6.421.

41. See, for example, Garry L. Hagberg, "The Instructive Impossibility of Being John Malkovich," in *The Philosophy of Charlie Kaufman*, ed. David LaRocca (Lexington: The University Press of Kentucky, 2011), 169–89.

42. See Ohad Landesman, "Holy Motors: Metameditation on Digital Cinema's Present and Future," in *Metacinema: The Form and Content of Filmic Reference and Reflexivity*, ed. David LaRocca (Oxford: Oxford University Press, 2021), 173–87.

43. See Villarejo, *Film Studies*, 32, 39–40, 58, 131.

Part IV

Plastics in Literature

12

Polymeric Thinking: Allison Cobb's
Plastic: An Autobiography

Lynn Keller

As I write, in September of 2020, I am among billions of people across this planet attempting to shelter in place in order to control the spread of the novel coronavirus, SARS-CoV-2. This virus is generally thought to have entered the human population through close contact with bats, perhaps with wildlife-livestock acting as intermediaries. Such contact has been made inevitable by the dramatic expansion of human populations and the consequent encroachment of human settlements into other species' territories. Enacting the increasingly complex entanglements of human and more-than-human realms and revealing the high costs of often only partially anticipated anthropogenic transformations of the planetary environment, the current global pandemic is a phenomenon distinctly of the Anthropocene. In this context, the approach to autobiography that Allison Cobb takes in her hybrid text *Plastic: An Autobiography* (Nightboat Books, 2021) makes powerful sense.[1] For, unlike most Western autobiographies, hers is not the story of individual development within an exclusively human society. Implicit in its organization and contents is the recognition that a person can no longer imagine her/his/their life story as the tale of a relatively autonomous, self-determining being. Instead, as Cobb's title and her methods indicate, we in the twenty-first century—a period I have elsewhere dubbed the self-conscious Anthropocene[2]—must understand our life stories and our identities as inextricable from the histories of the objects and species around us, and also inextricable from human technology, which is bound up in military and imperial histories as well as changing environmental conditions. The life stories of even the most privileged and protected humans are intertwined with the movements of vulnerable refugee populations, with the suffering of oppressed groups, and with the adaptability or non-adaptability of more-than-human species. We need to understand our lives as thoroughly entangled with those of the creatures whose environmental conditions humans have in recent decades so drastically altered, and with the substances as well as the machines that humans in the industrial age have invented. Among the most crucially influential, toxic, persistent, and pervasive of those substances is plastic. Whether we understand Cobb's title to mean that she traces the life of plastic from an inside perspective or that the story of her life *is* the story of plastic (or if both meanings seem operative), the invention of plastic, its current production, its consequences as waste, and a particular plastic car part are central to Cobb's "autobiography."

Plastic, especially plastic's polymeric form, possesses figurative as well as literal significance in this book. Polymers, a category that includes DNA and cellulose and many non-synthetic materials as well as synthetic ones, are large molecules made of chains of small molecules (monomers). Their forms may be branching, unbranched, or cross-linked, but they always involve covalently bonded chains or networks. Their architecture as extended chains developed from multiple units that bond together provides, I contend, a model for thinking about literary form, about the nature of the self, about social relations and responsibilities, and about ethics more generally. While attending to the damaging environmental entanglements of plastics Cobb tracks, this essay will focus on exploring the affordances of what I am calling *polymeric thinking*—thinking that's alert to or reaches for extending chains of connections that are demonstrated and advocated in this unconventional, particularly timely autobiography.

Polymeric form

Cobb's achronological work is composed of seventy-four titled sections of prose which sometimes shift to short-lined free verse. (Several titles, such as "Work," "Joy," "Zero," and "Desire," repeat.) These are gathered under three numbered headings—"I. THE THING," "II. REFUSE," and "III. THE LIVES," framed by an introductory section titled "GIFT: THE THING" and a coda, "LEGACY: 'THAT'S YOURS.'" The text is followed by notes containing scholarly citations as well as a "Selected Bibliography" with more than 150 entries. My assertion above that plastic is "central" to *Plastic: An Autobiography* may not be exactly apt, for the book lacks the single core and defined perimeter the term may imply. Even a synonym like "integral" may suggest a structure more neatly bounded than this autobiography. Ranging surprisingly in subject matter, it frequently shifts unexpectedly to a new focus or suddenly returns to an earlier one. While the book feels shapely—its shape depending in part on frequent returns to some apprehension of plastic—we are invited to understand this text as a fragment of something perhaps infinitely extending, part of an always expandable network of "kaleidescopically interwoven" connections.[3]

Even Allison Cobb herself, at least as the self is usually understood, may not be "central" to this disjunctive autobiography, which reveals nothing of her education or professional development and depicts little of her personal life. Instead, because she is the daughter of a physicist who spent his career at Los Alamos National Laboratory, where the atomic bomb was developed, her recounting foregrounds the bomb's development, a tale that is inextricable from the life stories of Jewish refugee mathematicians and physicists such as Stan Ulam and Edward Teller. Developed partly through branching research into the military and technological past, her story also incorporates histories of others who were not part of the Manhattan Project, including Jiro Horikoshi, who designed the Zero fighter plane; American PBY pilot Elwyn Christman, who was shot down by a Japanese Zero; William Perkin, who produced in the mid nineteenth-century a fashionable purple dye from coal tar; and August Kekulé, who at about the same time intuited in a vision the ring shape of the benzene molecule, the bonding ability of which is crucial to the creation of plastic.

Plastic: An Autobiography does not open by dramatizing some foundational event in Allison Cobb's life; nor does it start by depicting an obviously foundational moment in the history of plastic. Instead, it recreates a moment from Stan Ulam's autobiography when he tells his wife that he *"found a way to make it work."*[4] How better to convey an expansive interdependence essential to one's identity than to begin one's autobiography with a scene taken from another person's autobiography—in fact, a scene that was added as a postscript by that person's wife, making it her story as well? The moment with which Cobb opens occurred in 1951, decades before her birth. It is the history-changing day when Ulam figures out how to make "the Super"—the hydrogen bomb—work.[5] Cobb's curiosity in this section focuses less on the discovery than on the mathematician's eyes, which he turned toward his wife as he reported his breakthrough—eyes his wife says stared "unseeing" out the window. Only much later in Cobb's book, after intermittent sections have fleshed out Ulam's history in Lwów, New York City, Madison, and Los Alamos in the years before, during, and immediately following World War II, does the reader learn that what Ulam discerned that day depended on a new plastic, polyethylene, to ensure the hydrogen bomb would ignite. That is, not until the penultimate entry of "REFUSE" is the critical link between the developing technologies of plastic and of nuclear weapons apparent. (Only later, too, will the reader recognize seeing and not-seeing as a recurring thematic motif in Cobb's book.)

As this example suggests, the interconnections among the book's multiple story threads, reaching across time and space, are not always immediately evident, though they are real. The book's form, as I understand it, echoes the molecular structure of plastic. Cobb herself, attempting to map the book's themes and connections early in her work on it, "tried to make the map look something like a complex molecule."[6] Apparently, her map wasn't much of a success: "I have no skill for drawing," she writes.[7] Nonetheless, the book's form can readily be understood as *polymeric*, enacting the material as well as the social entanglements of our time, as it requires the reader to trace the sometimes subtle or unexpected chains linking its units—that is, to engage in polymeric reading. The author's prior tracing of these chains yielded the substance of the book, but her tracking process was far from linear; importantly, her representation remains true to the nonlinear complexity of both her investigations and the histories they expose. Hence the book's many-stranded structure, more weaving than collage, in which sometimes surprisingly interrelated narratives or topics disappear and resurface in irregular patterns.[8] The form dramatizes entanglement.

Environmentalists and environmentally inclined writers have often sentimentalized interconnectedness within the biosphere and sweetened the science of those interrelations, ecology.[9] Plastic, however, puts the brakes on such idealizing impulses, and often the interconnections Cobb reveals are deeply disturbing. While her narrative is shadowed always by the development (and the testing or deployment) of the atomic and hydrogen bombs, the primary object Cobb traces through the book is a piece of trash that blew into her yard: a black plastic car part that she determines, after considerable effort, comes from the fender of an early Honda Odyssey. This part has been designed to cover over the ugliness of "the car's raw metal underbelly, which betrays its brute machine birth and pierces the illusion of speed and ease the shining surfaces impart,"[10] and its polymerically expanded story is not a pretty one.

Contemplating the curved shape of this large plastic car part, "folded in half like a wing at its narrowest point," Cobb sees the image of "an albatross carcass bursting with plastic."[11] She is recalling a particular young bird, dead of starvation, from whose stomach the photographer Susan Middleton extracted—and then photographed—more than five hundred pieces of plastic that had been fed unwittingly to the chick in the fish eggs and other food gathered by its parents from the Pacific. Its fate is a common one among young albatross for whom adults, flying on the longest wings of any birds, forage far over the ocean. The accumulation of plastic waste in the environment, and particularly its incorporation into ocean gyres, has meant its dangerous integration into food chains. While part of her would prefer to turn away from her vision of plastic's interconnections in the biosphere and be numb to its horrors, the persistence of this disturbing image of plastic's deadly entanglements hooks into her and pushes Cobb to follow their course. By the time she writes the book, she has grasped its inseparability from herself: "Me and this car part, its dirty carapace curled around me. There is no gap between us, no other 'out there' to access, by microscope or imaginative vision. Here we are. Together. And the industrial chemicals we share, the resonating molecules of our bodies."[12]

Among the plastic fragments in the albatross's belly, the earliest datable is a Bakelite fragment from the equipment, perhaps a bombsight, of World War II bomber squadron VP-101, a fact that leads Cobb to the story of Elwyn Christman (1915–1945), who flew—and died—in that squadron. In Cobb's polymeric work, the terrible violence of modern warfare is multiply interconnected with the development of plastic. The hydrogen bomb, for which the crucial part, as noted above, was plastic, provides one profound example. Radar technology, so important to the Allies' bombing campaigns, provides another: that technology depended on the development of insulating polyethylene, while carrying radar on airplanes required lightweight polyethylene cables. Moreover, radar, which dispensed with the need for a bomber to see his target and enabled continuous attacks regardless of cloud cover, "cemented the strategy of terror bombing cities, an approach for which the atomic bomb turned out to be the ultimate weapon."[13]

Pursuing the origins of her plastic car part as well as a better understanding of the impact of plastic production on human lives eventually leads Cobb to make several trips to areas of Texas and Louisiana where plastics are currently manufactured and to the communities, primarily communities of color, that are being razed to make way for today's vast production plants or being sickened by the chemicals the plants discharge. These often impoverished "sacrifice communit[ies]"[14] are the human equivalent of the albatross, Shed Bird, poisoned in capitalism's "global networks of consumption, waste, and pollution."[15] Much of the third section of the book, "THE LIVES," explores these issues of human environmental injustice, tracking what Cobb learns as she becomes involved with the environmental activists of Freeport, a town of 12,000 people on the Texas Gulf Coast, 65 percent Latinx and 11 percent Black, surrounded by nearly a dozen industrial plants. Among those plants is the complex of Dow Chemical, "the world's leading supplier of polyethelene," identified by the local chamber of commerce as "the world's largest integrated chemical manufacturing site":[16]

The Dow plant is called a "cracker." It takes molecules of ethane from natural gas and heats them to high temperatures, "cracking" them apart to form ethylene for plastic. It was the first in a $150 billion boom of chemical plants along the Gulf Coast, as companies looked to profit from transforming the oil and gas freed by fracking into products for the global economy.

The "cash cow" among those products, Cobb notes, is single-use plastic, the source of nearly inconceivable amounts of plastic waste piling up in landfills, accumulating in oceans, and washing onto beaches and shores.[17]

The most common chemical air pollutant in Freeport, Texas is ethylene, from which polyethylene is manufactured. Interacting with sunlight and other materials in the air, ethylene produces smog that consists mostly of ozone. Freeport's ozone levels have never met federal standards, and the asthma widely suffered in the population is only one of its health effects. An assessment by the Texas Department of Health of cancer levels in Freeport between 2000 and 2015 found that "[t]he number of all-age liver and intrahepatic bile duct, lung, nasopharynx/nose/nasal cavity and middle ear, and stomach cancers was above the range expected."[18] Dioxin, another toxic byproduct of plastic production, contributes to "cancer, heart disease, diabetes, endometriosis, early menopause, reduced testosterone and thyroid hormones, skin, tooth, and nail abnormalities, damage to the immune system."[19] It harms the central nervous system of developing fetuses. Cobb does not concern herself with the valuable uses of plastic, but with the multiple levels of harm to human and nonhuman bodies that have resulted from plastic's proliferation and its environmental entanglements.

From trans-corporeality to the extended polymeric self

In her books *Bodily Natures* (2010) and *Exposed* (2016), environmental humanities scholar Stacy Alaimo develops a concept of trans-corporeality, applicable to issues of environmental health and environmental justice, "that traces the material interchanges across human bodies, animal bodies, and the wider material world."[20] Material interactions that preoccupy Allison Cobb, such as the absorption by human and animal bodies of carcinogenic or endocrine-disrupting chemicals leached from plastics into the environment or ingested in water and food, or the absorption of nuclear radiation released by the deployment or testing of atomic weapons, are among those interchanges. Alaimo observes that, "[a]lthough trans-corporeality as the transit between body and environment is exceedingly local, tracing a toxic substance from production to consumption often reveals global networks of social injustice, lax regulations, and environmental degradation."[21] Cobb's tracing of plastic, prompted by a particular Honda part, nicely demonstrates Alaimo's point. Indeed, given that trans-corporeality is a "sense of the human as perpetually interconnected with the flows of substances and the agencies of environments," and that tracing trans-corporeal interchanges "reveals the permeability of the human, dissolving the outline of the subject,"[22] it's clear the term captures important dimensions of Cobb's orientation in *Plastic: An Autobiography*.

Some of Cobb's explicit meditations on the redefined nature of the self in our plastic-saturated world align strikingly with Alaimo's concept. One of these is the section "White Whale," where Cobb admits to having identified her title before knowing what it meant, and where she narrates a turning point in her thinking about the autobiography. This turning point occurs before the plastic car part enters her life and shortly after a beach clean-up expedition to Kamilo Point, Hawaii, where, encountering first-hand the plastic debris that washes up "twenty-four hundred miles from the closest continent," she is overwhelmed by the scale of the project she has taken on.[23] Not taking into account that the tale of plastic was already (trans-corporeally) her own, Cobb despairingly imagines she would need to travel to see all the plastic in the world in order to tell its story from the inside. Given that plastic, in addition to being constantly produced, is "impervious to flame, corrosion, electricity, water, decay, or other destructive force," its amounts are continually growing.[24] She thinks that in order to write *Plastic: An Autobiography* she would have to visit "every junk beach on the planet" where plastic accumulated, descend by submersible into sea canyons and beneath ice floes to witness the plastic collecting in the sea's depths, and even somehow become microscopic in order to slide as a tiny particle of plastic into the gut of a lugworm in beach sand. Feeling "incapable" of writing the book, she finds herself "paralyzed." Then, stepping outside her own door and picking up a weathered plastic ring that her dog seems to have chewed, she understands that she is already in it, that particles from that ring have made their way into the soil of her yard and will be incorporated into the peas she will grow and eat: "In there, along with whatever molecules make a pea, there might be a few broken free from the plastic bits, and whatever else has washed this coast in its sixty years as suburban tract: particles of soot from car exhaust, bits of mercury fallen with rain drops, asbestos slivers from the house shingles." Realizing that the plastic inside lugworms is already inside her, so that effectively she is inside the worm and the worm inside her, "sloshing molecules back and forth," she rejects her earlier fantasies of "piloting around, the watcher peering out from her safe suit of self."[25] Cobb is determined to strip off that suit of bounded selfhood, which is more like the emperor's new clothes: a socially shared delusion. This section fits neatly with Alaimo's conceptualization of trans-corporeality, focusing as it does on bodily permeability, on the body's absorption of materials in the environment, and on a radical opening of the allegedly discrete self into a network of material agencies.

Yet Cobb's recounting of another turning point, when her twenty-year partnership with her wife unravels, reveals differences between Cobb's thinking and Alaimo's, as Cobb explores an expanding sense of relationality that seems to me better interpreted through the metaphor of the polymer than through metaphors of porosity. In this section, "Seed," Cobb announces, "The basic unit of existence is not the individual, but the relationship." She attributes this notion to the physicist Karen Barad—whose concept of agential realism has influenced many of those, including Alaimo, who are exploring new materialist thinking—and immediately quotes a passage from Barad's *Meeting the Universe Halfway*: "To be entangled is not simply to be intertwined with another, as in the joining of separate entities, but to lack an independent, self-contained existence . . . Individuals do not preexist their interactions; rather, individuals emerge through and as part of their intra-relating."[26]

Announcing that "All that exists is merging and overlapping phenomena," Cobb begins to define (polymerically) how she herself exists, starting with bodily connections but including relations that are also immaterial.

> I am of this: my mother, in cell and bone and breath. How could I not be? One nervous system takes shape enfolded within another. I am plastic blister pack, calculus, computer. I am nuclear weapon, and car part, and war. I am—deeply—of my father.
>
> I am of Jen also, this "we" out of twenty years twined together. But something happened. Our entanglement started to fray and falter. Loss. Jen and I began to split.
>
> And then I was—what?[27]

What makes this autobiography so compelling is a capaciousness evident in this passage: The writing pursues Cobb's interest in what is—or comes to be—materially part of cell and bone and nervous system, including not just one's genetic heritage but also chemicals or nuclear radiation absorbed into human or animal bodies from anthropogenic materials released into the environment. Her interest in these bodily conditions draws her into environmental justice concerns involving the lives of those most directly affected by plastic production. Yet she is no less invested in emotional connections and in interconnections across time and space that are far less directly material—including Cobb's intangible connections to centuries-old mathematical disciplines like calculus, to Stan Ulam and Jiro Horikoshi, and including semantic links between words' current meanings and their etymological roots. This range of relationality is more readily captured in the figure of the polymer—of extending chains of linked entities—than in conceptions based in the permeable body.

Representing polymeric relationality: Cobb and Jiro Horikoshi

To demonstrate how such polymeric relationality functions in the book, let us consider the example of Cobb's entanglements with Jiro Horikoshi (1903–1982), whose story occupies four sections about midway through *Plastic: An Autobiography*. Presumably, the chain that led to her researching his life began with the fragment of plastic from the World War II bombing squadron found by Susan Middleton in Shedbird's stomach; from there, inquiries on a veterans' Listserv led Cobb to the nephew of a man whose journal entries recounted flying in that squadron under the command of Elwyn Christman, who in turn was shot down by the Japanese Zero, an astonishingly light and maneuverable plane designed by Horikoshi.[28] Only a few lines into the first of the sections about Horikoshi's life, titled "Grief," Cobb attributes something she recounts to his autobiography. As she did with the book's opening depiction of Stan Ulam that enfolded his autobiography into hers, Cobb thereby creates an expansion of self through mirroring reflection.

Thematically, Horikoshi's dreaming of flight links him to multiple flyers appearing in Cobb's book, from albatross, to bomber pilots like Christman, to the Al Qaeda suicide bombers of 9/11. Like Ulam and those working on atomic weapons, this aircraft designer was pushing the boundaries of technology for military advances. And, as was true for his American counterparts, much of what his remarkable inventions enabled was utterly horrifying. His first breakthrough in fighter planes, the Type 96, made possible the Rape of Nanking[29]—though bearing no responsibility for the preexisting motivations for that massacre and mass murder (vengeance, xenophobia, etc.). His next design triumph, the nimble long-range fighter aircraft, Type 00, dubbed the Zero, for a while gave Japan a tremendous advantage in the air war. Cobb, while highlighting the links between technological advances and violence or harm, pursues other threads as well. The aircraft's number prompts meditations on zero that extend ideas introduced in early sections' ruminations on zero (e.g., "ratios cannot be made with zero. Zero consumes all relations"[30]) and on nothing. The etymological roots of the word "nothing" intertwine a negation ("absence, a lack") and an affirmation ("vital force, long life").[31] That no/yes combination captures Cobb's own relation to ongoing environmental degradation, as at once a preference for numbness, "the uniform distanceless" (an allusion to Heidegger on technology in his essay "The Thing") and a desire to touch and be touched by the painful shards of the world.[32] Her meditation on zero in connection with Horikoshi also furthers the book's interest in circular forms, particularly rings in motion, things that bend and come round again—as in the benzene molecule and the recurring figure of oroboros. Horikoshi's character and life story take shape, then, via elaborate entanglement in the intellectual web of this book.

From polymeric thinking to ethical responsibility

One of the most important ideas to emerge from Cobb's pursuit of Horikoshi's story within her own comes from her polymeric investigative reading in the disputed history of the Rape of Nanking. In a "beautiful essay" by Simon Han, she finds the following: "Perhaps in a world that tells us how to feel about our past, a way forward is to ask a different kind of question—not how a scar came to be, but how it hurt. How it continues to."[33] (The emphasis on continuance may bring to mind the endurance of plastic, how "[i]t fails to disappear," further reinforcing concerns with the ongoing character of historical damage.[34]) Han's sentences subsequently become a kind of refrain in *Plastic: An Autobiography*. They resonate powerfully with the kanji, presented at the close of "Remember," used by Japanese speakers to designate what English speakers call "ground zero": three characters that mean "blast / heart / place."[35] For while the English phrase points only to a location, the Japanese one speaks also of a bodily organ that is subject to both physical and emotional wounds. The three kanji descending in dramatic boldface alongside their English translations remind the reader that the scar most salient to Cobb's history comes from atomic weaponry. Inclusion of the word *heart* in the sign system of the bombs' victims may be a reminder, too, of somatic vulnerability, of trans-corporeality. Yet "the way forward" requires an attention to ongoing suffering for which the notion of scarring is more metaphorical than literal.

Cobb does not treat this as an either/or choice. Clearly, she wants her readers to remember the dreadful wounds inflicted on humans and the environment during World War II and in the subsequent Great

Acceleration—damage that is in crucial ways material, somatic. Thus, in late sections of the book, Cobb provides often horrifying information about the testing of atomic weapons by the US on Enewetak and Bikini Atoll in the Marshall Islands, about the colossal amounts of radioactivity generated on the land and sea there, and about the US government's treatment of the Marshallese as another sacrifice community. Recognizing harm to bodies and to ecosystems is crucially important to Cobb. At the same time, her book's structure insists that the ways we are consequentially entangled with others are not only material. Its many polymeric threads emerge from her painful awareness that her upbringing, with its experiential and educational privileges—material that might be covered in a more conventional autobiography but is largely elided here—was supported entirely by her father's professional involvement with nuclear weapons. Cobb's sense of transgenerational entanglement in an unfathomably violent and destructive history, which accounts for her incorporating Ulam's fateful discovery and Horikoshi's inventions as parts of her own story, does not depend on the bodily permeability of trans-corporeality. Moreover, the aim of her autobiography is neither self-flagellation nor the denigration of physicists and engineers; nor is it simply to remind readers of the horrifying and lasting material consequences of atomic radiation or plastic production. Simon Han's sentences bring into focus her overriding aim: to move forward from past harm via an ethical understanding of the thoroughness of our interrelation, and of our consequent responsibility for and to one another's pain.

Alaimo and Barad both speak to the ethical implications of the versions of interconnected being in which they are invested. Alaimo writes, "trans-corporeality as an ethical practice requires not only that citizens seek out information . . . about risks to their own health but also that they seek out information about how their own bodily existence—their consumption of food, fuel, and specific consumer products—affects other people, other animals, habitats, and ecosystems."[36] Barad closes her massive volume with a grand and powerful plea for an ethics appropriate to our essential intra-relatedness, worth quoting in full:

> If we hold on to the belief that the world is made of individual entities, it's hard to see how even our best, most well-intentioned calculations for right action can avoid tearing holes in the delicate tissue structure of entanglements that the lifeblood of the world runs through. Intra-acting responsibly as part of the world means taking account of the entangled phenomena that are intrinsic to the world's vitality and being responsive to the possibilities that might help us flourish. Meeting each moment, being alive to the possibilities of becoming, is an ethical call, an invitation that

is written into the very matter of all being and becoming. We need to meet the universe halfway, to take responsibility for the role that we play in the world's differential becoming.[37]

Cobb, too, seeks an ethical enactment of the ways in which we are bound to others—to people and actions from the past as well as to human and more-than-human beings in the present. She states directly what she has come to understand: "by remembering across generations and without refusal our pained entanglements, and by being responsible, answerable to that pain, we can carry each other into a past that might make a living future."[38] This is the most hopeful potential affordance of polymeric thinking and a polymeric understanding of the self.

Perhaps paradoxically, in limited ways both Cobb's father and Horikoshi begin to model such a taking of responsibility, by acknowledging harm in which they have been directly, or through chains of relations, implicated. Her father, who arrived in Los Alamos a quarter century after bombs were dropped on Hiroshima and Nagasaki, "spent most of his career at Los Alamos in nonproliferation, working to stop the spread of nuclear weapons," even serving as Director of Threat Reduction.[39] Horikoshi revealed in his 1970 book about the development of the Zero fighter how he had wished to include "at least a paragraph of protest" in a solicited essay in praise of the kamikazes that he reluctantly produced in 1945; at the time his muted objections to the focusing of human ingenuity on the development of weapons were registered in his statement: "We have reached the limits of human intelligence and have selfishly tried all kinds of methods to make effective adjustments to our limited human and material resources so that new arms could emerge."[40]

While such steps aimed at preventing repetition of past errors represent a beginning, they are insufficient means of confronting how, in Han's terms, the scar hurt and continues to hurt. With the help of her Japanese friend Yukiyo Kawano, Cobb comes to appreciate authentic apology as a fuller resource for confronting past suffering in which one is complicit and for facing the ongoing damage. Apology has an important place in Japanese culture, a form of politeness that grows from empathy and respect;[41] Cobb does not try to unravel its cultural significance, but through interactions with Yukiyo, a third-generation *hibakusha* (survivor of either of the atomic bombings) whose maternal grandfather also probably fought in the Battle of Nanjing, she comes to take very seriously the ethical force of apology in a world of relational entanglements. Apology is particularly powerful, for both the entity making and the entity receiving it, when it becomes a "way of being."[42]

Prompted by Yukiyo first to consider the possibility that a particular white poet in Portland, working on a project "regarding the poisoning of the Willamette River," might apologize "to the river, to the Chinook salmon, to native communities, to all harmed by white colonizers" and then the possibility that Cobb herself, if she were performing in Los Alamos, might apologize "to all those displaced and harmed by the nuclear lab," Cobb initially resists. She thinks of the impossible scale of the apology she would have to give "for everyone harmed all around the world by nuclear radiation, to all beings."[43] That perception of impossibility echoes the moment when she had felt paralyzed by the scale of what she imagined she would have to be responsible for in narrating the autobiography of plastic. And the resolution on this occasion similarly involves a kind of scaling down via a response to what is immediately present. In this case it's an apology she finds herself making specifically to one link in that vast chain of irradiated beings:

> "Yukiyo," I said, "I'm sorry. Your family has suffered so much from nuclear weapons and war. You've lost your mother, your uncle, your aunt, and had such hurt and fear in your family, and in your own life." Yukiyo—she looked in pain. Tears came up in her eyes and spilled down her cheeks. I put a hand on her arm, tears came up in me also. I didn't know fully what to say. I fumbled—it's not as clear as I'm writing it here—but the words came out, from my body, my heart and my guts, not my head. I felt it all through me, that state. The state of being sorry. It hurt, because it required really seeing Yukiyo, and knowing and feeling how she has suffered. It required seeing myself also, or feeling myself, and the harm my privilege carries. It felt like release, like something broke free in me.[44]

It's worth observing that there is a "myself" in the passage that is a particular privileged individual, just as there is a defined Yukiyo with a distinct family history. The self is not dissolved in this perspective but exists in a state of polymeric entanglement, analogous to a monomer molecule—one that can react with other molecules—bonded with others in a polymeric chain. If we think of autobiography as necessarily a document of memory, the point seems to be that if the writing is to be ethical, the memory it relies on must expand across time and space beyond the merely personal. Cobb's autobiography has to remember her painful multigenerational nuclear entanglements as well as her participation in the "consume-and-dispose violence" of plastic.[45]

Not seeing the self as isolated yields an expanded sense of agency, which means that responsibility, too, is not neatly bounded. In "Wind" Cobb presents a sharp critique of the romanticized vision of Horikoshi's life created in Miyazaki's 2013 feature film *The Wind Rises*, because of its avoidance of responsibility. The animated movie features "gorgeous dream sequences of Horikoshi flying fantastical airplanes that resemble birds or sea creatures." It transforms the obsessively dedicated engineer often hobbled by ill health into a heroic, if dreamy, figure. One of his invented exploits, in which he rescues a young girl and her governess from the burning city of Kanto, carrying the governess on his back, might bring to mind Aeneas rescuing Anchises and Ascanius from burning Troy. Yet, Cobb notes, for a hero he is also rendered "strangely passive"; "[t]hroughout the movie, the plane designers are depicted as lovers of knowledge and beauty swept along by impersonal forces." Aircraft are presented as "destined to become tools for slaughter and destruction."[46] Repeatedly in the movie rising wind is used to signal the fate governing human destiny. Cobb's critique is largely implicit, but clear: the responsibility for what Horikoshi's airplanes did lies with people and the choices they make, not with their machines or some controlling fate.[47]

Throughout the book, her presentation of individuals' choices is compassionate. Her narration makes clear that many of those most directly responsible for the atom bomb had suffered terrible losses due to Hitler; a Jew like Ulam, many of whose family members had been murdered by the Nazis, had good reason to fear Axis victory and devote himself to developing the awful weapon to defeat them. "Fear drove them," Cobb observes of the refugee scientists.[48] But that understanding does not prevent her from taking issue with Ulam's retrospective attempt in his own autobiography to avoid moral responsibility by separating his theoretical work on the bomb from the political contexts and historical consequences with which it is in fact entangled.[49] The authentic apology that she learns from Yukiyo to value acknowledges agency and takes responsibility, and the more it encompasses, the more it registers the thoroughness and sweep of an agent's entanglement in the lives of other beings.

Honda and Dow Chemical offer no apologies. Far from it; instead, they enact the "hierarchy of values in global capital: stuff comes first; consumers—the people who buy the stuff and keep profit flowing—come second. The lives of those who might interrupt this flow have negative value. They are obstacles for removal."[50] To expand ports for commerce, Dow and other chemical companies force land sales, breaking communities already rendered vulnerable by long histories of racial oppression; they use their wealth and power to avoid restrictions on pollution to air and water, and to avoid responsibility for the lives and bodies they damage. These corporations do everything they can to generate among consumers a sense of endless unsatisfied desire so that the public will consume (and discard) more and more plastic products. But even as Cobb reveals their consequential abuses, she discourages adoption, at least by middle-class readers, of a sense of righteous removal from corporate wrongdoings. This is a crucial contribution of polymeric awareness; it requires recognition that our entanglements extend in every direction. We Western consumers cannot understand ourselves as apart from Dow Chemical any more than we exist apart from the albatross that is its victim; if we are white, we cannot imagine ourselves disentangled from the violent structure of white supremacy that allows companies like Dow readily to sacrifice poor Black and brown lives.

The important question, of course, concerns our collective course for the future. Partly through her interactions with African American activists, for whom mere words like "I'm sorry" coming from a privileged white woman like Cobb are meaningless, she comes to understand apology as "a long-term commitment, an ongoing relationship," with three key components: regret, responsibility, and remedy. She recognizes that the kind of apology needed in our "white supremacist, heteropatriarchal culture, built and organized around the notion that certain lives are disposable" will require cultural transformation "from the root."[51] When at the book's end the tour leader at the Honda plant refuses to take back the plastic car part, telling Cobb's partner's daughter, "That's yours," the woman speaks a polymeric truth.[52] Of course, it is most definitely Honda's as well, even if Honda refuses to acknowledge that responsibility, but the plastic in our world affects all of us in our polymeric entanglement and becomes a massive responsibility we must all embrace.

Polymeric thinking doesn't in itself provide answers to the problems we face, but it does provide appropriate frameworks for understanding them and for seeking solutions. The implications of this vision—of a perceptual lens that brings into focus the thorough entanglement of the present and future with the past, of the violence of war with patterns of capitalist consumption, of industrial development with racial inequities, of human bodies with those of other species, and of all living bodies with the anthropogenic chemicals and radiation strewn through the environment—go well beyond literary form or genre. As new materialist scholars in the environmental humanities have recognized, an understanding that humans exist interactively with the non-human realm denies human exceptionalism and undercuts arrogant pretenses to human mastery and control. A polymeric understanding of the self as more defined by its interrelations and thereby more expansive than has been previously recognized in Western thought may enhance our ability to grasp the realities of our environmental situation. It's a disconcerting irony that plastic polymers, even as they litter and poison our material world, suggest a fruitful ontology and an appropriate ethics for our environmentally precarious time.

Trans-generational remembrance

Having begun this essay positioning myself, I will end it, oroboros fashion, by doing so again and at somewhat greater length. For what I have not yet acknowledged is that, like Allison Cobb, I am a daughter of Los Alamos, although I never lived there. Like Cobb's father, mine earned a doctorate in high-energy physics. His degree was supervised by J. Robert Oppenheimer at UC Berkeley in the late 1930s. I believe he was working at Columbia in 1941 when Cobb describes Edward Teller and Enrico Fermi there, strolling in Manhattan and considering the possibility of an atomic bomb that could ignite thermonuclear fusion.[53] Later in the war, my father joined the scientists at Los Alamos working on the bomb. He died of cancer at age fifty-nine when I was eighteen, and though I recall his deep loyalty to Oppenheimer (who, I've been told, charmed me as a child by blowing smoke rings), I don't remember him ever talking about his time at Los Alamos. My father was a modest person, and it may have been he who led me to think of him as occupying a lowly position there, similar to the women who served as computers, working with pencil and slide rule—always in his jacket's inside breast pocket—to produce endless calculations on pads of lined paper. Allison Cobb herself, when I wrote her acknowledging this connection between us, tracked down his badge photo at the Lab and informed me that he was, at least briefly, a leader at the Lab working on the hydrodynamic lenses for the plutonium bomb. I knew him as someone, like her own father, invested in nuclear nonproliferation. Yet I have always carried, as an albatross around my neck, a burdensome awareness of entanglement in atomic warfare and the nightmarish possibility of its future recurrence.

There are passages in Cobb's book that enfold my own biography. One occurs in the section titled "Lament," where Cobb recounts a reading of ecopoetry she gave at the University of Hawai'i in 2014—a visit, she notes, that "marked [her] place in a long lineage: white settlers imbued with authority to speak as protectors and defenders of the ecologies they helped wreck." As a white ecocritic from a major research university whose father helped develop the atom bomb, I have acquired a comparable guilty authority. So I find it particularly meaningful that Cobb goes on to remind her readers of the range of things that bind us, the no and also the yes:

The word "complicit" comes from a root that means to fold, or weave together. I stood before students and colleagues in Hawai'i, threaded to them by the violence I carry, in my ancestry, in my body, in my every step and breath. Other threads also tied us together—the care and concern we shared.

"Concern" comes from a root that means to sift or sieve, plus "con," the root for together. To sift or sieve together. It suggests a mixing in which the constituent parts retain their integrity. A weaving that displays each original color. This mix of students, poets, teachers—all of us in this place with our particular heritages, our histories, our unequal suffering, our stakes. It was our care—whose earliest meaning was mourning—that held us there.[54]

Material things bind us as well, and plastic is one of them. Much of the rest of that section of *Plastic: An Autobiography* treats the history of anthropogenic threats to albatross, including the threat posed to albatross and seabirds worldwide by plastic. Linking birds to humans, it ends with information about the environmental impact of plastic on humans:

A 2020 study found invisible plastic particles suspended in the air and raining down everywhere the researchers looked, so much plastic they kept rechecking their results. They concluded plastic fallout exists in "every nook and cranny" of the planet. Like birds, people eat and breathe it; on average the weight of a credit card in plastic goes into people's bodies each week. Scientists also found plastic particles in placenta that nourishes human fetuses. No one knows what this means for us—we all, the concerned, threaded together.[55]

What Cobb's readers *have* come to know is that remembrance, including the recollection of past harm and suffering and injustice, will be necessary to the generation of a survivable future.

My father is similar to numerous men in Cobb's book in having chosen not to talk about his involvement in war or damaging invention. Her own gentle grandfather, for instance, did not speak of having, in the war, killed at least one Japanese man with just a knife. We may sympathize with their desire to leave painful or morally fraught memories behind, but the polymeric thought that Cobb practices and advocates insists that we will move into a better future only by recognizing the past and its shards in the present. By giving an expansive polymeric form to her autobiography, Allison Cobb is breaking what Miss Jessie of Freeport, Texas calls "the generational curse": "'The older generations didn't talk about their lives,' she said. 'But if you don't know what happened to your mother and your grandmother, and the one before her, and the one before her, how can you change the future?'"[56]

Notes

1. Allison Cobb, *Plastic: An Autobiography* (New York: Nightboat Books, 2021). I was given advance access to the manuscript, for which I wish to thank Stephen Motika, Director and Publisher of Nightboat Books, and Allison Cobb.
2. Disengaged from debates about the dating or validity of the Anthropocene as a geological epoch, the self-conscious Anthropocene names our era of widespread, often anxious awareness of the profound impact industrialized humans are having on planetary systems. It designates a cultural reality, not a stratigraphic one. For elaboration of this concept and explanation of its dating from 2000, see the opening pages of Lynn Keller, *Recomposing Ecopoetics: North American Poetry of the Self-Conscious Anthropocene* (Charlottesville: University of Virginia Press, 2017).
3. Cobb, *Plastic*, "GIFT: THE THING," 4–5. Cobb's having earlier published, through Essay Press, a much shorter version online may itself demonstrate the work's flexible boundaries.
4. Cobb, *Plastic*, "Work," 9.
5. Ibid. 10.
6. Cobb, *Plastic*, "Refusal," 99. The exercise reveals to Cobb her own erasure of women as she has traced "histories of plastic, of the atomic and thermonuclear bombs, World War II airplanes, European canonical writings. White men dominated these accounts" ("Refusal" 100). Anger at silences that she herself had accepted and repeated led her to temporarily abandon the project in order to write the poems of *After We All Died*. The final version of *Plastic: An Autobiography* incorporates extended histories of individual women scientists whose contributions to knowledge have been erased or downplayed—Ida Noddack, Lise Meitner, Mary Tsingou (Menzel)—as well as discussion of the work done at Los Alamos by groups of women, often scientists' wives, acting as computers, crunching numbers on calculators, or programming the digital computer, once it came on line. To my chagrin, few women figure in my own abbreviated analysis here, so that I have repeated the erasure that Cobb, in a deliberate effort to remember women, worked to correct. Later parts of her book delve into activism by people of color whose communities on the Gulf Coast are most directly harmed by plastic production in the United States.
7. Cobb, *Plastic*, "Refusal," 99.
8. Donna Haraway's tentacular thinking may come to mind here; see Donna J. Haraway, *Staying with the Trouble: Making Kin in the Cthulucene* (Durham, NC: Duke University Press, 2016). But while Cobb, like Haraway, would reject notions of the bounded individual or thinking in terms of the human individual plus context, the notion of the polymer seems to me more appropriate than that of the tentacular organism because of Cobb's ties to atomic and molecular science and her focus

on the plastics developed with increased human ability to manipulate polymers. Cobb acknowledges what she terms both the no and the yes involved, as polymers are not inherently good or bad, while Haraway wants to advance the tentacular as a solution to the wrong thinking of the past. Later parts of this essay will flesh out additional reasons for the appropriateness of the polymer as explanatory model here.

9. See Dana Phillips, *The Truth of Ecology: Nature, Culture, and Literature in America* (New York: Oxford University Press, 2003) for a sharp critique of ecocriticism's uses of ecology.
10. Cobb, *Plastic*, "Car Part," 29.
11. Cobb, *Plastic*, "GIFT: THE THING," 4.
12. Cobb, *Plastic*, "Desire," 75.
13. Cobb, *Plastic*, "Radar," 137.
14. Cobb, *Plastic*, "Risk Factor," 202.
15. Stacy Alaimo, *Exposed: Environmental Politics and Pleasures in Posthuman Times* (Minneapolis: University of Minnesota Press, 2016), 113.
16. Cobb, *Plastic*, "Freeport," 189.
17. Cobb, *Plastic*, "Lights," 196, 197.
18. Cobb, *Plastic*, "Risk Factor," 161.
19. Cobb, *Plastics*, "Tired," 235–6.
20. Alaimo, *Exposed*, 112.
21. Stacy Alaimo, *Bodily Natures: Science, Environment, and the Material Self* (Bloomington: Indiana University Press, 2010), 15.
22. Alaimo, *Exposed*, 112.
23. Cobb, *Plastic*, "Gyre," 52.
24. Cobb, *Plastic*, "Infinity," 62.
25. Cobb, *Plastic*, "White Whale," 54–5.
26. Cobb, *Plastic*, "Seed," 129. The ellipses appear in Cobb's text.
27. Ibid. 157.
28. Cobb's recounting of the aircraft's development makes clear that, had Allied officials been less arrogant and less racist, they might have anticipated the threat posed by Japanese ingenuity; a retired pilot "repeatedly tried to warn US, British, and Australian officials about the Zero. They ignored his reports. No one believed the Japanese could create such an advanced machine. They concluded the defeats in China must have resulted from lack of skill among Chinese pilots. To Western powers, the Zero remained a blank, invisible" (Cobb, *Plastic*, "Job," 139). Had their sight been less occluded, perhaps Christman and other US airmen would have been spared.
29. The march of Japanese troops into the capital city of Nationalist China, Nanking (or Nanjing), on December 13, 1937 was preceded by aerial bombardment that "emptied the skies of Chinese fighters." For more than a month the Japanese soldiers then "went on a rampage, raping and murdering hundreds of thousands of people" in that city. (Cobb, *Plastic*, "Grief," 128).

30. Cobb, *Plastic*, "Zero," 14.
31. Cobb, *Plastic*, "Nothing," 15.
32. Cobb, *Plastic*, "Work," 20.
33. Cobb, *Plastic*, "Grief," 129.
34. Cobb, *Plastic*, "Refuse," 86.
35. Cobb, *Plastic*, "Remember," 131–2.
36. Alaimo, *Exposed*, 127.
37. Karen Barad, *Meeting the Universe Halfway: Quantum Physics and the Entanglement of Matter and Meaning* (Durham, NC: Duke University Press, 2007), 396.
38. Cobb, *Plastic*, "To Live," 150.
39. Cobb, *Plastic*, "The Thinker," 156.
40. Cobb, *Plastic*, "Job," 142.
41. An earlier draft of Cobb's book, which included longer treatment of Horikoshi, told of his making an apology: when an early version of the Zero disintegrated in a 1940 test flight and the pilot was killed, he rushed by train to the airfield and apologized to the director of flight testing.
42. Cobb, *Plastic*, "Sorry," 170.
43. Ibid. 170–1.
44. Cobb, *Plastic*, "Sorry," 171.
45. Cobb, *Plastic*, "GIFT: THE THING," 6.
46. Cobb, *Plastic*, "Wind," 143–4.
47. Concerned with critiquing the use of notions of fate or destiny to absolve leading aircraft designers Caproni and Horikoshi of responsibility for their inventions, Cobb is not here exploring how people's choices may be genuinely constrained by the power of the states in whose interests wars are waged.
48. Cobb, *Plastic*, "Adventure," 89.
49. Cobb, *Plastic*, "Boy," 266.
50. Cobb, *Plastic*, "Heartland," 204.
51. Cobb, *Plastic*, "Life," 239.
52. Cobb, *Plastic*, "LEGACY: 'THAT'S YOURS,'" 294.
53. Cobb, *Plastic*, "Adventure," 91–2.
54. Cobb, *Plastic*, "Lament," 111.
55. Cobb, *Plastic*, "Lament," 115.
56. Cobb, *Plastic*, "Remember," 217.

13

Plastic City: Temporality, Materiality, and Waste in Vanessa Berry's *Mirror Sydney*

Emily Potter and Kirsten Seale

This chapter thinks through the ubiquitous and dynamic life of plastic as it contributes to the imaginary and material landscapes of Australia's largest city, Sydney. Using *Mirror Sydney: An Atlas of Reflections*, a book-length work of "creative cartography" by writer and artist Vanessa Berry, as our guidebook to Sydney,[1] we theorize plastic as a symbolic and material actor whose dissonant temporal presence simultaneously supports and challenges subjective and objective visions of Sydney as progress-oriented and capitalist-efficient. Considered through new materialist terms, plastic brings realities into being by actively informing and shaping human practices and modes of inhabitation. Plastic is the ultimate agent of modernity. In this guise it appears as a "futural form"[2] indexical to and bringing about the world to come. At the same time, plastic also enacts multiple and at times dissonant temporalities and realities that are coexistent rather than linear.[3] It is a material constituent of many of today's disposable commodities, which will inevitably be consumed or thrown away. As deferred trash, plastic is an anachronistic presence in its refusal to integrate into a forward-focused sense of time.

In a similar way, Berry's encounters with tangible artifacts of an urban past that have been thrown away produces a counter-narrative to the hyper-modern story of Sydney as global city. Responding to what she sees as "Sydney's drive towards reinvention,"[4] Berry's concern is to uncover and document the complexity and ephemerality, but also the dense histories, of urban places through practices of wandering, mapping, and narrating localized situations of inhabitation in contemporary Sydney. These situations and Berry's "wandering" of sub/urban places belong to a performative methodology of "subversive mappings"[5] of city space, where dominant or official narratives of place are disrupted through the unpredictable encounters that the urban context affords. Place stories are "found" and recollected through the embodied mobility of the writer moving through vernacular spaces and activating the unprogrammed aspects of place.[6] Berry's methodology evokes the *flâneuse* or the Situationist *dérive* as a tactic of intervention in urban imaginaries and speaks to her ultimately political project, which is to invest the matter of urban memory with questions of social and environmental justice.

Sydney: plastic city

Jennifer Hamilton describes Sydney as a "settler colonial capitalist imposition on unceded Indigenous Country; the property market is outrageously dominant, both as a mediator of place relations and a keystone aspect of a financial system that governs everyone's life and work."[7] In Sydney, a city built on the violent dispossession of the Gadigal people, the weight of forgetting is heavy. Paradoxically, this heaviness is performed through the lightness of temporal excision: this is a city that prefers to look forward. Sydney's prevalent imaginary is focused squarely on the future and provides little imaginative or literal space for divergence. As Hamilton writes, Sydney's "developers and governments are currently peddling a spectacularly homogenous future-vision of fancy cafes, green space, and luxury apartment life."[8] This abstracted vision of globalized cosmopolitanism renders the local generic, while still loading it with the marketable specifics of Sydney's renowned natural beauty and lifestyle.

Outwardly, Sydney presents itself as an icon of global modernity, a constellation of soaring skyscrapers around a glittering blue harbor. For this reason, Sydney has come to be associated with rapacious development and a certain emphasis on superficiality. While this association does not always match lived and experiential realities of Sydney, the cultural privileging of the surface does model the spatial planarity that colonial-capitalist visions rehearse. A stable ground cleared of impediments to progress is a developer's dream.[9] However, this frequently involves the displacement of communities and their stories who do not conform with or actively resist the shape of imagined futures.[10]

In Sydney, as in other cities, the material city does not exist separately to the imagined city: it is a formation that is shaped by the stories told about it. James Donald recognizes this when he writes that "cities are best understood as complex 'imagined environment[s] composed of the layered narratives told by their various residents, visitors, and observers.'"[11] This is an "imaginary" that manifests a conceptual framework underpinning a particular way of knowing and existing in the world: that is, it is "constitutive of, and constituted by, ontic and epistemic commitments."[12] These, in turn, assert and put into material practice a certain way of seeing the world.[13] Just how the city is narrated, what stories are brought into discourse in the telling of its collective history, will inform how it is inhabited and made inhabitable on a spectrum of erasure and inclusion. Urban imaginaries are therefore political. The urban imaginaries that dominate are ones that are told by those in power. This has the effect of promoting certain narratives as normative, suppressing others, and sustaining entrenched systems of power.

On the other hand, material forms are also crucial in bringing into view what Alison Young calls "other cities within 'the city.'"[14] Plastic—the synthetic, or semi-synthetic, polymer and celebrated modern invention—is a key material that literally and figuratively mediates urban multiplicities. Since its development and exponential spread into every aspect of twentieth-century life, plastic has become a normalized presence across bodies and life forms, littered through landscapes and structuring daily life. As Katie Schagg writes, "We eat, drink, touch and absorb plastic polymers."[15] We sit in its forms, transact through its affordances, and inhabit worlds of play and movement brought into being by its reality. Plastic's ubiquity speaks to its absolute integration as a force of ontological making in contemporary life despite the persistent connotation that plastic is "wholly unnatural." This "plastic ecology"[16] disturbs assumed boundaries between the "natural" and synthetic worlds, while also refuting any assumption of passive materiality in the hands of human agents.

There are three ways in which we understand plastic as we consider Berry's and *Mirror Sydney*'s interests and interventions. The first is plastic as the cheap, disposable material of a modernity that is degraded and trashed easily. Plastic is the ultimate detritus, signaling the swift evacuation of value when desired products become empty, and crumpled packaging and the fantasy of single use curdles into rifts of garbage. It is also an icon of modernity's quest for efficiency and fast consumption without consequence. Hawkins describes the signifying role of plastic in the rise of consumer modernity, a material that practically enabled mass consumption and rapid manufacturing growth, as well as construction practices, and also spoke figuratively as a harbinger of "better futures": "plastic was represented as a revolutionary material triggering a dazzling array of new objects, practices and experiences that had no precedent. . . . Modernity had arrived."[17] Temporality, as Hawkins contends, was crucial to the material ascendancy of plastic. Plastic could create new realities, and in doing so displace "natural time and natural limits,"[18] generating its own temporalities based on capitalist norms of end use and disposability. The plastic city, in this frame, writes its own stories of the future through its material facades. However, by looking to the past to recognize the disposability of the present and the ephemerality of the future, Berry's work is a more accurate portrait of the plastic city, as she asserts its dissonant capacity for reinvention and remembering, for stretching and remaking, but also for multiplicity, for cities within the city, than the dominant place-myths[19] about Sydney allow. Plastic materiality, in this light, takes on a complex role in her text.

This connects to our second consideration of plastic in Berry's text: how Sydney's futurity is a plasticity, a malleability and ability to be shaped, that is based on a fiction of *terra nullius* which illuminates the paradoxical strength of historical narratives in a future-focused city. This is unceded land of traditional owners, Berry constantly reminds us. It is unsettled country. Yet the imaginary of a "tide of history . . . [that] has washed away"[20] the connections and entitlements of the Gadigal people to their land persists, a forgetting that conveniently enables developers' visions. The paradox of the Australian city—as with many cities elsewhere—is that it claims colonized land, land with deep histories, as ground zero for the starting point for the making of the modern nation and its landmarks. Still, the capacity of the city to also dictate urban life and its possible worlds through its material and imaginative forms weaves these questions about justice, access, and entitlement into the fabric of postcolonial cities. For Henri Lefebvre, after all, the plasticity of the city, its capacity to be made and remade over and over again, is the very condition of the urban.[21] This was picked up by radical geographer David Harvey, who argues that:

> The right to the city is far more than the individual liberty to access urban resources: it is a right to change ourselves by changing the city. It is, moreover, a common rather than an individual right since this transformation inevitably depends upon the exercise of a collective power to reshape the processes of urbanization. The freedom to make and remake our cities and ourselves is, I want to argue, one of the most precious yet most neglected of our human rights.[22]

However, Berry recognizes the limits to this freedom in her episodic wanderings through the city and its environs, noting the continued transformation of the city in the pursuit of a single, seemingly intractable goal. "The city as I knew it was being overwritten as fast as I could chronicle it . . . I saw the shadow city that I had for my whole life been drawn to also disappearing."[23]

Our final approach to the plastic city is to consider Sydney as a city of the surface: what Watson and Murphy elaborate as "superficial Sydney, glitzy Sydney, shiny Sydney. The surfaces of the opera house sails glisten in the sun, elegant yachts skim the bright harbor waters, cafes spill people onto the scorched city streets while chatter clatters off the hard brittle walls of Sydney restaurants . . ."[24] This is Sydney as a place of artifice, of light, that eschews the shadows. As environmental philosopher Val Plumwood notes in her treatise on "shadow places," capitalism relies upon the evacuation of the shadows.[25] It maps and surveys to take in its resources; it monetizes and instrumentalizes life in the pursuit of profit as an unquestioned goal. The capitalist veneration of the surface is complicit in the colonial projects of extraction, but also speaks to its fear of the unknown and the resistant that might undermine the goal of economic pursuit and ultimate reward. However, as Plumwood theorized, capitalism relies upon the creation of externalities. The surface is only clear because the shadows have been displaced elsewhere. Potter et al. write on this point: "shadow places are capitalism's externality: they are its 'disregarded places of economic and ecological support' [whose] neglect or forgetting continues to perpetuate the fantasy of untouched western lives in abstract circuits of production, consumption and disposal. Shadow places bear the marks of these processes; they wear them; they are them."[26] Shadow places are thus material sites that manifest what the capitalist vision excludes; they are everywhere but rarely made visible. Berry's *Mirror Sydney* seeks to do just that, as she plays with the association of a one-dimensional surface image, welcoming narcissistic preoccupation. As Watson and Murphy point out, even the surface city has potential for multiplicity: "There is no one Sydney, just many surfaces reflecting different lives, different images and different lenses."[27]

The cities within the city

As we have argued, Sydney is a plastic city endlessly open to remolding itself, throwing away the past to remake anew. Complicating this image, Berry's book deliberately refuses the dominant imaginary of the plastic city that is made over by smoothing over, by rupturing the impulse to forget. A work of creative nonfiction, *Mirror Sydney* began life as a blog that constituted a performative component of Berry's PhD thesis. Here, she recorded her journeys across Sydney, almost always by public transport, to various hinterlands of the bright and shiny CBD (central business district), journeys that involve very localized encounters generated through walking and lingering in place. These narrative reflections on place are accompanied by detailed "story maps,"[28] visualizations of the sites in focus through a revision of the cartesian plane. Rather than reduce space to an abstract view from above, these maps materalize place stories through the tangible remains of objects, ephemera, and fragments of the embodied environment, often accompanied by scraps of text that speak to these and evoke the particularity of the place story told through the map's form and shape. Berry calls these "subversive mappings" because they do not match up with "textbook cartography."[29] They are memory maps, including Berry's own, but also memories shared by or belonging to others that she doesn't necessarily have access to. Fragments of the past do not always disclose themselves, and it is the uncertainty of what the complex detritus of our urban environments means, where its history leads to, that leaves a charge for Berry. In the plastic city, memories of urban place are disposable, yet for Berry they are enduring. Berry's inversion of the formula contains an implicit critique of the temporality of modernity's most iconic material. Plastic may be enduring and future-facing, yet it is easily trashed and thrown away.

However, as Hawkins points out, plastic's multiple temporalities complicate this, proving stubborn and enduring, potentially for thousands of years.[30] Berry's maps become a mode of rematerializing stories where they have been forgotten or cleared away. At the same time, she doesn't seek to "know" these places in their totality. In part, it is the personal experience of place, over time, that precludes this totalized knowing. The city is inhabited multiply and diversely, and each inhabitant carries different histories and memories of place. Berry acknowledges this by interweaving her own biography, her own memories of Sydney, with those of others. This again points to the plasticity of the city, where inhabitants "possess his or her own map of the place, a world of amenities, amours, transit routes, resources and perils, radiating out from home."[31] These micro stories of inhabitation become the stuff of broader place stories: they weave the living fabric of the city, and intersect with the "big" narratives of culture and history. Walking is key to Berry's practice, a mode of both encounter and notation that brings the mirrored city into view. The city begins to unfold around Berry as she moves through it, immersed in its spaces, generating moments of both familiarity and strangeness. This is the city "on the cusp of change"[32] but also full of the past, pressing in, tripping us up, arresting our movement. These moments give rise to place-based encounters through which, as Berry tells us, "stories emerge,"[33] and many Sydneys become apparent.

Berry's chapter on the forgotten theme parks of Sydney underscores this point through the broken and eerie grounds of the Magic Kingdom amusement park in the southwestern Sydney suburb of Lansvale. A relic of the "amusement park boom" of the 1970s and '80s, Magic Kingdom ran for forty years and now sits as a "suburban wasteland,"[34] awaiting its next reinvention. When Berry climbs unauthorized through a hole in the fence, most of the rides have been sold off, "leaving the giant slide, a giant concrete shoe, a few buildings, and the ghosts."[35] Rubbish litters the ground, both old and more recent: "At the foot of the slide is a pile of plastic bread delivery trays used in place of mats to ride down the slide by those who came to visit the park after dark. The Kingdom has never ceased to be a playground for some."[36] What might appear as wasted artifacts—the abandoned park infrastructure, and the park itself, a strange island in the middle of suburbia—and thus out of place in the single-storied account of Sydney, the city of progress, are in Berry's eyes much more complex. This is a place whose use has varied, rather than ended. It is a place with multiple investments by different communities over time. In Berry's urban imaginary, these kinds of "wastelands" are imbued with other lives and realities, quietly pulsing with this charge, and always there to be reactivated through engagement. They are also, in their decay, sites of potential and reimagining. They are "places to dream in": "Their neglected buildings and overgrown grounds become somewhere to slip away from the present day into the past or the future, or at least into a different version of the present."[37]

Not all Berry's "mirror" sites are literal wastelands. Although they might be outside the shiny city's view of itself, some of these places are actively inhabited, frequented and used. They are other living cities in the city. One of these is the Penrith Arcades, situated some 30 miles west of the CBD, in the edgeland suburb of Penrith. These arcades, which date from the 1960s and '70s, are a retail warren off the main street that pre-date the monolithic shopping mecca, Westfield, which is the alternative retail space. There are seventeen arcades in total, filled with the niche or unprofitable shops that do not belong at the mega-mall. Unlike the "Victorian era galleries that might come to mind," these arcades are "secret passageways ... paved in brown tiles"[38] and, fifty years after their heyday, are worn out and down at heel. They seem a world away from the Parisian arcades that Walter Benjamin wrote about; however, to Berry there are affective overlaps. Both the arcades of Paris and those of Penrith transport the visitor to "past atmospheres."[39] These atmospheres bring forth diverse histories and times: they recollect Sydney as it was, but also as it still is for the many people who move through these arcades daily. This is Sydney against the grain of the future, steeped in active memories.

Poetically announcing this is the "Memory Mall" arcade, which triggers Berry's reflections on the malls of her own past: "malls aren't generally regarded as places of memory. Memories do inhabit malls though, often childhood and teenage ones."[40] Transported to the malls in her memory, Berry recalls, "there was something about arcades that was almost like houses in their mixture of public and private, interior and exterior. Yet they were exciting in a way that houses usually weren't, with endless promises of new things to contemplate ... sometimes I imagined being locked in one at night and having it as my domain."[41] These memories fold into Berry's arcade experience; they map onto the comings and goings of shoppers through an eclectic range of businesses such as "One Stop Cake Decorations," "The Cottage Lane," and the "Prima Ballerina" ballet shop. These shops are tactically contrasted with the generic, globalized retail chains of the Westfield down the road. Hidden away, some of them ill-frequented and decaying, such establishments are invested with the resonance of Benjamin's arcades, redolent with the capacity to transport to other lives and places: "anachronistic, full of mysterious signs, strange objects, antiquated trades and vacant stores with secretive atmospheres."[42] Ultimately, in Berry's hands, these are sites of imagining: an anti-futurist site of dreaming both forward and backwards. The Penrith arcades reclaim agency in the process of urban placemaking from the developers' totalizing vision: "People walking down High Street disappear into the arcades ... the For Lease signs ... make them remember what used to be there, or imagine what might be in the future."[43]

Unsettled country

While Berry's focus is the journeys of others, and the density and diversity of human experience in place, she is also attendant—as part of this complexity—to the environmental histories which interlace the fabric of contemporary Sydney. These histories are invariably sites of repression across the "case studies" of *Mirror Sydney* and are also imbricated in the long process of Indigenous dispossession that has accompanied the colonial-capital city's claim on the land. Writing against this, Berry is attentive to the deep histories of human occupation that still mark the land, and the nonhuman environment that has supported human life here, pointing to radically different modes of inhabitation that contrast First Nations peoples with the dispositions of extraction capitalism. "Excavating St. Peters" opens with this reflection: "For thousands of years, like much of what is now known as Sydney, this area was forest ... Gadigal people walked the track that ran across the top of the ridge, and hunted kangaroos in the forest. When the land was invaded, the British stripped its trees and turned the earth inside out ..."[44] The rich clay soil, it turned out, was perfect for brickmaking, and the industrial life of this place was brought into being. Yet the extractive mentality of the colonial city, and its efforts to erase native ecology and transform its materials into units of transaction, could not ultimately render the land flat and secure. It could not erase the past. What is discarded, Berry reminds us, lingers behind, for generations. There is much beneath the surface of a razed ground, laid open for development. "In 1896," Berry relates, "when the creek was being widened into a canal, workers discovered fragments of stone tools and the skeleton of a dugong with cuts to its bones. . . . Thousands of years ago people were hunting dugong here in the warm seas. Their gestures, preserved by these bones, return in a moment of overlapping time."[45]

The mirror city seeks to tame the nonhuman world, to constrain it to its ends. In "The Sydney Underground" Berry tells the story of the Tank Stream, whose post-invasion history is a parable of the plastic city. For many thousands of years a key source of fresh water to the Gadigal people, it was a primary reason for the decision to site the city of Sydney where it stands—but soon after colonization, it was too polluted to drink from. It was eventually contained in pipes and built over, and to this day it runs beneath the city: "a mythical presence" that Sydneysiders and visitors can tour (with admission allocated via ballot due to its popularity) and actually walk through. Visiting the Tank Stream provides a link to the past and to the environmental predicates that make place habitable, that make life possible, and that the city tends to ignore. Standing with water running over her boots, Berry reflects: "I have a sense of being at the city's source, a vital place where elements combine."[46] Despite the constraints of the pipe and the concrete layers above, there is a recognition threaded through *Mirror Sydney* that the city is on borrowed time. The innate power of the land remains. Magic Kingdom's ruins offer another point for reflection on this: these environs were once swampland, Berry narrates, and as shaping human hands have retreated, leaving the strange structures of the amusement park to quietly decay, the water has returned. The giant concrete boot, painted with the faded figures of "the Old Woman Who Lived in the Shoe and her many children," is now a "fitting symbol for the futility of attempts to control the sodden land."[47] While the city destroys much in its rapacious growth and capital-driven imaginary, it is also reclaimed when its guard is down. The chinks in a uniform imaginary give way to other life forms, other ways of being returning.

"In former industrial zones," Berry continues, "new developments arise . . . flattening the environment into tidy functionality. . . . The real estate floodlight has come to illuminate every shadow, so no place is spared reduction to its economic value."[48] At the same time, as she also journeys through the detritus of what urban development leaves behind, Berry's intervention is to point to the ultimate plasticity of the city, made and remade through constant encounters between physical form and ephemeral memory, and to refuse the forgetting that this might otherwise entail. Her interest in documenting what is fast disappearing is also—in a different temporality—about recalling and recording what has made this place and gives it its current shape and energy. This coexistent reality unsettles the uniformity of the wealthy global city. These stories and lives haven't been wiped away, but remain across the city, to be encountered still. Berry uncovers the tracks that lie beneath other tracks, the echoes that are ephemeral, the stories that linger despite transformation: "The Sydney I know best is one of undercurrents and weird places, suburban mythologies and unusual details. . . . These are the city's marginalia, the overlooked and the odd. . . . They form an alternative city."[49]

Shadow places of modernity

In *Mirror Sydney*, Berry explicitly states that her Sydney is a "shadow city," an "uneven landscape of harmonious and discordant places"[50] that doesn't yield itself up to the light. Instead, one has to go looking, to walk through its hidden labyrinths. Through her wanderings, Berry generates alternative and incomplete maps of a far from shiny city marked by "the new." Instead, she uncovers what Patricia Yaeger calls "a rubbish ecology,"[51] full of traces and ephemera, broken and scattered things. Berry is a gleaner, modeling a kind of ragpicking[52] through memory and the material and textual refuse of history. What she finds is value-full, far from useless or degraded. Even in the ruins of abandoned theme parks and long-closed shop fronts, Berry locates an active and affective energy that counters the logic of its mirrored other. There is an ethics to this practice, an ethics of the shadows. "Recording the city in this way is to contemplate and care for its histories, its inhabitants, its ecology and its future," she writes.[53]

Berry journeys to and through the abject spaces of capitalist production and consumption, when the party—that is, the focus of capitalist privilege and value—has moved on. Her reflections in "Excavating St. Peters" illuminate the forward-focused city through a study of a place through time. From the early days of colonization, this was an industrial suburb—a brick works followed by a rubbish dump, much of which was converted into a much-loved and -used public park, still surrounded by post-industrial detritus. Now it is being wrought into a travel corridor which will "convert much of the eastern half of the suburb into a motorway exchange, a coil of overpasses and feeder roads."[54] At a protest to "Save Our City, Save Sydney Park" that Berry documents, traditional owner Uncle Ken Canning makes the point that the WestConnex motorway eating up the suburb "comes from the same colonial mentality that invaded this land 228 years ago."[55] An urban imaginary of "progress," of the "old" giving way to the new and of relentlessly pushing on forward, underscores the colonial-capitalist city in both its past and contemporary iterations.

This capacity for rupture, for upturning the ground of an exclusionary and violently dispossessing city, is where Berry's poetic practice invests its own ethical imagination. She does not fantasize about untainted landscapes or utopian communities. Her mirror Sydney, the Sydney of the shadows that she tracks through many different wanderings and encounters amongst the city's range of spaces and structures, is as far from a pastoral imaginary as is possible. She is upfront about the pollution, the dirt, and the decay. This is, in line with Yaeger's reading of the postmodern concern with the polluted and tainted, "the power of waste at the center of contemporary literature . . . [where] waste managers and garbage haulers are its poets and purveyors, its historians and makers."[56] If Berry is such a manager and hauler, a writer of waste, hers is not simply an aesthetic pursuit. Rather, it is the material density of urban modernity that Berry seeks to capture, its rich multidimensionality that creates the conditions of our living. In the midst of the industrial suburb St. Peters, for instance, she recalls: "I turn into congested Canal Road, where the air is thick with fumes so strong I can taste them. . . . Above me a plane shrieks from the airport. . . . As I trudge through it I remind myself that this confusion of traffic, industry and noise is as much Sydney as any other place within the city's boundaries."[57] This material density results in profoundly uneven socioeconomic outcomes and successive removal of communities—like the residents of St. Peters, soon to be evicted from their homes to make way for a giant motorway, and the suburb's artists, who occupied the disused warehouses and industrial spaces, "the kinds of underground cultures nourished by discards and margins."[58]

Berry reanimates urban detritus, alerting us to its affective afterlives and its eruptive potential. Once artifacts and places have been "dropped," as Tim Edensor writes, from their stabilizing networks of meaning and value, they radiate a particular charge[59]: "The object's abrupt loss of the magic of the commodity . . . seems to confirm Julian Stallabrass's observation that 'commodities, despite all their tricks, are just stuff'. . . . And this loss allows us to reinterpret and use them otherwise."[60] More significantly, these "dropped" artifacts and places point to a lapsed future. For example, Berry tells the story of the now disassembled Sydney Monorail, a piece of public infrastructure that was always out of step with the dynamic requirements of the city's transport needs. A similar reality emerges from Berry's discussion of

Sydney's "memorial stores": shops that have closed and remain "as objects of contemplation long after they had stopped trading,"[61] their facades and windows still dressed as they were, congealed in time. Berry is drawn to shop fronts that once gleamed with the future-facing promise of the commodity, but are now the shadow places which reveal capitalism's dirty secret: that commodity and refuse are two sides of the same coin. On the façade of one store: "The advertisement in the center, for the long-defunct *Daily Mirror* [newspaper], has been painted over in white, but the name can still be seen faintly. The suburbs are full of such shadows, which persist until something comes along to cover them over."[62]

The Mary-Louise Salon is another memorial store, originally run by brother and sister hairdressers who became well known in the 1980s after winning first-division Lotto. Well patronized over the years, the Mary-Louise Salon sat with its "pink-and-mauve façade"[63] on the main street of the Sydney suburb of Enmore, its interior covered in photos of the famous siblings and lined with "trays of curlers and bottles of blue rinse, swatches of different hair tints"; "a row of hair dryers on pedestals, their domed heads like giant snowdrops."[64] When the siblings died, the store remained closed but looked after, visited regularly by patrons who would stare in the windows at the "time capsule"[65] display of pink and mauve objects: an "unofficial museum, in memory of people and times that have gone."[66] This kind of commemoration—local, vernacular, and very situated in place—resists the momentum of the modern city, which positions closure, or endings, as the drive for new beginnings. Yet in the city it cannot ever totally escape capture by capitalism. The Mary-Louise Salon was eventually sold, and with it the faded interior that attracted so much nostalgia and interest: "Mary-Louise Memorabilia Sale. Grab Your Piece of Inner Waste History. Sticky Beaks Welcome." Berry is there, and chooses her own Marie-Louise souvenirs: "a giant novelty comb from World Expo '88 and . . . a pink business card for 'George and Nola Mezher, Leading Sydney Hairdressers."[67] Not long after, the Salon was remade through adaptive reuse into a high-end restaurant.

Conclusion

Discarded and enduring, plastic performs a present that is "endlessly replaceable"[68] but is always undone by its material recalcitrance. Plastic—like memories—endures. To return to Hird, the resistant presence of what we discard, even as we turn away from it and face an imagined future, calls on us "to remember . . . [to] bear . . . witness to the waste we want to forget."[69] It is in these marginal spaces, the spaces where memories gather and are active, that the image of a more inclusive and generous city emerges, spaces where lives entangle and times overlap. In Berry's hands, Sydney, the plastic city endlessly remaking itself, encounters its own limits in the echoes of its past. As much as the city drives forward, it is also undone, and the temporal paradox of plastic makes visible this tension. Witnessing a transforming Sydney through her work, Berry laments: "I was right in suspecting that many of the anachronistic or neglected places I was documenting would not be around much longer."[70] And yet her project pushes against the smoothness of the plastic city vanishing and asserts its politics in contrast, offering stories of remaking through time as an active, radical act. Here, the possibilities for belonging in and to the city expand, refusing the exclusivity of the shiny, surface urban environment with its repressed shadows at the margins. In turning to these shadow places, Berry's advocacy is to re-thread marginal or forgotten histories into the fabric of Sydney, to re-seed its atmosphere with nearly-lost memories. Urban detritus illuminates the multiplicity of place, its histories and narratives. As rubbish theorist Myra Hird writes, "when people have lived for millennia in the same place, experience is all about paying attention to landscape variability."[71] Berry's wandering, mapping, and writing pays such attention, opening up new urban imaginaries in the process that enable us to dwell in the refuse, diversity, and complexity of the modern city as it relentlessly pursues the future.

Notes

1. Vanessa Berry, "Spatial Experiments: Autobiographical Cartography," in *New and Experimental Approaches to Writing Lives*, ed. Jo Parnell (London: Red Globe Press, 2019), 117; Vanessa Berry, *Mirror Sydney: An Atlas of Reflections* (Artarmon, NSW: Giramondo, 2017).
2. Amanda Boetzkes, *Plastic Capitalism* (Cambridge, MA: MIT Press, 2019), 214.
3. Gay Hawkins, "Plastic and Presentism: The Time of Disposability," *Journal of Contemporary Archaeology* 5, no. 1 (2018): 91–102.
4. Berry, "Spatial Experiments," 10.
5. Berry, *Mirror Sydney*, 117.
6. Emily Potter and Kirsten Seale, "The Worldly Text and the Production of More-than-Literary Place: Helen Garner's *Monkey Grip* and Melbourne's 'inner north,'" *Cultural Geographies* 27, no. 3 (2020): 367–78; Kirsten Seale and Emily Potter, "Wandering and Placemaking in London: Iain Sinclair's Literary Methodology," *M/C Journal* 22, no. 4 (2019).
7. Jennifer Hamilton, "Rewriting Redevelopment: The Anti-Proprietorial

Tone in Sydney Place-Writing," *JASAL: Journal of the Association for the Study of Australian Literature* 18, no. 1 (2018): 1, https://openjour nals.library.sydney.edu.au/index.php/JASAL/article/view/12404.

8. Ibid. 5.

9. Rebecca Solnit and Susan Schwartzenburg, *Hollow City* (London: Verso, 2001); Kirsten Seale, *Markets, Places, Cities* (Abingdon: Routledge, 2016).

10. Seale, *Markets, Places, Cities.*

11. Sarah Harrison, *Waste Matters: Urban Margins in Contemporary Literature* (Abingdon and New York: Routledge, 2016), 9–10.

12. Helen Verran, "Re-Imagining Land Ownership in Australia," *Postcolonial Studies: Culture, Politics, Economy* 1, no. 2 (1998): 239.

13. Seale and Potter, "Wandering and Placemaking"; Potter and Seale, "The Worldly Text."

14. Alison Young, "Cities in the City: Street Art, Enchantment, and the Urban Commons," *Law and Literature* 26, no. 2 (2014): 145.

15. Katie Schagg, "Plastiglomerates, Microplastics, Nanoplastics: Towards a Dark Ecology of Plastic Performativity," *Performance Research* 25, no. 2 (2020): 14.

16. Ibid. 14.

17. Hawkins, "Plastic and Presentism," 95.

18. Ibid. 95.

19. Rob Shields, *Places on the Margin: Alternative Geographies of Modernity* (London: Routledge, 1991).

20. Justice Brennan in Ann Genovese, "Turning the Tide of History," *Griffith Review* 2 (2003), https://www.griffithreview.com/articles/tur ning-the-tide-of-history/.

21. Henri Lefebvre, *Writings on Cities* (Oxford: Blackwell, 1996), 79.

22. David Harvey, "The Right to the City," *New Left Review* no. 53 (2008): 23.

23. Berry, *Mirror Sydney*, 10–11.

24. Sophie Watson and Peter Murphy, "Sydney: City of Surfaces," *City* 1, no. 5–6 (1996): 38–46.

25. Val Plumwood, "Shadow Places and the Politics of Dwelling," *Australian Humanities Review* 44 (2008): 139–50.

26. Emily Potter, Fiona Miller, Eva Lövbrand, Donna Houston, Emily O'Gorman, Jess McLean, and Clifton Evers, "A Manifesto for Shadow Places: Re-imagining and Co-producing Connections for Justice in an Era of Climate Change," *Environment and Planning E: Nature and Space*, December 2020.

27. Watson and Murphy, "Sydney," 6.

28. Robert McFarlane in Berry, "Spatial Experiments," 122.

29. Berry, "Spatial Experiments," 117.

30. Hawkins, "Plastic and Presentism."

31. Rebecca Solnit in Berry, *Mirror Sydney*, 119.

32. Berry, *Mirror Sydney*, 11.

33. Berry, "Spatial Experiments," 122.
34. Berry, *Mirror Sydney*, 88.
35. Ibid. 81.
36. Ibid. 84.
37. Ibid. 88.
38. Ibid. 52.
39. Ibid. 53.
40. Ibid. 53.
41. Ibid. 53.
42. Ibid. 52.
43. Ibid. 57.
44. Ibid. 135.
45. Ibid. 138.
46. Ibid. 171.
47. Ibid. 86.
48. Ibid. 12.
49. Ibid. 3.
50. Ibid. 11.
51. Ibid. 329.
52. Walter Benjamin, *Charles Baudelaire: A Lyric Poet in the Era of High Capitalism* (London: Verso, 1991).
53. Berry, *Mirror Sydney*, 11.
54. Ibid. 136.
55. Ibid. 143.
56. Patricia Yaeger, "Editor's Column: The Death of Nature and the Apotheosis of Trash; or, Rubbish Ecology," *PMLA* 132, no. 2 (2008): 331.
57. Berry, *Mirror Sydney*, 139.
58. Ibid. 141.
59. Tim Edensor, "Waste Matter: The Debris of Industrial Ruins and the Disordering of the Material World," *Journal of Material Culture* 10, no. 3 (2005): 313.
60. Berry, *Mirror Sydney*, 320.
61. Ibid. 160.
62. Ibid. 179.
63. Ibid. 153.
64. Ibid. 154.
65. Ibid. 157.
66. Ibid. 154.
67. Ibid. 159.
68. Hawkins, "Plastic and Presentism," 3.
69. Myra J. Hird, "Waste, Landfills, and an Environmental Ethic of Vulnerability," *Ethics and the Environment* 18, no. 1 (2013): 105.
70. Berry, *Mirror Sydney*, 10.
71. Hird, "Waste, Landfills," 116.

14

Better Learning through Plastic?:
The *Moby-Duck* Saga and Pedagogy

Donna A. Gessell

Numerous books, films, and even a song and poem have been produced to tell the story of the 1992 rubber duck spill. Many accounts have explored the event as a curiosity, while others have focused on teaching about the problem of plastic in the oceans. This essay examines a variety of the accounts of that singular incident of plastic waste entering the ocean and how its results have influenced efforts to educate people—especially children—about the connections between plastic consumerism and plastic ocean pollution. To become more effective, education attempts must focus on integrative learning techniques, ones which at once acknowledge the complexity of ocean plastic pollution and connect it with people's everyday lives, particularly the decisions they make about their personal plastic use.

The problem

The media is awash with reports on ocean plastic pollution. For instance, on 11 September 2020, a newsfeed reprinted an article reporting on "a growing island in the North Pacific—one that consists solely of trash," one that "presents the shocking reality of the magnitude and composition of the Great Pacific Garbage Patch."[1] Highlighting the increasing mass and density of the island, its name shortened to GPGP, the article links the effects with the cause: our everyday use of plastics. It begins "Single-use plastics plague our modern daily lives. They are cheap to produce, convenient, and sanitary—it's no wonder why we produce a little over 300 million tons of plastic globally every year."[2] Revealing the environmental impact of the disposal of plastic waste, the article explains plastic's effects on the oceans:

> Once plastic reaches the ocean, it is subject to waves, winds, and currents that break it down and transport it in various directions. Depending on the size, density, and location of these plastic pieces, they may wash ashore,

sink to the ocean floor or get trapped in oceanic currents. Through a combination of environmental processes, circulating ocean currents form massive vortexes called gyres. Since the rise of plastic production, gyres have become garbage hotspots for the accumulation of marine debris.[3]

It then details how the GPGP, estimated to be "1.6 km—more than twice the size of Texas," was studied, using two planes and eighteen vessels trawling 652 nets over three months.[4] Because the various plastics found in the GPGP originated in a variety of countries, "plastic pollution is a global problem."[5] Over the last forty years, the GPGP has expanded exponentially. The study generated a model to understand better how the GPGP has moved and grown. The reason for the study was the hope that "Understanding the size, growth rate, and composition of the GPGP should alert communities and nations worldwide to more efficiently manage their waste, while incentivizing a reduction in plastic production."[6] The article warns: "[i]t is also essential that behavioral changes around single-use plastic happen at the consumer level. Reduce, reuse, and refuse plastic—the health of our oceans is a global responsibility, and its protection depends on support from individuals to governments."[7]

Despite such studies and the proliferation of media attempts to raise awareness about the problem, few people associate their daily habits with ocean pollution. After all, the GPGP seems innocuous at such a distance, its composition of different-sized plastics swirling in a gyre almost as outlandish and remote as any of the islands depicted in Jonathan Swift's *Gulliver's Travels*. How, then, can a compelling story be made to alert consumers to the dangers of plastic waste? As improbable as it may seem, a flotilla of "rubber duckies" has become part of the solution, after accidentally spilling into the North Pacific Ocean. From that story, people continue to learn about the effects of plastic pollution on the ocean. However, even with all of its success, the effects of this pedagogy are questionable. The association of ocean plastic pollution with cute rubber ducks normalizes the situation, potentially leading it to be ignored. It raises the question: What will it take to lessen our use of plastic?

The history and depth of the problem

Two books especially used the yellow ducks to raise the awareness of the adult public to the increasing problem of plastics: Curtis Ebbesmeyer and Eric Scigliano's *Flotsametrics and the Floating World: How One Man's Obsessions with Runaway Sneakers and Rubber Ducks Revolutionized Ocean Science* (2009), and Donovan Hohn's

Moby-Duck: The True Story of 28,800 Bath Toys Lost at Sea and of the Beachcombers, Oceanographers, Environmentalists, and Fools, Including the Author, Who Went in Search of Them (2011). Each tells the story of the spill and then describes how its aftermath contributed to scientific discovery of the mechanics of ocean currents and gyres, as well as raising awareness of the enormity of the amounts of plastic waste polluting the ocean.

During a 1992 storm in the Northern Pacific Ocean, falling containers from a ship launched 28,800 plastic bath toys. Almost no notice of this event occurred at the time, yet Donovan Hohn exuberantly reports the results of his energetic investigation. He reports the longitude, "44.7° N, 178.1° E, south of the Aleutians, near the international date line, in the stormy latitudes renowned in the age of sail as the Graveyard of the Pacific, just north of what oceanographers, who are, on the whole, less poetic than mariners of the age of sail, call the subarctic front"; the date, "January 10, 1992—but not the hour"; and the ship's name, which he researched "by process of elimination" using microfiche records from the *Journal of Commerce*: the *Ever Laurel*.[8] Hohn compulsively fills in the specifics, detailing the size of each of the twelve containers that spilled overboard: "eight feet wide and either twenty or forty feet long."[9] He claims:

> We know that as the water gushed in and the container sank, dozens of cardboard boxes would have come bobbing to the surface; that one by one, they too would have come apart, discharging thousands of little packages onto the sea; that every package comprised a single shell and a cardboard back; that every shell housed four hollow plastic animals—a red beaver, a blue turtle, a green frog, and a yellow duck—each about three inches long; and that printed on the cardboard in colorful letters in a bubbly childlike font were the following words: THE FIRST YEARS, FLOATEES. THEY FLOAT IN TUB OR POOL. PLAY AND DISCOVER. MADE IN CHINA. DISHWASHER SAFE.[10]

Of the 28,800 PVC toys, one quarter—or 7,200—were yellow ducks.

If Hohn's rhetoric seems overblown in its detail, then studying what happened next puts his account into perspective. As flotsam, the toys began to wash up on shores across the northern Pacific. The first sighting was at Sitka, Alaska, according to Ebbesmeyer and Scigliano. They detail further sightings.[11] Although these include Japan, Indonesia, Australia, New Zealand, and South America, it was the sightings in the North Pacific that most interested Hohn. Over a sixteen-year period, the yellow ducks were found along the Aleutian Islands, northward through the Bering Strait, and then across the Arctic Ocean, then east into the Atlantic Ocean, where they landed in Canada, New England, Iceland, and even Scotland.

Ebbesmeyer and his colleagues used the data to make predictions not only for the flows along the ocean currents but even within the gyres themselves: "three kinds of orbits: Some toys drifted once around the Aleut Gyre's periphery. Some made repeated orbits around the periphery—five in fifteen years. And others made loops—suborbits—within it. The research showed that the Aleut Gyre (and by implication any gyre) is not a simple circle but a system of wheels within a great wheel, somewhat like a planetary gear."[12]

Initial awareness: scientific and mainstream popular cultures

The spill is variously called "The Rubber Duckies Spill" or "The Great Ducky Spill," even though only a fourth of the toys were yellow ducks; for that reason, some call it "The Bathtub Toy Spill" or "The Spill of the Friendly Floatees." No matter the name, the spill has had large ripple effects for both scientific research and the popular imagination. As the number of spottings of the PVC bath toys on far-flung beaches increased, so did stories in the popular press. The first sighting of the ducks was reported in the *Daily Sitka Sentinel*, which ran classified ads to find out more information about them.[13] Over the sixteen years of their strandings, the toys made news. The stories were published in newspapers and magazines and were broadcast on radio and television. Publicizing their movements, the stories also included calculations of future movements, which prompted beachcombers and the general public to search for them. By the time they were forecast to wash up on North Atlantic shores, "The First Years, Inc. offered a $100 reward for the first confirmed Atlantic duck."[14]

Books for adults proliferated, including the two already mentioned, which each feature the yellow duck bath toy on the cover; however, not The First Years toy. Ebbesmeyer and Scigliano's has a single duck in relatively calm waters with longitude and latitudinal lines imposed to show the global impact; Hohn's has one large duck foregrounded against large breaking waves, with two more ducks riding a wave's crest. Books published later mention the yellow duckies spill and subsequent events, but focus on the larger problem of ocean plastic. Charles Moore and Cassandra Phillips published *Plastic Ocean: How a Sea Captain's Chance Discovery Launched a Determined Quest to Save the Ocean* in 2011, and Michel Roscam Abbing published *Plastic Soup: An Atlas of Ocean Pollution* in 2019. Neither book cover features a duck, even though each features an array of plastic, the earlier one a beach deposit, and the latter a stylized arrangement of plastic in the shape of a fish ready to swallow the earth. The evolution of book covers suggests that the dialogue has moved past cute bathtub toys and amateur beachcombers working with scientists. Instead, the topic has become dire: saving the oceans from plastic pollution that threatens the entire planet.

At least three children's books have also been inspired by the spill and the subsequent spotting of "rubber ducks." Each prominently features a bright yellow plastic duck on its cover; the first, *Ducky* by Eve Bunting, published in 1997, features a single duck, while Eric Carle's 2005 book *10 Little Rubber Ducks* features ten ducks, and Janeen Mason's 2012 book, *Ocean Commotion: Caught in the Currents*, features seven.

Several films about the event, including animations, are featured on YouTube. Emma Dobken created the animation "The Epic Journey of the Plastic Ducks" to tell the story of the 29,000 floating toys, 19,000 of which went south to Australia, Indonesia, and Chile, while the other 10,000 had "other plans."[15] Although anthropomorphized, the ducks on their journeys illustrate details of the ocean current movements, including the observation that the North Pacific gyre takes three years to return the ducks to where they started. After detailing the trip through the Bering Strait and the Arctic Ocean to the US and Scotland, the film ends by discussing how many of the ducks are trapped in the "Big Plastic Soup" in the Pacific Ocean. Another animation, "Ducks Overboard," by Christian Dorian, uses no voiceover yet shows vividly the conveyer action of the ocean currents as each duck enters and is moved to a different destination.[16] The film, part of the *How the World Works* series, ends with the message "So, if you find a rubber duck, give it a safe home." Of the non-animated films, two are most informative. The first, "How 29,000 Rubber Ducks Helped Map the World's Oceans," is also an ad for Skillshare, an education company. However, it includes important observations.[17] Despite saying that the ducks were made of rubber, the film points out that the ducks continue to float because of the absence of holes in their bottoms. Tracing the routes of the ducks, the film connects them to Ebbesmeyer's research and proves their model's predictions. It discusses how the sheer number of ducks involved made this spill scientifically productive: scientists could never ethically release so many items into the sea to track, especially because, "after all, humans already dump over 8 million tonnes of plastics into the oceans each year."[18] The other non-animated film is "When Thousands of Rubber Ducks Were Accidentally Dumped into the Ocean."[19] Made by the Science Channel, this film is part of the *What on Earth?* series, and its associated comments indicate that it is often shown in science classes. Hohn, the *Moby-Duck* author, tells the story, explaining how the spill provides information on the North Pacific Ocean currents with real-time data. Here, the rubber ducks are described as "almost indestruct-ible," washing up on beaches some twenty years later, still providing scientific data.

At least one song has been written about the spill and its legacy. "Yellow Rubber Ducks" (2011) by Rich Eilbert is accessible on YouTube complete with a video; together they supply the story in detail, even if some of its details have been altered to fit the medium.[20] For instance, the song refers to 29,000 rubber ducks, not 28,800 plastic floatees. However, it does accurately portray the movements of the ducks, as well as the increasing interest in them by Ebbesmeyer and other scientists. It folds in details of climatology, including global warming.

The legacy of the ducks even appears in poetry. In what one reviewer labels "a tiny epic poem," Kei Miller, a Jamaican poet, includes the "rubber duck" story in his collection *The Cartographer Tries to Map a Way to Zion*.[21] Dave Coates describes the poem "When Considering the Long, Long Journey of 28,000 Rubber Ducks" as "ennobling 'them who knew to break free from dark hold of ships . . . to them / that pass in squeakless silence over the *Titanic* . . . who instruct us yearly on the movement of currents; / those bright yellow dots that crest the waves / like spots of praise: hail.'"[22] Although the details are distorted, the "heroism" of the ducks becomes a metaphorical celebra-tion of freedom, as well as a recognition of their scientific impact even as such small, insignificant objects.

The proliferation of stories of the spill in all of their formats has produced a variety of effects. The "rubber duckies," already popular in the global cultural imagination, have enjoyed an even larger repu-tation; the scientific community, through the resulting crowdsourced evidence, has proven new theories of oceanography; and plastic ocean pollution has become regularly monitored. Perhaps most importantly, popular imagination has created strong pedagogies for children to connect plastic in their daily lives to plastic ocean pollution.

"Rubber duckies" and the popular imagination

Rubber ducks have been around since the late 1800s, developed after Charles Goodyear vulcanized raw rubber to stabilize it. Too heavy to float, the original duckies were meant to be chew toys. However, by the 1940s, the iconic duck was being made out of plastic, and it could float. Its stylized shape was colored bright yellow and it sported an orange bill, white eyeballs, and black lines for other facial details. Although most toy bath ducks are now made out of plastic, the toy's name nearly always reflects its original material: "rubber duckie." The ducks have become symbols of the comfort and fun of taking a bath. The Museum of Toys, into which the rubber duckie was inducted in 2013, describes its appeal:

> Rubber ducks naturally inspire water play that develops muscle strength and coordination. With their bright color, smooth texture, and (for some) squeaky or quacky sounds, rubber ducks sharpen toddlers' senses. Their presence in the bathtub soothes youngsters' fears of water and water immersion and makes good clean fun of the routine hygiene they're learning.[23]

The article then explains how the toy came to "have been recognized as the quintessential bathtub toy since 1970, because of Burt and Ernie, two of the Muppets from *Sesame Street*, Public Television's educational television show for preschoolers. Ernie sang 'Rubber Duckie,' at bath time to his toy, declaring 'Rubber duckie you're the one / You make bath time lots of fun / Rubber duckie I'm awfully fond of you.'"[24]

A part of contemporary global popular culture, floatable yellow ducks with brightly colored eyes and bills have become so iconic that their classic use as bathtub toys has been modified. Now many people around the world consider them keepsakes, displaying them in personal collections. Ebbesmeyer and Scigliano report that "one collector has amassed a mind-boggling nineteen hundred toy ducks."[25] As collectibles, they have reached the epitome of consumerism and materialism.

Because rubber duckies make cheap giveaways and are ready made in various guises, they have become popular party favors. The iconic design is altered to fit themes ranging from graduation parties to holiday events to breast cancer awareness; they can represent occupations, from construction worker to lifeguard to chef; they come in costume: dragon, fairy, mermaid, space alien, cat, pineapple, and carrot; they can be customized for a sports team; or they can be pre-engraved with messages of all sorts. They come with black eyes or blue, many with sunglasses. They come in a variety of sizes; and they can even glow in the dark. Often, they are hand sculpted and painted to create celebrity ducks, the celebrity faces blended into duck faces. Their celebrity appearances include sports figures, movie stars, and politicians. Even the likeness of the venerable face of Queen Elizabeth II of England has been rendered onto rubber ducks. Perhaps this is fitting, as she is rumored to display in her bathroom a rubber duck with an inflatable crown.[26]

Ducks and the science of currents and oceans

Scientists had previously tried to track ocean currents, but never had access to such good data. They had predicted models of the current flow, but they had difficulties proving their predictions with the small numbers of data points they collected. Previously, they would drop 1,000 bottles with messages in them, called "drift bottles." Only a small number were found and returned. However, finding the duckies and their other floatee friends became so popular that Ebbesmeyer comments on his nevertheless high rate of recovery: "I received wash-up reports for a full 3.3 percent of the toys spilled, a 30 percent higher rate than the sneaker elicited—even though media attention turned the orphaned mascots into collectibles, making finders loath to part with them."[27]

What made the toys such good "drift bottles"? Ebbesmeyer and Scigliano explain how the ducks differ from other ducks, called "false canards":

> The *Ever Laurel* critters, designed by the prominent child psychiatrist T. Berry Brazelton for The First Years, Inc., toy company are unique among all the toys adrift. The ducks, made of hard plastic rather than rubber, scarcely resemble the Disneyesque duckies of popular lore and décor. Their shape, like those of the other three critters, is angular and unexpressive, simplified rather than stylized, cubist rather than cartoonish.[28]

As they explain, even though the plastic bath toys were not as well equipped to be tracked as drift bottles (which carry directions and return address cards), nor as valuable as the Nike sneakers that he had been tracking, they "inspired even more public enthusiasm than the shoes. That seems to reflect in part the peculiar popular fixation on so-called rubber duckies, which have migrated from infant bathtubs to adult fashion and design. The yellow ducky is an icon of whimsy, nostalgia, childhood innocence, and pop-cultural kitsch."[29]

Organized communication among those collecting the ducks tracked the spill's progress, causing oceanographers to revise their modeling of ocean currents. Instead of separated and individual, the data showed the currents as one huge, interrelated system. Instead of seven seas, each with its own set of individually contained currents, the dispersal of the plastic toys demonstrated an interlinkage not previously evident. Ebbesmeyer and James Ingraham used the data they collected to "refine their magnum opus: a computerized ocean current model called Ocean Surface Current Simulator (OSCURS)."[30] Using the coordinates of the spill, "OSCURS got huge amounts of new data. Best of all, Kiddie Products printed the duckies' calculated twisty

15,000-mile route on the back of every new bathtub-toy package. 'By far our best publication,' says Ebbesmeyer."[31]

The pedagogy of plastics

As sightings of the "Friendly Floatees" continued to capture the public imagination, others who were monitoring plastics in the ocean, particularly the huge garbage patch east of Hawaii, used the phenomenon—which had morphed into rubber duckies in popular imagination—to raise awareness of the immensity of the plastic problem. Three broad areas of pedagogical materials for children have emerged from the friendliness of the beloved childhood icon: children's picture books; online science lessons; and museum-sponsored educational programs.

Children's picture books

Three of the children's books that have been written about the event each focus on the ducks in a different way to tell their story. These rhetorical differences affect not only the connection with the books' targeted audiences, but also the apparent reason for telling the story.

Ducky by Eve Bunting, published in 1997 and illustrated by David Winiewski, Caldecott Medal winner, was the first of the children's picture books.[32] It tells the story of a single duck, yellow with orange facial feature details, lost overboard. After the introductory pages, with pictures depicting a container being hoisted from a dock and another of a container ship leaving harbor, the next picture shows a single detailed duck, surrounded by other plastic toys, none like it, sweating. The first words of the story set the emotional condition: "I am a yellow plastic duck and I am in great danger."[33] After mentioning that yesterday was "safe," it quickly relates the situation: "A storm came. Our crate was washed overboard."[34] The next page depicts the crate rupturing, and toys—indistinctly unfocused in their depiction, other than vivid reds, yellows, blues, and greens—bubbling out upwards. The text is laid out to reflect its meaning: "DOWN DOWN DOWN it went," sinking until we find out that "We tumbled around inside, yellow ducks, green frogs, blue turtles, and red beavers."[35] The "BUMP" and "CRASH!" disrupt the text and we are told that after the crate hits the bottom and breaks open, the toys "bobbed like colored bubbles to the surface."[36] The danger is not over, as the next several pages make clear. First, they are lost at sea—described as "big, big, big."[37] Although the duck is surrounded by "bathtub friends," most are indistinctly drawn among the red waves in the red sea against a red sky. The duck reacts: "Oh, I am scared!" However, on the following pages it "go[es] from scared to terrified" after being confronted by staring fish, a sea snake, and a shark with "A great monster head."[38] In keeping with *Jaws*-inspired popular culture, the shark is terrifying, and as we see the small duck inside the large shark's mouth, we are told about his lack of agency: "I wish we could swim and get away. But all we can do is float."[39] The shark leaves (we are not told how or why) and the following pages describe how the toys become separated from one another in the huge ocean, until the plastic duck is alone. Then there is a long journey of floating, again without agency, but with loneliness. Pictures show the ocean in all seasons and at all times of day. One image is of "nights of constellations" with "a moon, and another and another."[40] We are told "The water must be colder now. Ice nudges me," and the duck wonders, "Will I float in this ocean forever and ever?"[41]

This is the turn in the book when the duck next lands on a rocky shore and "Someone is shouting. 'I've found one! I've found one! It's a duck!'"[42] At once the duck is on solid ground, found by a human and associated with its lost bathtub friends, at least intrinsically. The following pages build community. The human is a young boy, who has a friend; the friend knows of a scientist who is tracking the toys. The duck's self-worth grows as he realizes he is data, "For science? Me?" and as he sees "so many of my friends again" who have also been collected.[43] After visiting the scientist, who notes his data on a map of Alaska with red data points arrayed across its southern shores, the boy proudly announces that the duck is "coming home with me," where he is put into a bathtub, which the illustration depicts filled with bubble bath, bubbles everywhere.[44] The duck exclaims, "Oh, I am so happy! I am a bathtub duck, fulfilling my destiny. How wondrous it is to be able to float!"[45] The moral of the story suggests accepting fate and trusting that, by relying on the ability to endure, it is possible to fulfill one's destiny.

The last page, with yet another plastic ducky in bubble bath illustration, contains the "Author's Note," which briefly gives the real story of the toys being lost overboard. Then it summarizes where they have been found in the past five years:

> Hundreds of the toys have since been found, beached on the eastern coast of the Gulf of Alaska. Scientists are checking findings and sightings to learn more about currents, winds, and tides. Using computers, they plot the track of the remaining toys. Some, they believe, will go into the Arctic Ocean and ride a course toward the North Pole. Some could sail around the top of the world to the North Atlantic. One lone duck has been found off Washington State.[46]

This account ends the same way as the story: "How wondrous it is to be able to float!"[47] It is not made clear whether the note is intended for the children who are the book's target audience, or the adult reader. "Ducky," named only once in the eponymous title of the book, overcomes danger and fear without exerting agency, simply by floating.

In 2005, the celebrated children's author Eric Carle—best known for *The Hungry Caterpillar*—published his picture book based on the spill and dispersion of the ducks, named *10 Little Rubber Ducks*.[48] In the front of the book appears a snippet of a newspaper article story about the spill, titled "Rubber Ducks Lost at Sea" and dated July 2003, along with a photo of the author with a signed caption stating, "I could not resist making a story out of this newspaper report."[49] The jacket of the hardcover book represents the "exciting voyage of discovery" by describing how readers can "follow the little ducks as they float to all parts of the globe."[50] The educational intention of the book is made clear: "young explorers can see for themselves the meanings of directional words, and learn simple math concepts such as counting and the use of cardinal and ordinal numbers."[51] The directional words are employed to describe how each duck drifts: *west, east, north, south, left, right, up, down, this way* and *that way*.[52] The book jacket also claims the story offers "a very simple first view of biology and geography" because each of the ducks meets a different creature, "seen in its own habitat," that "behaves in a true-to-life manner."[53] The exception may be for the tenth duck, which meets a mother duck and her nine ducklings, and is adopted by them. The book closes as the rubber duck "floats along with" the other ducks to their home, enlarging the number of that group of ducklings to ten.[54]

Although inspired by the original Friendly Floatees story, Carle takes license with the details beyond providing an education in directions, counting, biology, and geography. In his version, the plastic ducks become rubber ducks; instead of being only yellow, the ducks are painted with red bills and blue eyes, a white circle around the blue, making the toy recognizable as a rubber ducky. The spill is reduced from a container holding 28,800 toys in four varieties to a single box holding ten toys that are all alike. The changes no doubt help to ensure appropriateness for the imagination of the book's stated audience, that of two- to six-year-olds.

The results of the adaptations not only help to normalize the appearance of the ducks, but they also normalize the spill of plastic in other ways. The rubber ducky is transformed by being recognized to what has become so familiar in global popular culture, and the environmental impact of the spill is lessened in importance by decreasing the numbers involved. By changing the story into an educational experience and locating the tenth rubber duck into a family, both the manufactured duck and the wild ducks are anthropomorphized. Most significantly, the adoption of the plastic duck normalizes ocean plastics by insisting on the plastic duck's unlikely inclusion in a family of wild ducks.

Another picture book appeared in 2012: *Ocean Commotion: Caught in the Currents* by Janeen Mason.[55] This book appears to be aimed at a slightly older audience, with its apparatus beyond the story including endpapers featuring world maps with depictions of major ocean currents; bolded words in the text linked to an illustrated glossary at the end of the story, cross-linked to a map showing the locations of place names; a representation of the Pacific Ocean showing the equator and international date lines, again depicting major ocean currents; and a pictorial summary of the actual event. The last printed page is a page-long note to readers arguing that "[t]hese little yellow plastic ducks are so much more than toys" and providing an explanation of the environmental impact presented by plastic waste in the ocean.[56] Mason details its effects on ocean animals and our food chain, as well as a list of activities anyone can pursue to "make a difference" in our environment. She suggests that readers may have environmental careers when they grow up. Furthermore, she traces the travel of the ducks beyond when the story ends, stressing that "the currents of our oceans are all connected."[57] Finally, she links breathing to air quality to photosynthesis by plankton and algae, ending the note with the statements: "If plastic is hurting the plankton on the other side of the world, that affects us all. These ducks have proven that."[58]

The story is more sophisticated as well, staying closer to the facts. Mason repeatedly emphasizes the factual, even though she fictionalizes her account. In an "Author's Note" before relating the story, she locates us:

> This story is based on fact. On January 10, 1992, a container ship left China bound for the United States. At 44° N and 178° E, the waves reached ninety feet high, and 28,800 bathtub toys were lost overboard. They were yellow ducks, red beavers, green frogs, and blue turtles. I have employed artistic license to simplify the story here. You will find only yellow ducks on these pages. I have changed the names and imagined some events, but the science is accurate.[59]

She includes a cargo container that falls in the storm, but she focuses on the yellow duck, which she illustrates with painted orange bills and black eyes with a white dot. Once she assures the reader that "[t]he armada of ducks righted, turned in the current, and sailed into history,"[60] she turns the focus of the story to a human one, focusing on a young boy, John Dunning, who lives in the Aleutian Islands of Alaska. While beachcombing with his newspaper reporter father, he finds several of the ducks and wonders about their origin. The father can only guess, but he mentions them in his column that evening. On the next pages, the young daughter of the captain of a fishing trawler discovers more ducks in the Bering Sea, reminding her father that they read about them in the paper. The captain is skeptical of the connection, musing that "rubber ducks could never make it this far down the island chain."[61] The next pages trace their journey into the Arctic Ocean, when some are "squeezed into ice floes" and the rest drift "endlessly toward the east and the Atlantic Ocean."[62] The book then backtracks to note those that have been caught in "the North Pacific gyre and swept farther south."[63] With a clever display of notes, photos, clippings, and other ephemera, the next two pages provide a pictorial summary of the event. Prominent is a *Time* magazine cover with the yellow duck on it, as well as charts and maps. Geography is key to this picture book, which even mentions Fiji, Tonga, and Tahiti as duck destinations. Also key is John Dunning, who keeps "a scrapbook filled with articles about duck sightings," then becomes an oceanographer, and eventually exposes his children to the legacy of the ducks, revealing the marvel of the ocean currents that would cause "the toys to float 15,000 miles around the entire North Pacific and back again."[64] The story ends with the observation from an old captain: "This mighty ocean keeps a lot of secrets caught in her currents."[65]

Each of these three books capitalizes on the original story. However, the degree of agency for both the ducks and the audiences increases with each telling, so that the final story takes place over a larger portion of the globe with people who are able to react to what they have learned, not only in the moment, but in their futures as adults who can make informed choices.

Children's books and promoting the development of scientific thinking

In her article "Promoting the Development of Scientific Thinking," Ruth Wilson uses a wide variety of research to argue for how science should best be introduced to young children.[66] Arguing for "Science as Active Exploration," she urges that "Young children should be involved in 'sciencing' versus the learning of scientific facts presented by others. . . . Sciencing is a verb and suggests active involvement. Such involvement should be both hands-on and minds-on in nature."[67] To achieve this goal of "action-oriented and inquiry based approach to science," she cites Lind's argument that "the best way to learn science is to do science."[68] What that involves is "asking questions, probing for answers, conducting investigations, and collecting data," or, in other words, to use Lind's again, "the focus is on the active search for knowledge and understanding to satisfy students' curiosity."[69] She recommends a constructivist approach "based on the understanding that knowledge is constructed by children versus being given or transmitted to them."[70] She details the necessary approaches: "Science viewed as active exploration; Teacher viewed as facilitator; Areas of study set by child interest; Individual and small group investigations; Evaluation based on multiple criteria; Content connected to children's experiences; Content of study open-ended; Multiple ways to collect and record data; [and] Science integrated with other curricular areas."[71]

Using these parameters to evaluate the children's books written about the ducky spill, one could argue that they could be used to establish child interest, begin children's experiences, establish multiple ways to collect and record data, and integrate science with other curricular areas, notably language arts.

Online science lessons and "the Great Ducky Spill"

The saga of the rubber duckies has even spawned online science lessons for sixth graders, aligned to professional teaching standards.[72] One of many, on the *Better Lesson* website, "Currents and the Great Ducky Spill" is part of the unit *Oceanography*, created by Melodie Brewer, who teaches at Pinnacle Peak Elementary in Scottsdale, Arizona. The stated objective is to "identify the primary causes for ocean currents, and explain the effect of currents on human activity in the ocean," and the "Big Idea" is that "Currents are the driving force behind many ocean phenomena. This lesson discusses the cause of currents and how they impact life in the oceans."[73]

To begin the 55-minute lesson, Brewer engages the class with ocean scenes caught by webcams at *Explore: LiveCams*.[74] Because of the popularity of these scenes with the class, Brewer keeps them displayed during many of the activities, explaining that not only do the students enjoy seeing the sea life, but the scenes "are relaxing and keep the students calm as well."[75]

Next, students explore ocean currents. First, they complete a lab which asks them to "predict what will happen when they place an ice cube in a glass of warm-hot water."[76] Then they "perform the experiment and record their results/observations both by writing and drawing" and explain their results.[77] Although Brewer admits that the students' "explanations will probably be very vague and may not take into account any changes in density or temperature between the ice and the water," she will have them watch a *Bill Nye the Science Guy* video on "Ocean Currents."[78] Then they read more about currents and conduct additional research so that they can "draw a diagram that shows the path of the water and how it is caused."[79] Next they explain how their new understanding affects what they originally wrote about the ice melting in hot water experiment.

Next they learn about the ducky spill and how it allowed scientists such as Ebbesmeyer to collect data to support his models of ocean currents. To do so, they complete a lesson developed by Brenda Paul, which meets three of the professional standards for sixth-grade science: "a. Explain that a large portion of the Earth's surface is water, consisting of oceans, rivers, lakes, underground water, and ice. c. Describe the composition, location, and subsurface topography of the world's oceans. e. Explain the causes of waves, currents, and tides," as well as the Ocean Literacy Standard 1: "The Earth has one big ocean with many features."[80]

After the students read "What can 28,000 rubber duckies lost at sea teach us about our oceans?" they summarize the article, then "discuss which aspects relate to ocean currents and the paths taken by the rubber duckies."[81] They list details and then are given maps of ocean currents, on which they chart paths that the rubber duckies could travel. There is no one correct answer: "So long as their path travels along ocean currents, their path may take the rubber duckies anywhere!"[82] Then they write down a description of the path they drew, listing oceans, ocean currents, and countries.

They also listen to the "Ducky Spill" podcast by National Public Radio.[83] Finally, they evaluate by completing the "Nike Shoe Investigation," which is similar to what they have been learning about but involves even more specificity of vocabulary and concepts.[84] It focuses on the spill Ebbesmeyer investigated before the ducky spill. Students read a brief account of the spill and its aftermath of spotting 80,000 Nike shoes; then they plot the sightings onto a map using longitude and latitude; then they analyze their data using "gyre," "current," and "eddy" to explain the pathway they plotted. They are given maps of shoe locations and surface currents.

Using Wilson's criteria, this lesson, which combines several lessons, is indeed "sciencing" because of its "active involvement, [which is] both hands-on and minds-on in nature."[85] Students "do science" by "asking questions, probing for answers, conducting investigations, and collecting data."[86] Without fully knowing students' backgrounds or interests, we can say it meets a majority of the nine criteria: "Science viewed as active exploration; Teacher viewed as facilitator; Individual and small group investigations; Evaluation based on multiple criteria; Content of study open-ended; Multiple ways to collect and record data; [and] Science integrated with other curricular areas."[87] Through the myriad of activities, students do meet the intended outcome: "identify the primary causes for ocean currents, and explain the effect of currents on human activity in the ocean."[88] However, one could be critical of the lesson, questioning whether it also helps students explain the effects of human activity on the ocean, and wondering how students can associate their everyday habits with the plastic pollution of the oceans.

To better connect the individual with the larger problem, another major learning outcome must be considered: the transfer of knowledge from one situation to another, which is a major impediment to learning at all levels of education.

Teaching "integrative learning"

The Association of American Colleges and Universities (AAC&U) defines integrative learning as "an understanding and a disposition that a student builds across the curriculum and cocurriculum, from making simple connections among ideas and experiences to synthesizing and transferring learning to new, complex situations within and beyond the campus."[89] Integrative Learning has five components: "Connection to Experience: Connects relevant experience and academic knowledge"; "Connections to Discipline: Sees (makes) connections across disciplines, perspectives"; "Transfer: Adapts and applies skills, abilities, theories, or methodologies gained in one situation to new situations"; "Integrated Communication"; and "Reflection and Self-Assessment: Demonstrates a developing sense of self as a learner, building on prior experiences to respond to new and challenging contexts (may be evident in self-assessment, reflective, or creative work)."[90]

The book *Plastic Soup* by Michiel Roscam Abbing (2019) devotes two pages to "Rubber Ducks."[91] Its dedication spells out its larger purpose: "This book is dedicated to children everywhere, in the hope and conviction that they will live in a world in which plastic soup is no longer on the menu."[92] The headnote to the entry on rubber ducks, which includes a map of the ducks' journeys, explains the reason for the entry in the book: "The ones that were found turned out to be incredibly valuable for oceanographers studying the ocean currents. As a result, the yellow rubber duck has become one of the symbols of plastic soup."[93] The entry also includes the photograph of a large yellow duck, one of many such installations, which the photo tag explains: "To symbolize plastic pollution, the Dutch artist Florentijn Hofman blew up a yellow bathtub duck, familiar throughout the world, to a monstrous scale."[94]

However, most notable in the book is the section on "Creative Reuse."[95] Among the various ways ocean plastics have been reused is through exhibitions such as *Gyre: The Plastic Ocean*, hosted by the Anchorage Museum. This exhibit, featuring hands-on activities, multimedia presentations, and artworks, ran from 7 February through 6 September 2014 and is still available electronically. Situated at "the intersection of art and the environment," it was part of a larger project that featured an expedition by scientists and artists "to document and collect trash along Alaska beaches."[96] Combining the science of ocean systems with art made from ocean plastics, the exhibit explored "the complex relationship between humans and oceans in a contemporary culture of consumption."[97]

Part of the lasting archive of the exhibit is "Gyre Tools for Teachers," a set of seven lesson plans for students in grades four through eight, correlated with Common Core Standards, Alaska State Standards, and Next Generation Science Standards. Each of the seven lessons combines an exploration of environmental science concepts such as ocean systems and plastics with students' personal connections to consumer culture, resulting in individual projects that include flowcharts, designs, and visual comparisons as well as artworks such as sculptures and masks.[98] Objectives include "Lesson 1: 'They will create a flowchart of plastic movement to the ocean, detailing how physical factors power this movement and how people can stop it'; Lesson 2: 'They will connect their personal use of plastics with the problems of marine debris, and design ways to minimize the improper disposal of plastics'; Lesson 3: 'They will connect their personal consumer choices with the effects of marine debris on the marine environment, and design and evaluate ways to mitigate the ecosystem effects of marine debris'; and Lessons 5 and 6: 'They will create a collaborative sculpture from marine debris representing the surface of the ocean choked with intertwined debris to help people understand and connect with issues of marine debris.'"[99]

These lessons not only teach "sciencing," they teach integrative learning. Children use reflection and self-assessment through the integrated communication of art projects to transfer knowledge learned about science to other disciplines, making connections from their experiences to other experiences.

Unintended consequences: normalization of the problem (or is it legitimization?)

Teaching about ocean plastic pollution can have unintended consequences. Many "baby boomers" may remember the Keep America Beautiful movement of the 1960s as a precursor to Earth Day and other subsequent green activities. However, a closer examination of this program and the people behind its creation tells another story. Rather than working to keep America beautiful through calling attention to "litterbugs" who would throw waste anywhere, the program promoted the use of "litterbags" to collect the trash until it could be properly disposed of. However, while purporting to normalize litterbugs and litterbags, the movement was really a part of the plastic industry's attempts to normalize the use of disposable plastic. As one blogger explains, "The pure genius of this highly emotive campaign was that it bought a social license for mass production of disposable packaging, by championing action to clean up the pollution it led to."[100] The deviousness of the campaign deserves more attention than this space permits; however, it helps raise an issue with yellow ducks that must be examined: normalization.

In his *Discipline and Punish: The Birth of the Prison* (1975), Michel Foucault describes the power of normalization in human society to control people's actions with a minimum of exertion.[101] If an action is made "normal," then people will work to maintain that norm, disciplining themselves and others to abide by it. The question of normalization then, when applied to the rubber duck spill, is whether their association with the problem normalizes plastic pollution in the ocean. For instance, do they make the issue so "comfy" that song lyrics can make light of 29,000 ducks diving "into earth's biggest bathtub"?[102] Do sixth graders understand the problem as one of pollution, or do they consider the yellow ducks helpful in normalizing something as complex as the unity of the ocean into one large current?

Rubber duck releases conducted as fundraisers are events that suggest how normalized dumping plastics into water has become. Ebbesmeyer and Scigliano report that the Great American Duck Races company "rents yellow ducks for charity 'races' worldwide," which he terms a "variation on both racing and roulette": participants pay to rent a numbered duck which they bet will be the first to float past a finish line after all the ducks are released. The money raised goes to charity after the winner is awarded a prize. Ebbesmeyer and Scigliano note that "In 1994, two years after the spill, Great American supplied 1.5 million ducks to 127 races."[103] A YouTube video, "Windy City

Rubber Duckie Derby" shows one such race on the Chicago River in 2017 to raise money for the Special Olympics Illinois.[104] A huge dump truck backs up and, in about ten seconds, dumps sixty thousand ducks into the river. For about twenty minutes, they float down to the finish line, sprayed from behind to keep them moving. The lucky duck that wins also wins a brand new car for its sponsor. However, even though there are booms set up to contain the ducks and boats patrol the raceway, some do get away. Ebbesmeyer and Scigliano assert that "With such volumes, some inevitably escape the post-race cleanup crews and reach the open sea."[105] The ducks' contribution to the ocean plastic problem is considered normal.

One final example demonstrates that even scientists have normalized the release of plastic ducks in their research.[106] When scientists in Greenland lost a piece of equipment being used to measure glacial flow, they "released the flotilla of rubber ducks, each labeled with the words 'science experiment' and 'reward' in three languages, along with an e-mail address." Their hope was for someone in Baffin Bay to point the direction for finding their missing equipment after recovering a plastic duck.

Notes

1. Mary Schoell, "Garbage Island: The Great Pacific Garbage Patch," *Yale Environment Review*, June 11, 2019, https://environment-review.yale .edu/garbage-island-great-pacific-garbage-patch.
2. Ibid.
3. Ibid.
4. Ibid.
5. Ibid.
6. Ibid.
7. Ibid.
8. Donovan Hohn, *Moby-Duck: The True Story of 28,800 Bath Toys Lost at Sea and of the Beachcombers, Oceanographers, Environmentalists, and Fools, Including the Author, Who Went in Search of Them* (New York: Viking Press, 2011), 9.
9. Ibid. 10.
10. Ibid. 11.
11. Curtis Ebbesmeyer and Eric Scigliano, *Flotsametrics and the Floating World: How One Man's Obsessions with Runaway Sneakers and Rubber Ducks Revolutionized Ocean Science* (New York: Harper, 2009), 78–85.
12. Ibid. 87.
13. Ibid. 78.
14. Ibid. 87.

15. Emma Dobken, "The Epic Journey of the Plastic Ducks," *YouTube*, January 31, 2015, https://www.youtube.com/watch?v=AvchlWftt80.
16. Christiane Dorion, "Ducks Overboard!," *YouTube*, December 12, 2014, https://www.youtube.com/watch?v=fjxLIMF2Fq0.
17. "How 29,000 Lost Rubber Ducks Helped Map the World's Oceans," *You Tube*, March 15, 2018, https://www.youtube.com/watch?v=_UjA xuSuLIc.
18. Ibid.
19. "When Thousands of Rubber Ducks Were Accidentally Dropped into the Ocean," *What on Earth? Science Channel, YouTube*, November 14, 2018, https://www.youtube.com/watch?v=_uuMpVf2R8E.
20. Richard Eilbert, "Yellow Rubber Ducks," *YouTube*, https://www.you tube.com/watch?v=-RPUmRmdcjw.
21. Dave Coates, "Kei Miller—The Cartographer Tries to Map a Way to Zion," *Dave Poems*, July 28, 2014, https://davepoems.wordpress.com /2014/07/28/kei-miller-the-cartographer-tries-to-map-a-way-to-zion/.
22. Ibid.
23. "Rubber Duck," *The Strong National Museum of Play/National Toy Hall of Fame*, https://www.toyhalloffame.org/toys/rubber-duck.
24. Lyric Find, "Sesame Street: Ernie and His Rubber Duckie," *Sesame Street Songs*, October 10, 2020, https://www.google.com/
25. Ebbesmeyer and Scigliano, *Flotsametrics and the Floating World*, 81.
26. Stephen Bates, "Going Quackers at the Palace," *Guardian*, October 4, 2001, https://www.theguardian.com/uk/2001/oct/05/monarchy.ste phenbates.
27. Ebbesmeyer and Scigliano, *Flotsametrics and the Floating World*, 81.
28. Ibid. 81.
29. Ibid. 80.
30. Charles Moore and Cassandra Phillips, *Plastic Ocean: How a Sea Captain's Chance Discovery Launched a Determined Quest to Save the Oceans* (New York: Avery, 2012), 53.
31. Keving Krajick, "Message in a Bottle," reprinted from *Smithsonian Magazine*, June 30, 2001, courses.washington.edu/ocean200/Read ings/Ebbesmeyer.pdf.
32. Eve Bunting, *Ducky* (New York: Clarion Books, 1997).
33. Ibid. 4.
34. Ibid. 4.
35. Ibid. 6.
36. Ibid. 6.
37. Ibid. 8.
38. Ibid. 11.
39. Ibid. 13.
40. Ibid. 20.
41. Ibid. 20.
42. Ibid. 22.
43. Ibid. 25, 27.

44. Ibid. 29.
45. Ibid. 30.
46. Ibid. 32.
47. Ibid. 32.
48. Eric Carle, *10 Little Rubber Ducks* (New York: Harper Collins, 2005).
49. Ibid. n.p.
50. Ibid., book jacket.
51. Ibid., book jacket.
52. Ibid. n.p.
53. Ibid., book jacket.
54. Ibid. n.p.
55. Janeen Mason, *Ocean Commotion: Caught in the Currents* (Gretna: Pelican Publishing, 2012).
56. Ibid. n.p.
57. Ibid. n.p.
58. Ibid. n.p.
59. Ibid. n.p.
60. Ibid. n.p.
61. Ibid. n.p.
62. Ibid. n.p.
63. Ibid. n.p.
64. Ibid. n.p.
65. Ibid. n.p.
66. Ruth Wilson, "Promoting the Development of Scientific Thinking," *Earlychildhood NEWS*, Excelligence Corporation, published 2008, http://www.earlychildhoodnews.com/earlychildhood/article_view.aspx?ArticleId=409%20.
67. Ibid. n.p.
68. Ibid. n.p.
69. Ibid. n.p.
70. Ibid. n.p.
71. Ibid. n.p.
72. Melody Brewer, "Currents and the Great Ducky Spill," *Better Lesson*, Creative Commons, https://betterlesson.com/lesson/resource/3288799/rubber-duck-ocean-currents.
73. Brewer, "Currents and the Great Ducky Spill," n.p.
74. "Tropical Reef Aquarium," *Explore Webcams*, https://explore.org/livecams.
75. Brewer, "Currents and the Great Ducky Spill," n.p.
76. Ibid. n.p.
77. Ibid. n.p.
78. Ibid. n.p.
79. Ibid. n.p.
80. Ibid. n.p.
81. Ibid. n.p.
82. Ibid. n.p.

83. Terry Gross, "'Moby-Duck': When 28,800 Bath Toys Are Lost at Sea," *Fresh Air*, NPR, March 29, 2011, https://www.npr.org/2011/03/29/13 4923863/moby-duck-when-28-800-bath-toys-are-lost-at-sea.
84. "Teacher's Guide: The Nike Shoe Investigation," COSEE Ocean Systems, University of Maine, October 10, 2020, http://cosee.umaine. edu/climb/resources/nike_invest1.pdf.
85. Wilson, "Promoting the Development of Scientific Thinking," n.p.
86. Ibid. n.p.
87. Ibid. n.p.
88. Brewer, "Currents and the Great Ducky Spill," n.p.
89. Association of American Colleges and Universities, "Integrative and Applied Learning VALUE Rubric" (2009), https://www.aacu.org/value /rubrics/integrative-learning.
90. Ibid.
91. Michiel Roscam Abbing, *Plastic Soup: An Atlas of Ocean Pollution* (Washington, DC: Island Press, 2019), 22–3.
92. Ibid. n.p.
93. Ibid. 23.
94. Ibid. 23.
95. Ibid. 72–7.
96. "Exhibit Overview," *Gyre: The Plastic Ocean*, Anchorage Museum, https://www.anchoragemuseum.org/exhibits/gyre-the-plastic-ocean /exhibit-overview/.
97. Ibid.
98. Center for Alaskan Coastal Studies, "Lesson Plan Summary and Standards," *Gyre: The Plastic Ocean*, Anchorage Museum, http://gyre .anchoragemuseum.org/pdfs/LessonPlanSummaryandStandards.pdf.
99. Ibid.
100. Chris Rose, "A Beautiful if Evil Strategy," *Plastic Pollution Coalition*, October 26, 2017, https://www.plasticpollutioncoalition.org/blog/20 17/10/26/a-beautiful-if-evil-strategy.
101. Michel Foucault, *Discipline and Punish: The Birth of the Prison* (London: Vintage Books: 1995), 215.
102. Eilbert, "Yellow Rubber Ducks."
103. Ebbesmeyer and Scigliano, *Flotsametrics and the Floating World*, 81.
104. Joel McFarlin, "60,000 Rubber Ducks Spill into the Chicago River," *YouTube*, August 4, 2017, https://www.youtube.com/watch?v=NlUm _kBCW9g.
105. Ebbesmeyer and Scigliano, *Flotsametrics and the Floating World*, 81.
106. Deborah Zabarenko, "Can Rubber Ducks Help Track a Melting Glacier?" *Reuters*, September 21, 2008, https://www.reuters.com/artic le/us-climate-glacier-idUSN17465140200080921?feedType=RSS&feed Name=environmentNews&pageNumber=1&virtualBrandChannel=0.

Part V

Plastics and the Future

15

Disposable: The Dirty Word in Medical Plastics

Patrick D. Murphy

Americans under the age of forty, and most under the age of fifty, will not be able to remember a time when dentists and dental technicians did not routinely wear nitrile gloves during examinations, fillings, and root canals. The same can be said for many visits to their family physicians. When I had the end of my finger sewn back on after a sledding accident in the winter of 1957, our hometown doctor did not wear gloves, nor did he a few years later when he closed up the gash on my forehead incurred during backyard tackle football. For more than half of the American population younger than me, disposable single-use plastic items just form part of their everyday medical world rather than a relatively recent change in standard medical protocols.

The HIV/AIDS crisis in the United States beginning in 1981 led to a relatively sudden and intense concern with the transmission of infectious diseases from doctor to patient unparalleled since the end of the nineteenth century.[1] Although other infectious diseases, such as hepatitis, were a cause of concern in the United States, no other illness matched the fear and anxiety prompted by AIDS/HIV. Within a few years, the invoking of anti-discrimination regulations about people with AIDS led to increasing concern about safe and effective means of treating such people in medical settings, as well as such people safely treating others. By 1989 the federal Occupational Safety and Health Administration (OSHA) proposed regulations for the wearing of personal protective equipment (hereafter PPE) in dental settings, and in December of 1991 it finalized those regulations over the opposition of the American Dental Association.[2] Making the use of such equipment standard procedure across all types of medical settings led to a dramatic increase in the use of disposable plastic-based supplies in the treating of contagious and potentially contagious people.

While not old enough to remember bare-handed dentists, most Americans are, however, old enough to remember the "syringe tide." As in the previous year, but more extensively in 1988, beaches were closed in New Jersey, Connecticut, and New York due to medical waste washing ashore. As an *NJ.com* article recently reminded viewers,

> The images were disturbing: hypodermic needles, syringes and vials of blood washed ashore miles of beaches along the Jersey Shore. In early June of 1988, beaches on Island Beach State Park and Ortley were closed after vials of blood tested positive for AIDS and hepatitis B virus. The trash kept coming and the beaches kept closing, creating what became known as the "syringe tide." It delivered a massive blow to the tourism industry along the Shore: $1 billion in lost revenue in 1988, tourism officials said at the time.[3]

Eventually, it was determined the trash came from New York City's largest landfill on Staten Island. That meant that hazardous medical waste was actually being mixed in with regular household trash. But why did *NJ.com* run an article on this thirty-year-old story? Because the same mismanagement of waste has occurred again: "Thirty years later, more than a dozen beaches in Monmouth County were closed July 19 after garbage, including hypodermic needles, washed ashore. But state officials are pointing to a different primary source for the recent debris: Combined sewage system overflows in New York City and North Jersey."[4] It will become clear a little later in this chapter how and why such medical disposables ended up in municipal landfills and sewage systems, rather than being disposed of properly as hazardous waste.

The container on the wall of an examining room marked with a red biohazard symbol for the disposal of single-use syringes no doubt draws a patient's attention, but it is unlikely that the much larger hamper in the room or down the hall for disposing of gloves, gowns, and other PPE similarly catches the eye. But in either case, how often do the people caught up in concerns about their health, or fears that they have contracted the latest contagious disease, whether a sexually transmitted one or just the annual round of flu or meningitis, consider where those syringes and PPE items end up? Unlike cloth gowns, these items are surely not just cleaned and reused, but are any of them recycled in any way? If not, where do they go and just what kinds of quantities are we considering? Just how much of these items that comprise the medical industry disposability problem consist of some type of plastic? And what exactly should be included under the label of medical plastics when analyzing their role, quantity, and increasing presence, replacing various other types of materials such as glass and latex?

While much of the plastic-based equipment used in hospitals and clinics is reused, such as bed components, most tubing, machinery housing, and technological items such as monitors and keyboards, an increasing amount of medical supplies are single-use disposables that end up in landfills or incinerators rather than being recycled. Also, single-use disposables tend to come individually wrapped in plastic film, which also tends to be discarded. The coronavirus pandemic, of course, has dramatically increased the percentage of potentially reusable items that have now become disposable ones to avoid cross-patient contamination and to protect healthcare professionals: Tyvek coveralls, polyester gowns, face shields, synthetic masks, stethoscopes, thermometers, and ventilator tubing, for instance. And, while there are various efforts underway to develop bioplastics from renewable materials, these pilot projects have had no impact on the quantity of medical disposables, nor do they appear likely to represent a significant source of feedstock for the manufacturing of such items at any time in the near future.

Syringes and gloves

The two most ubiquitous disposable medical devices have to be gloves and syringes. Even before the coronavirus pandemic, single-use gloves had become standard procedure in virtually every type of medical office. The exception would be optometrists' and ophthalmologists' offices, where practitioners need only wear gloves if they are likely to come into contact with bodily fluids, such as when treating corneal abrasions or weeping wounds of some kind. Thus, in medical offices, hospitals, rehabilitation centers, and many mental health clinics one sees nurses changing gloves between each patient who receives an injection, is examined in some way, provides a urine sample, or has blood drawn.

For years after World War II latex gloves were the disposable standard, made of natural rubber rather than petroleum-based synthetics (petrosynthetics). Gradually, however, the problem of allergic reactions, including fatal ones, by both patients and practitioners—particularly those people who had to wear them for long periods of time day after day—led to the substitution of nitrile, vinyl, and butyl for latex.[5]

The nitrile glove market is mostly discussed in terms of global sales figures across the various industries that use them, rather than per unit use or sales in the medical industry alone. That makes it difficult to determine quantities of gloves consumed. One hospital in Canada, however, which did an ecological footprint review some years ago, does provide some idea, prior to the pandemic, of the typical use of disposable gloves in such a setting: 1.75 million pairs of gloves per year.[6] There are nearly 6,200 hospitals in the United States,[7] and, if they consume disposable gloves at a rate similar to the one in Canada, that would mean an American consumption rate of over 11 billion pairs of gloves per year for hospitals alone. According to the CDC, as of 2016 survey data, there were nearly 900 million physicians' office visits per year separate from trips to the hospital.[8] If we assume that in addition to the doctor there is at least one nurse attendant who will also need to wear gloves, we can add another 1.8 billion gloves to those used in hospitals. Then there are the innumerable other sites requiring the wearing of disposable gloves for contact with bodily fluids, such as mental health clinics and corporate random drug testing facilities, for which there is no reliable consumption rate data. Glove use in medical research and school facilities is beyond the scope of this chapter but would constitute another significant area of medical disposable glove consumption.

The consumption of disposable gloves for health care occurs mainly in hospital and clinical settings. When looking at the widespread use of disposable, single-use syringes, however, one finds that literally billions of them are used by individuals at home and in other non-medical locations, including alleys, stairwells, and motor vehicles.[9] As the medical supplies distributor Omnisurge puts it, "there are few medical tools so commonplace, and yet so indispensable, as the plastic disposable syringe."[10] Diabetics taking insulin at home comprise the group using the largest number of these syringes outside medical establishments. Recent estimates peg the number of diabetics, whether having type I or type II, needing to take insulin at approximately 8.3 million.[11] Even subtracting the 350,000 or so Americans who use insulin pumps instead of taking injections, there are still nearly 8 million using single-use syringes or preloaded plastic pens for injections averaging two or more per day, resulting in a non-medical setting consumption rate of nearly 6 billion syringes and pens per year by this group alone. The number of Americans with diabetes, particularly type II, and those requiring insulin injections has been steadily increasing and is projected to continue to increase.

There has been a substantial uptick in the use of disposable needles over the past two decades, not only because of the increasing unhealthiness of the American public and their drug abuse habits,[12] but also due to efforts to reduce infectious disease transmission. As one market analysis report has noted, "Disposable syringes market in the US has significantly benefited from the implementation of US Needle Stick Safety and Prevention Act of 2000, which advocates usage of safer medical devices to eliminate or minimize exposure to pathogens from needlestick injuries."[13] According to a 2011 press release by Jenny Schumann of the Coalition for Safe Community Needle Disposal,

Americans discard 7.8 billion needles every year, all of which were initially attached to a disposable syringe.[14] Given the rising rate of diabetes in the US, it is reasonable to assume that even more needles, and therefore more plastic syringes and pens, are being disposed of in 2020 than were in 2011.

Not only are there the disposable syringes used for adjusted insulin injections, for example, but an increasing number of syringes are being marketed pre-filled with a particular medication. By 2011, more than twenty companies were marketing some fifty drugs this way.[15] Pharmaceutical companies now advertise on television and social media such devices to treat a variety of chronic illnesses, such as Humira (adalimumab) for arthritis and Emgality (galcanezumab-gnlm) for migraines. In 2017, one research firm suggested that these syringes will account for more than $18 billion in business.[16]

In addition to Americans discarding nearly 8 billion needles and syringes per year outside of medical care settings, one medical industry online forum estimates that hospitals and medical offices dispose of another two million syringes per day, or about 800 million per year.[17] Setting aside any upsurge in use during the pandemic, one can safely say that in the US alone disposable syringes are discarded at a rate of about 25 million per day. These are not recycled in any way, shape, or form but either dumped into the nearest landfill if disposed of outside of medical settings, incinerated when treated as biohazard waste, or autoclaved to make them nontoxic and then dumped in landfills.

While hospitals and medical offices have plastic biohazard boxes for collecting used syringes, which are then discarded, box and all, as biohazard waste, there is no equivalent mechanism for household use collection, although some meager efforts exist to collect needles. Even if there were mechanisms for collecting all of these syringes, they would probably not end up being recycled since they are generally made of polypropylene, recycling number 5. Although it can be recycled by being melted down into plastic blocks that could be sold if a waste management company were able to find a buyer, less than 3 percent of all of the products made from polypropylene are actually recycled due to technical difficulties.[18] It is doubtful that medical waste, given the restrictions on its handling and processing, comprises a significant portion of that 3 percent.

The personal pregnancy test, the urine collection cup, and the prescription container

Into the 1960s, for a woman to determine what she frequently intuited—that she was indeed pregnant—required a visit to the doctor's office, the submission of a urine sample, and a two-week wait for results. In 1969 a female researcher gained a patent for an at-home pregnancy test, and seven years later the first FDA-approved tests went on sale in the United States.[19] These simple devices are made almost entirely of plastic. Given their relatively low cost of ten dollars or less, the privacy they afford, the quickness of their results, and their high degree of reliability, it is no surprise that the two most popular brands, Clear Blue and First Response, sold a combined 13.1 million of these in 2019 in the United States.[20] Of course, they are not recycled, but invariably tossed into the nearest trash container to make their way to the local landfill as household waste.

If a woman does want to confirm the findings of her at-home personal test, she can visit a medical office and there provide a urine sample in a disposable plastic container. These specimen cups, however, are not only used for pregnancy testing, but used by both men and women for myriad tests. Drug testing by clinics and businesses may produce their highest rate of consumption, but numerous other types of tests, such as checking for signs of prostate cancer, also involve the use of urine specimen containers. Almost all of these are disposable and non-recyclable, not because of the type of plastic used but because they house bodily fluids. Even a potentially biodegradable one for sale on the market is unlikely to have much impact on the production of plastic waste because of the specific requirements for its degradation.[21]

The specimen cup appears pretty much everywhere these days, since many companies in addition to medical facilities routinely conduct drug tests. The most common form of such tests is urinalysis, since it is cheap, quick, and relatively noninvasive—no needle required. Such tests, however, do not involve only single-use specimen cups. Plastic dipsticks are often inserted in the urine to obtain a variety of information. Subjects of these tests may also be required to suck on a plastic saliva swab as well. All of these items are single-use disposables. Millions of them are used across the country every day, in workplaces as well as medical facilities, and add to the mountains of medical waste growing around the country.

Estimates of the number of prescriptions filled each year in the United States vary because some data is obtained only from pharmacies, while other organizations also include in-hospital prescriptions. According to KFF.org, pharmacies filled 3.8 billion prescriptions in 2019, while Medscape.com published an article stating in 2018 that 5.8 billion prescriptions had been filled nationwide.[22] While the dispensing of prescription drugs in hospitals intravenously involves a disposable plastic drip bag, dispensing in pill form frequently does not involve a disposable plastic container except for the one in which the pills are originally packaged. The opposite is the case for pharmacy prescriptions. Whether the prescription is for a pill, a liquid, or an ointment, or the previously mentioned preloaded syringes, the container is almost always made of plastic. Although the former two types of containers are made of recyclable PET plastic, few of these containers actually get recycled. Likewise, there is the even larger amount of plastic containers for over-the-counter drugs (OTCs). With American consumers spending over five billion dollars on OTC analgesics (aspirin, acetaminophen, and the like) alone in 2019,[23] it is easy to imagine the billions of small plastic containers in which these and other OTC medications are marketed piling up in landfills.

Since almost all of these OTC and prescription containers do have a recycling number on them and could theoretically be recycled, some people will shrug at the notion that they constitute a significant waste management or pollution problem. Few of these containers, however, get recycled since they require voluntary personal recycling initiatives that few people are willing to undertake, even in communities that offer curbside recycling. Yet with refill bottles, if not the OTC containers as well, if a program were initiated through pharmacies to provide incentives for individuals to drop off empty prescription containers when they pick up their refills, these objects would be easy to collect in much larger numbers. And, unlike the complex plastic processing required of something like disposable contact lenses, most of these containers are simple #2 plastics, which are already processed by recycling centers. The caps, however, are likely not a desirable type of plastic for reprocessing and will still end up being discarded.

Masks and gowns

The pandemic has caused widespread use of all kinds of masks, with their diversity perhaps only matched by the variety of ridiculous arguments made against wearing them. With the majority of homemade masks being made from cotton and paper fabrics of one kind or another, many people do not realize that the majority of medical masks are actually made from plastic, with some surgical masks being made from a combination of cotton and plastic. Companies such as DuPont, BYD Care, McKesson, and others use high-density polyethylene as the primary feedstock. Merrow Manufacturing, a textile company in Massachusetts, shifted to PPE production during the pandemic and uses a type of nylon intended for tents and travel gear.[24] A critical care patient's space will be visited by a combination of hospital staff, doctors, nurses, aides, and cleaners, fifty to eighty times per day—with each visit requiring a new mask.[25] The scale involved here can be sensed from the fact that BYD Care, a wholly owned subsidiary of BYD Ltd., a Chinese company in which Warren Buffet's Berkshire Hathaway owns a 25 percent share,[26] is currently making 50 million facemasks daily, ratcheting up production from its pre-pandemic rate of 5 million per day.[27]

These are figures for worldwide sales and not just for the US market, but BYD is also only one of the companies manufacturing face masks, which they do market in the US through the drugstore chain CVS among other outlets. The majority of BYD masks are the earloop surgical ones. 3M is the largest US manufacturer of N95 masks, the ones with two elastic bands. They are the type of mask most people would have recognized prior to the pandemic and are often used outside medical settings, such as by painters, house remodelers, and other types of hardware store customers, in addition to hospital staff. 3M ramped up production for this model to 50 million per month.[28] 3M's N95 shell is made from polyester and polypropylene, which are not easily recycled.

During the pandemic, 3M stopped shipping these to retail stores such as Home Depot in order to reserve production for hospitals only. They were rarely seen being worn by the general public after a few months into the pandemic. Single-use earloop face masks, like the ones made by BYD, are the ones most likely to be seen in 2020 forming a component of street litter. It is virtually impossible to estimate the total number of disposable face masks being discarded daily in the US during the course of the pandemic, since sales figures do not distinguish between masks being used and discarded and those being purchased for future use or stockpiled for resale. Most of these face masks are made from a combination of synthetic fibers of different types of plastic or a mixture of synthetic and natural fibers, making them virtually impossible to recycle due to the complexity of separating the materials and the high cost of doing so even when the technology exists for that purpose.

Some reusable gowns are made from cotton fabric, but many of them are actually polyester blends. Disposable isolation gowns in turn are made from various petrosynthetics, including chlorinated polyethylene (CPE).[29] It is difficult to get a sense of the volume of petrosynthetic plastic gowns being used and discarded at the present time, because the pandemic has skewed production and sales figures. Some sense of the scale, though, can be gleaned from Merrow Manufacturing. When they refitted their factories to make gowns for pandemic utilization, they did so to achieve a production rate of 700,000 gowns per week, or about 36 million per year.[30]

Contact lenses

Certainly in terms of physical volume, disposable contact lenses account for a tiny proportion of the medical waste stream. Nevertheless, they are worthy of attention for three reasons: one, the high number of people who use them and their behaviors in regard to their disposal; two, the characteristics of the type of pollution caused by improper disposal; three, they are an example of a medical disposable unaffected by the pandemic in terms of sales or disposal. According to the *CDC Fast Facts* on contact lenses, approximately 45 million Americans wear them, and roughly 93 percent of these people buy soft contact lenses.[31] Soft lenses are the disposable ones that are designed either to be worn for thirty days and then discarded, or to be replaced daily. Angela Lashbrook writes in *The Atlantic* that approximately 14 billion contact lenses get discarded every year in the United States, 3 billion of which are flushed down the toilet or washed down the drain, according to Rolf Halden, director of the Biodesign Center for Environmental Health Engineering at Arizona State University.[32] That means, says Halden, that nearly fifty tons of contacts could end up in American water supplies each year.[33]

Here the pollution aspect of these lenses becomes magnified. Like certain other kinds of plastic products, disposable contact lenses neither remain wholly intact nor completely dissolve. Rather, they degrade into microplastic particles that are then passed into oceans and rivers, from which some communities draw their water supplies, as well as becoming part of sewage sludge that is used as crop fertilizer.

Microplastics threaten marine life in ways different from that of larger plastic trash because they are easily digested by a wide array of animals, including fish and crustaceans. As Subhankar Chatterjee and Shivika Sharma note in their abstract for an article on microplastics,

> The microscopic size of these plastic fragments gets them easily available for ingestion by an array of marine habitants, causing adverse effects on their health. The potential of microplastics to absorb various harmful hydrophobic pollutants from the surrounding environment indirectly transfers these contaminants in the food chain.[34]

And eventually, either through that food chain or through river uptake of municipal water supplies, some of that microplastic ends up in American stomachs.[35]

Whether people flush their lenses down the toilet or throw them in the household trash bin, none of them get recycled. Yet getting a significant quantity of them to a recycling center would be a relatively simple process. Most people obtain these lenses through the mail via prescription subscription services. These services could provide a metered return address container for the used lenses and offer a small discount on future orders for their timely return. That would be the easy part. The hard part, then, would be finding manufacturers or recycling vendors willing to process this complex form of plastic in order to create new feedstock for the contact lens industry or some other commercial purpose. Even if not recycled, such a return to the manufacturer could reduce the number of lenses ending up in rivers and oceans, since the manufacturer would handle their landfill disposal or incineration.

IV and blood bags

As has often been the case, modern warfare produces advances in medical technology. In part these result from the need for a high degree of mobility in treatment centers and portability of equipment. Another contributing factor is the need for medical supplies to withstand the rigors of transport and the harsh conditions of battlefield proximity. For example, the ability to transport blood to field hospitals was seriously hampered by the use of easily breakable glass containers. During the Korean War, this problem was solved by the introduction of the PVC blood bag, which was much more flexible and resistant to breakage.[36] By 1970, two decades after the war, glass blood containers had virtually disappeared from hospitals. Setting aside the increase in blood transfusions and plasma therapy during the coronavirus pandemic, one still finds an enormous quantity of disposable PVC bags in use. The American Red Cross states that some 21 million blood components are transfused each year and that it collects nearly 14 million units of whole blood annually.[37] The Red Cross, though, accounts for only about 40 percent of the country's blood supply. Most of the rest comes from America's Blood Centers: "This is a network of more than 50 independent, local blood suppliers that supply about 50% of the nation's blood. Its member organizations manage more than 600 donation sites in 45 states. Two of the largest members are Vitalant (western United States) and Versiti (midwestern United States)."[38] These groups and the Red Cross, then, account for the use of more than 35 million blood bags per year, with contents that have a relatively short shelf life of six weeks for whole blood and only a few days for platelets.[39]

But this particular introduction of plastic into the medical supplies chain not only affected the storage and delivery of blood—it also led to the adoption of PVC bags for virtually all intravenous liquid delivery systems. The most common of these is saline solution, which both treats dehydration and serves as a transport medium for other medications. Some forty million saline IV bags are used per month in American hospitals and clinics. American reliance on this staggering amount of PVC bagged saline solution came to light as a result of Hurricane Maria's devastation of Puerto Rico.[40] A number 3 plastic, PVC is generally not recycled, and any of these bags considered to have been in contact with bodily fluids would need to be disposed of as biohazard waste, either incinerated or autoclaved and then dumped.

The plastic itself

As an *American Geosciences* report points out, "Ethylene and propylene are the two dominant petrochemicals: in 2016, the US produced over 26 million tons of ethylene and over 14 million tons of propylene." The former is mainly used to create polyethylene, which "comes in 10,000 different types."[41] Propylene is converted to polypropylene, which is then combined with chlorine to make PVC and CPE disposable medical supplies, such as blood bags and isolation gowns.

The only types of plastic readily recycled are items marked with recycling numbers 1 and 2. Many consumers, no doubt, imagine that recycling numbers have been placed on items for their benefit in practicing curbside or volunteer recycling. Actually, the numbers are designed to inform commercial processors of the type of resin used in a product. Even with the seven-number system, this labeling is not exhaustive. Rather, number 7 just means "other," and often describes an item made from a combination of different types of plastic. Basically, since its inception, plastic recycling has only processed in any large amounts plastic items labeled 1 or 2. Many recycling centers will not even accept anything made from the higher-number types of resin. Others will accept the items but if, as is usually the case, they cannot find a vendor to take these types of plastic, they will ship them off to the landfill. As E. Allison and B. Mandler, writing for the American Geosciences Institute, noted in 2018,

> Despite the many potential reuse options for plastics, only about 9.5% of plastic material generated in the US was recycled in 2014. Over 75% went into landfills, while 15% went into trash-to-energy plants that burn waste to generate electricity. A small but significant proportion of used plastic is not properly disposed of and ends up in the surface or marine environment, where it takes hundreds of years to decompose and can harm wildlife.[42]

The feel-good illusion of recycling

Four key dates in the history of medical plastic waste and its disposition would be 1970, 1976, 1991, and 2017. Not only was 1970 the year of the first Earth Day celebration, it was also the year in which the Clean Air Act was passed. Although this had no immediate impact on the handling of medical plastic waste, it did begin to raise consciousness in the United States about air quality and the major causes of air pollution, one of which was the burning of hazardous waste in municipal incinerators. E. Timothy Oppelt notes that despite growing awareness, "Hazardous waste incinerators were not regulated until the passage of the Resource Conservation and Recovery Act of 1976 (RCRA)."[43] The following year, the EPA distributed strict guidelines for toxic emissions that affected municipal waste incinerators.[44] These guidelines led to a near-death experience for that industry, with the number of municipal incinerators declining rapidly over the past forty years to the point that only seventy-two such incinerators were operating as of 2019.[45] As Zack Fishman reports, "In the US, 90% of medical waste is autoclaved. . . . The other 10% is incinerated and converted into energy. More than 90% of medical waste was incinerated in the US until 1997."[46] In 1991, however, medical waste lost the benefit of federal oversight, when the 1988 Medical Waste Tracking Act expired and oversight devolved to the varying regulations of individual states.[47] As a result the federal government could only regulate pollutants resulting from the burning of medical waste, not whether or not it was allowed to be incinerated in the first place.

Perhaps because of the relaxation of environmental regulations and standards under the Trump administration, medical waste processors have again been promoting the idea of incinerating medical waste as a way to address the problem of limited landfill space and the difficulties involved in opening new landfills. Stericycle, a major medical waste management company, basically argues that incineration is a necessary outcome of the United States having a strong healthcare system and allegedly strong pollution regulations: "With advancements in medical treatment technologies and increases in regulations for proper waste management, the amount of medical waste resulting from healthcare is expected to increase. Medical waste and the need for incineration is an outcome of the advanced healthcare system from which we all benefit."[48]

Also, with the pandemic dramatically increasing the amount of hazardous medical waste that has to be autoclaved and then dumped, the National Waste and Recycling Association has requested a relaxation of rules regarding the handling of such waste, which would likely include stepping up incineration.[49] This movement to increase incineration suggests that these companies do not anticipate a significant uptake in medical waste recycling in the foreseeable future. Keep in mind that "US health-care facilities generate about 14,000 tons of waste per day; up to 25 percent is plastic," or 3,500 tons of plastic per day, 1.277 million tons per year.[50] Greenhealth estimates that about eight percent of a hospital's total waste is biohazard material, or 280 tons per day.[51] This amount is the percentage most likely to be incinerated.

Countries such as the United States have made their environmentally conscious citizens feel good by allowing them in some areas to take higher-numbered plastic products to recycling centers. Most people do not realize that this plastic is rarely reprocessed into other products. Until 2017, the bulk of it was being shipped overseas, primarily to China. There it was burned, often with little attention to any environmental oversight. In that year, however, China announced that it would no longer import plastic waste beginning in 2018, and countries that had been using it as a dumping ground have since had to scramble to find other countries to serve as international garbage dumps.[52] Until the ban, few people realized that "95 percent of the plastics collected for recycling in the European Union and 70 percent in the US were sold and shipped to Chinese processors."[53] In other words, American recycling of plastic was never actually resulting in Americans recycling their discarded plastic into other products. Also, prior to the Chinese ban Americans were burning more plastic waste than the amount being recycled. With the increasing difficulty of exporting the so-called recyclable plastic that is not being recycled, pressure will rise, as Stericycle has already recognized, to incinerate more of it as a less expensive solution than dumping it in landfills.

An example of the feel-good illusion is "recycled polyester." Hearing this term, a person might easily be forgiven for imagining tons of PPE being autoclaved to render it non-infectious and then meltblown into new polyester threads for the manufacture of new gowns and masks. But such is not the case. "Recycled polyester" is not actually made from polyester at all. Rather, PET plastic, which is already recycled into numerous products, is melted down and then converted into fibers used to create polyester fabrics. That polyester fabric then becomes a single-use disposable piece of medical equipment, destined for the furnace or the trash heap. As a result, a type of plastic that could be recycled numerous times if it were used to make plastic water bottles ends up being reused once to make non-recyclable polyester.[54]

Pandemic intensification of a persistent problem

Americans generate and disperse more plastic litter than they have ever recycled. Yet, until the coronavirus pandemic, relatively little of that litter consisted of medical plastics. Generally using medical plastics inside their homes, people either threw them in the trash or flushed the smallest items down the toilet. Now, however, at least two types of medical plastic waste are joining water bottles, takeout containers, and plastic wrappers of all types in forming roadside and public space litter: disposable gloves and face masks.

In a report for CNN, Rob Picheta quotes John Hocevar, a Greenpeace campaign director: "Right outside my house there are discarded gloves and masks all over the neighborhood. ... It's been raining here for two days, so these are very quickly washed down into the sewer. Here in Washington DC, they end up in the Anacostia River, out in the Chesapeake Bay, and then the Atlantic Ocean." In addition to the problem of PPE breaking down into microplastics, Hocevar claims that the items pose special hazards to marine life before they begin to decompose: "Gloves, like plastic bags, can appear to be jellyfish or other types of foods for sea turtles, for example. The straps on masks can present entangling hazards."[55]

Limited alternatives and the problem of the oil glut

The Hospitals Plastic Recycling Council (HPRC) was founded in 2010 to bring together hospitals, research universities, and medical supply corporations to address the problem of plastic waste. On their website they claim that 85 percent of the 3,500 tons of plastic medical waste generated by hospitals each day is nonhazardous and therefore a clear target for recycling.[56] Setting aside the fact that these figures still leave 525 tons per day that is not recyclable, how realistic are their goals of significantly increasing disposable medical plastic recycling? After ten years of existence, the following statement still on their website provides a good indication of the answer to that question at this time:

> We recognize that there is not one single magic bullet that will solve the problem of clean, recyclable healthcare plastics ending up in landfills. There are product design features that inhibit recyclability, hospital staff lack training on what is and what is not recyclable, there is limited space and infrastructure within hospitals to aggregate recyclable materials and recycler demand and availability to collect these materials vary greatly by geography. All of these challenges combined, hinder the ability to recycle.[57]

As with so many other aspects of alternatives to the piling up of single-use plastics, in and out of the medical industry, the HPRC can mainly point only to pilot projects in terms of success in reducing medical plastic waste. The major roadblock seems not so much to be the problem of hospitals being unwilling or unable to take action, but the inability to interest corporations in designing products that do not rely on non-recyclable plastics and the challenge of getting recyclers to develop technologies to process plastic waste that does not primarily consist of PET plastic. Aside from their Chicago pilot project focused on hospitals, the HPRC's other major achievement has been developing a directory, as of 2019, for connecting hospitals with recycling vendors that might be willing to handle complex plastic items.[58]

There are various small programs underway to utilize the mixed plastic medical waste that is generated. One company makes plastic lumber out of medical waste plastic.[59] It can do so because its process allows for a mixing together of different types of plastics, rather than having to distinguish carefully between types 1 and 2 and 6 or 7. Such lumber, though, at this point in time is not a cost-effective substitute when a wooden board would serve the same purpose. A typical one-inch wooden board, eight feet long, costs one-third of its PVC equivalent. The main area in which such plastic lumber is selling well is where it is used for fencing. Demand seems too slight for this alternative to make a significant dent in the volume of medical plastic waste.

Another proposal for recycling mixed-plastic disposable medical waste calls for turning it into a type of biofuel.[60] The pyrolysis of plastic to convert it to fuel forms part of a new movement in the waste management industry being touted as an alternative to recycling through mechanical processing. Numerous corporations and university research teams have been developing pilot projects. Although the process itself is not new, previous attempts have not produced the level of efficiency and energy levels necessary to render it economically viable. But even if these technical limitations can be addressed to develop an efficient conversion mechanism for chemical plastic waste to biofuel or "syngas," is it really the path forward for plastic trash? An *EcoWatch* article on this subject casts doubt about pursuing such a path as a solution rather than a band-aid. Brigitte Osterath, author of "Do Fuels Made from Plastic Make Eco Sense?" for *Deutsche Welle*, states that Henning Wilts, director of the Circular Economy Division at Wuppertal Institute for Climate, Environment and Energy, warns that the process is not necessarily environmentally friendly: "If you break waste apart at a molecular level, you need a lot of energy, so the CO_2 savings are quite low. . . . If the energy needed comes from burning coal, then the whole thing is an environmental disaster."

Further, "If countries use the existence of chemical recycling as an excuse to stop any efforts in mechanical recycling, that will become a problem."[61]

Bioplastics as an alternative to petroplastics

Rather than trying to address the problem of disposable medical plastics waste after it has been produced, which is the focus of chemical recycling, other environmentally oriented organizations are seeking to develop "bioplastics," i.e., plastic-equivalent materials made from renewable non-petroleum resources. Ashlee Jahnke begins her article on bioplastics with the terrifying statement that "In 2016, the World Economic Forum made a prediction that there will be more plastic than fish in our oceans by 2050."[62] She then continues to promote a new type of plastic process developed by the corporation for which she works, Teysha Technologies. Since this company is just getting production underway, it remains to be seen whether or not it can really produce what it claims, and, even more importantly, whether the heavily invested petrochemical industries and medical supplies companies will adopt their product and its production technologies to replace petro-plastic.

Apparently, the Plastics Industry Association, which claims to support sustainability and the entire plastics industry supply chain simultaneously, views bioplastics as a promising growth area for plastics production:

> Globally, over 1.7 million metric tons were produced in 2014 and contributed to $4.4 billion and 32,000 jobs in the US. Bioplastics are one of the fastest growing sectors of the plastics industry, with an anticipated 20–30% annual growth. Biodegradable plastics also reduce the amount of trash that is sent to landfills.[63]

While clearly the development of bioplastics will prove to be an essential step in efforts to reduce petroleum consumption, they will only demonstrate their other purported environmental benefits if ways are found to guarantee that the products made from plant-based polymers, rather than petroleum polymers, can be placed in locations that enable them to biodegrade completely or to be recycled into new products multiple times—unlike, say, the one-off recycling that occurs with "recycled polyester."

Conclusion

There is certainly no end in sight for the use of plastic materials made from petroleum feedstocks in general, and even less likelihood of a reduction in the use of such plastics in the medical industry. That is particularly true in the near term, due to the enormous uptick in plastic consumption during the coronavirus pandemic. When this pandemic does eventually subside, a worldwide period of building up stockpiles for the next one should and probably will occur, despite the significant costs of building up such reserves. Thus, the dramatic increase in medical plastics production witnessed in 2020 should continue for some time, although the amount of waste will decline.

There seems little basis, then, for projecting a significant effort to substitute other sustainable types of raw materials as the starting point for making medical supplies in any kind of considerable quantities. These alternatives are either in the stages of initial experimentation and small quantity production, or merely conceptual projects. The first practical concern would be to insure a steady supply of crude oil suitable for the production of requisite feedstock for medical supplies. While that is not a problem in the United States given the short-term abundance of fracked and conventional oil, it certainly ought to be one for countries that are totally dependent on imported oil, such as Japan. The grindingly slow transition to electric personal and commercial transportation to the degree that it reduces the profligate burning of gasoline as fuel, then, has the unintended benefit of making more oil available for plastics production, which in turn facilitates the adequate production of feedstock for medical supplies. To the degree that oil remains a global commodity, though, and is increasingly being purchased for the benefit of high-population countries that are just converting to car culture patterns of consumption, casts a shadow over the life expectancy of adequate oil supplies.

That, however, in no way addresses the enormous environmental problem of disposability and the piling up of single-use medical materials made from plastic. The PPE worn in contagious wards will not be reused or recycled except in dire situations, and should not be so reused for patient and caregiver safety. Right now, that means burning it, since the high temperatures destroy contagious organisms, or autoclaving the materials to render them sterile. Any increase in medical plastics incineration calls for an equal, if not greater, attention to monitoring the environmental safety of hazardous waste disposal. That is largely a political issue rather than an economic or logistical one. Will the US and other governments expend additional funding to insure the safety of those who live downwind from medical waste incineration sites, or will they show the same laxness that all too often allowed the coronavirus and other health crises to spiral out of control?

The incineration of disposable medical waste or eventually its conversion into biofuel, lumber, or any other product that can utilize mixed plastic feedstock, however, relies entirely on hospitals and other care facilities collecting PPE and contaminated supplies and equipment and contracting for their disposal. That process then requires vendors who will accept the collected materials. This is one of the problems that the HPRC is attempting to address. There is, however, a much larger world of disposal medical plastics, including PPE, used by individuals in the home or in non-medical commercial settings. Masks, which are often made primarily out of polypropylene fabrics, are turning up in gutters and parking lots all over the country. Eventually someone will pick them up, perhaps after observing that they were discarded earlier enough that they can no longer carry the live virus, and throw them in a trashcan. Or they will blow from one location to another until they end up in some kind of waterway, joining the billions of tons of plastics fouling the world's oceans, rivers, and lakes.

A different type of PPE that is less injurious to the environment and does not rely on disposability is merely a glimmer in the eye of the futurist. As one sustainability researcher cogently observed:

> Once the pandemic is over, there will be time to reflect on the suitability of current systems and whether alternatives can be explored. Replacing single-use with reusable PPE that is cleaned between uses would reduce the amount of waste. However, the use of chemical cleaning may have other environmental impacts.
>
> To reduce the risk of infection, technology that sterilizes wastes and separation techniques that reduce the mixing of infectious waste with general waste could also be introduced. With more waste classed as non-infectious, more recycling options would become available. Because they require new systems and infrastructure as well as extra staffing, these options should only be considered during a time of reflection when the pandemic is over.[64]

Separate from contaminated PPE appearing outside of medical facility settings, where it is not safely handled in the waste stream, is the much larger quantity of prescription drug plastic containers, single-use insulin syringes, throwaway thermometers, and at-home pregnancy tests that end up mainly in household waste baskets and from there make their way to municipal landfills. Even though these items may be made of the kind of plastic that can be reprocessed, tens of millions of potentially recyclable pieces of medical plastic end up in garbage cans across the United States every day. The only hope for a change in this waste of products made from a non-renewable resource would be a seismic shift in consciousness regarding recycling, a commercial commitment to developing the technology to recycle medical plastics waste into other products, and a fundamental alteration in the methods for promoting and facilitating such sensible behavior on the part of consumers, producers, and processors.

Although there are clear medical reasons for having a certain amount of disposable medical materials and supplies, their ubiquity seems much more a matter of their production reflecting a consumer culture mindset cultivated since the end of World War II to convince people that personal convenience is more important than social responsibility or environmental sustainability. The problem of disposable medical plastics remains largely invisible to many people because the more general problem of disposability is not even perceived as a problem. While roughly a third of Americans actually engage in some degree of recycling, either via curbside programs or taking items to a recycling center, much of their medical plastics waste is not recyclable. As for the other two-thirds of Americans, recyclable or not, their personal medical waste ends up in the trash, which in turn trashes the planet.

At this point in time, the most immediately implementable action for reducing the production and consumption of disposable single-use plastic medical supplies is the one least likely to occur: fundamental change in American lifestyle practices, habits, and attitudes. The major diseases that lead to high levels of consumption of single-use medical supplies are mostly preventable, such as type II diabetes, hypertension, heart disease, and obesity. Rather than addressing what to do with medical waste to reduce its deleterious environmental effects once it is produced, Americans could dramatically reduce its production by becoming healthy individuals. That is not to say that the consumer alone is responsible. It is, however, individuals acting collectively who can pressure the government to coerce, or to encourage through incentives, corporations to find alternatives to disposable medical plastics, and to fund the research and pilot projects that will produce viable alternatives.

Notes

1. See "History of AIDS and HIV Overview," *AVERT*, October 10, 2019, https://www.avert.org/professionals/history-hiv-aids/overview.
2. Letter to Senator Sam Nunn, February 28, 1991, OSHA Archive, https://www.osha.gov/laws-regs/standardinterpretations/1991-02-28 -0. Regulations for medical and dental offices are updated regularly. See "Medical & Dental Offices: A Guide to Compliance with OSHA Standards OSHA 3187-09R 2003," https://www.osha.gov/Pub lications/OSHA3187/osha3187.html. On ADA opposition, see Celeste

Monforton, "From the Dentist's Chair, Remembering How They Cried Wolf," *The Pump Handle*, May 7, 2013, http://www.thepumphand le.org/2013/05/07/from-the-dentists-chair-remembering-how-they-cried -wolf/#.X2USAKjYphE.

3. Michael Sol Warren and Alex Napoliello, "More Syringes Are Washing up on Jersey Beaches. It's a Problem that Starts Miles Away," *NJ.com*, August 7, 2018, updated May 14, 2019, https://www.nj.com/news/20 18/08/more_syringes_are_showing_up_on_jersey_beaches_her.html. *NJ .com* is the state's largest provider of digital news.

4. Ibid.

5. "Latex vs. Nitrile," *QuickMedical*, February 12, 2013, http://www.quick medical.com/blog/post/latex-vs-nitrile.

6. Barbara Sibbald, "Hospitals Leaving Huge 'ecological footprints': Report," *CMAJ* 166.3 (February 5, 2002): 363, https://www.ncbi.nlm. nih.gov/pmc/articles/PMC99328/.

7. This figure is based on 2018 survey data in "Fast Facts on US Hospitals, 2020," American Hospital Association, https://www.aha.org/statistics/ fast-facts-us-hospitals.

8. "Ambulatory Care and Physician Office Visits," National Center for Health Statistics, https://www.cdc.gov/nchs/fastats/physician-visits.htm; *National Ambulatory Medical Care Survey: 2016 National Summary Tables*, https://www.cdc.gov/nchs/data/ahcd/namcs_summary/2016_na mcs_web_tables.pdf.

9. While most people would think of heroin addiction in terms of people "shooting up," the use of disposable syringes by drug addicts has increased in recent years due to the widespread production of methedrine and the intensifying opioid addiction crisis. See "Opioids Injection Drug Use," Minnesota Department of Health, https://www.health.state .mn.us/communities/opioids/basics/intravenous.html.

10. "The History of the Syringe," *Omnisurge*, May 5, 2015, https://omnisur ge.co.za/the-history-of-the-syringe/.

11. Tara O'Neill Hayes and Margaret Barnhorst, "Understanding the Insulin Market," *American Action Forum*, March 3, 2020, https://www. americanactionforum.org/research/understanding-the-insulin-market/.

12. For data on opioid drug use, see https://www.cdc.gov/drugoverdose/data /index.html.

13. "US Disposable Syringe Market to Cross $18 Billion by 2021: Pharmaion Consultants Report," *CISION PR Newswire Report*, February 15, 2016, https://www.prnewswire.com/news-releases/us-disposable-syringe -market-to-cross--18-billion-by-2021-pharmaion-consultants-report-56 8836051.html.

14. "Americans Discard 7.8 Billion Needles Every Year," *HCMS*, October 10, 2011, http://www.hcmsgroup.com/wp-content/uploads/2012/02/Ne edle-Utilization-Statistics.pdf.

15. Sagar Makwana, Biswajit Basu, Yogita Makasana, and Abhay Dharamsi, "Prefilled Syringes: An Innovation in Parenteral Packaging,"

International Journal of Pharmaceutical Investigation 1, no. 4 (October–December 2011): 200–6, doi: 10.4103/2230-973X.93004, https://www.ncbi.nlm.nih.gov/pmc/articles/PMC3465144/.

16. "Global Pre-filled Syringes Market 2017–202—Shift from Glass Pre-Filled Syringes to Plastic Pre-Filled Syringes," *CISION PR Newswire Report*, July 21, 2017, https://www.prnewswire.com/news-releases/global-pre-filled-syringes-market-2017-2021---shift-from-glass-pre-filled-syringes-to-plastic-pre-filled-syringes-300491731.html.

17. "How Many Needles Do Hospitals Use and How Do They Dispose of Them?" *MedPro Disposal*, October 17, 2016, https://www.medprodisposal.com/sharps-container-disposal/how-many-needles-hospitals-how-dispose/.

18. Emily Chasan, "There's Finally a Way to Recycle the Plastic in Shampoo and Yogurt Packaging," *Bloomberg Businessweek*, September 25, 2019, https://www.bloomberg.com/news/features/2019-09-25/polypropylene-plastic-can-finally-be-recycled.

19. Cari Romm, "Before There Were Home Pregnancy Tests: How Women Found Out They Were Pregnant When They Couldn't Just Pee on a Stick," *The Atlantic*, June 17, 2015, https://www.theatlantic.com/health/archive/2015/06/history-home-pregnancy-test/396077/.

20. M. Shahbandeh, "Unit Sales of the Leading Pregnancy Test Brands in the United States in 2019," *Statista.com*, September 7, 2020, https://www.statista.com/statistics/1052023/pregnancy-test-brands-unit-sales-in-the-us/.

21. "Eco-Friendly Spectainer™ Urine Collection Containers," *LabStorage Systems*, https://labstore.com/product/eco-friendly-spectainer-urine-collection-containers/.

22. "Number of Retail Prescription Drugs Filled at Pharmacies by Payer," 2019, *KFF.org*, https://www.kff.org/health-costs/state-indicator/total-retail-rx-drugs/?currentTimeframe=0&sortModel=%7B%22colId%22:%22Location%22,%22sort%22:%22asc%22%7D; Megan Brooks, "US Prescriptions Hit New High in 2018, but Opioid Scripts Dip," May 10, 2019, https://www.medscape.com/viewarticle/912864.

23. Elizabeth Doughman, "US Painkiller Market to Reach $5.9 Billion by 2023," *Pharmaceutical Processing World*, March 2019, https://www.pharmaceuticalprocessingworld.com/u-s-painkiller-market-to-reach-5-9-billion-by-2023/.

24. "Merrow Manufacturing Becomes Largest Producer Of USA-Sourced PPE," *Cision PR Newswire*, May 21, 2020, https://www.prnewswire.com/news-releases/merrow-manufacturing-becomes-largest-producer-of-usa-sourced-ppe-301063818.html.

25. Zack Fishman, "Plastic Panic in the Pandemic: How Single-Use Items Meant to Protect Us Will Harm the Planet," *Medill Reports Chicago*, June 20, 2020, https://news.medill.northwestern.edu/chicago/plastic-panic-during-the-pandemic-how-single-use-items-meant-to-protect-us-will-harm-the-planet/.

26. David Reid, "Buffett-Backed BYD Announces Electric Car Partnership with Toyota," *CNBC*, November 7, 2019, https://www.cnbc.com/2019/11/07/buffett-backed-byd-signs-up-to-electric-car-partnership-with-toyota.html.
27. "How BYD Built the Capacity to Make 50 Million Facemasks Per Day," *Innovation In Textiles*, June 1, 2020, https://www.innovationintextiles.com/how-byd-built-the-capacity-to-make-50-million-facemasks-per-day/.
28. "Coronavirus Outbreak Boosts the Sales of World's Top N95 Mask Manufacturers," *Technavio* blog, April 8, 2020, https://blog.technavio.com/blog/top-10-n95-mask-manufacturers.
29. See, for example, https://www.cov.care/cpe-disposable-gown.
30. "Merrow Manufacturing Becomes Largest Producer."
31. "Healthy Contact Lens Wear and Care," *CDC*, https://www.cdc.gov/contactlenses/fast-facts.html.
32. Angela Lashbrook, "Why Are So Many Americans Flushing Their Contacts Down the Toilet?" *The Atlantic*, August 19, 2018, https://www.theatlantic.com/health/archive/2018/08/contacts-down-the-drain/567850/.
33. Ibid.
34. Subhankar Chatterjee and Shivika Sharma, "Microplastics in Our Oceans and Marine Health," *Field Action Science Reports*, December 2019, https://journals.openedition.org/factsreports/5257.
35. Isabelle Gerretsen, "You Could Be Swallowing a Credit Card's Weight in Plastic Every Week," *CNN*, June 17, 2019, https://www.cnn.com/2019/06/11/health/microplastics-ingestion-wwf-study-scn-intl/index.html; Wijnand de Wit and Nathan Bigaud, Dalberg Advisors, *No Plastic in Nature: Assessing Plastic Ingestion from Nature to People*, World Wildlife Fund, 2019, https://awsassets.panda.org/downloads/plastic_ingestion_press_singles.pdf.
36. "The Blood Bag (1950)," British Society of Immunology, https://www.immunology.org/the-blood-bag-1950; "PVC—an Essential Component of Blood Donation," *PVC Med Alliance*, June 14, 2017, https://pvcmed.org/pvc-essential-component-blood-donation/.
37. "Blood Needs & Blood Supply," American Red Cross, https://www.redcrossblood.org/donate-blood/how-to-donate/how-blood-donations-help/blood-needs-blood-supply.html.
38. James Allen, "How Hospitals Get Blood for Transfusion," *The Hospital Medical Director*, a blog for and about medical directors, May 8, 2019, https://hospitalmedicaldirector.com/how-hospitals-get-blood-for-transfusion/.
39. "Blood Needs & Blood Supply."
40. Maryann Mazer-Amirshahi and Erin R. Fox, "Saline Shortages—Many Causes, No Simple Solution," *New England Journal of Medicine*, April 19, 2018, 378: 1472–4, doi: 10.1056/NEJMp1800347, https://www.nejm.org/doi/full/10.1056/NEJMp1800347.

41. E. Allison and B. Mandler, "Non-Fuel Products of Oil and Gas: Plastics, Fertilizers, Synthetic Fibers, Pharmaceuticals, Detergents, and More," Part 17 of *Petroleum and the Environment*, American Geosciences Institute, pp. 17-1 and 17-2, updated June 1, 2018, https://www.amer icangeosciences.org/geoscience-currents/non-fuel-products-oil-and-gas.
42. I, 17-4.
43. E. Timothy Oppelt, "Incineration of Hazardous Waste," *Journal of the Air & Waste Management Association* (formerly *JAPCA*) 37, no. 5 (1987): 559, doi: 10.1080/08940630.1987.10466245, https://www.tan dfonline.com/doi/pdf/10.1080/08940630.1987.10466245.
44. "Toxic and Priority Pollutants Under the Clean Water Act," EPA, https:// www.epa.gov/eg/toxic-and-priority-pollutants-under-clean-water-act.
45. Ana Baptista, "Is Burning Trash a Good Way to Dispose of It? Waste Incineration in Charts," *PBS Newshour*, June 23, 2019, https://www .pbs.org/newshour/science/is-burning-trash-a-good-way-to-dispose-of-it -waste-incineration-in-charts.
46. Fishman, "Plastic Panic."
47. "Who Regulates Medical Waste," EPA, https://www.epa.gov/rcra/medic al-waste.
48. "Discover the Importance of Medical Waste Incineration," *Stericycle*, https://www.stericycle.com/blog/healthcare-solutions/may-2015/dis cover-the-importance-of-medical-waste-incinerat#.
49. Louis Sahagun, "Medical Waste Industry Braces for Flood of Virus-Contaminated Trash," *Los Angeles Times*, March 30, 2020, https:// www.latimes.com/environment/story/2020-03-30/medical-waste-indust ry-coronavirus-contaminated-trash?.
50. Sarah Gibbens, "Can Medical Care Exist without Plastic?" *National Geographic*, October 4, 2019, https://www.nationalgeographic.com/sci ence/2019/10/can-medical-care-exist-without-plastic.
51. "Regulated Medical Waste," *Practice Greenhealth*, https://practicegreen health.org/search?keys=medical+waste.
52. Cheryl Katz, "Piling Up: How China's Ban on Importing Waste Has Stalled Global Recycling," *Yale Environment* 360, March 7, 2019, https://e360.yale.edu/features/piling-up-how-chinas-ban-on-importing -waste-has-stalled-global-recycling.
53. Ibid.
54. "Fabric Focus: What Is Recycled Polyester?" *United by Blue*, June 20, 2018, https://unitedbyblue.com/blogs/united-by-blue-journal/fabric-foc us-what-is-recycled-polyester; Alyssa Couture, "The Recycled Polyester Trend and Raised Questions about Its Sustainability," https://conscious magazine.co/recycled-polyester-trend-and-raised-questions-about-its-sus tainability/.
55. Rob Picheta, "Coronavirus Is Causing a Flurry of Plastic Waste. Campaigners Fear It May Be Permanent," *CNN*, May 4, 2020, https:// www.oceangrants.org/ocean-news/2020/5/4/coronavirus-is-causing-a-flurry-of-plastic-waste-campaigners-fear-it-may-be-permanent.

56. "Inspiring and Enabling Plastic Recycling Solutions in Healthcare," *HPRC*, https://www.hprc.org/about-hprc.
57. "A Value Chain Approach to Solution Development," *HPRC*, https://www.hprc.org/solutions.
58. Alison Bryant, "Help HPRC Launch New Recycling Vendor Directory!" April 22, 2019, https://www.hprc.org/post/hprc-launches-recycling-vendor-directory-project.
59. "Triumvirate Environmental Is Transforming Medical Waste into Plastic Lumber," *Waste Dive*, January 4, 2016, https://www.wastedive.com/news/triumvirate-environmental-is-transforming-medical-waste-into-plastic-lumber/411452/.
60. Taylor and Francis, "Disposed PPE Could Be Turned into Biofuel, Study Shows," August 4, 2020, https://phys.org/news/2020-08-disposed-ppe-biofuel.html.
61. Brigitte Osterath, "Do Fuels Made from Plastic Make Eco Sense?" *EcoWatch/Deutsche Welle*, May 18, 2020, https://www.ecowatch.com/plastic-fuel-2646025410.html?rebelltitem=1#rebelltitem1; see also "Chemical Recycling of Mixed Plastic Waste," *Biobased Press*, https://www.biobasedpress.eu/2019/12/chemical-recycling-of-mixed-plastic-waste/.
62. Ashlee Jahnke, "Could Bioplastics Reduce the Amount of Single-Use Plastic?" *Medical Plastics News* 16 (September 30, 2019): 51, https://www.medicalplasticsnews.com/news/breaking-it-down/.
63. Plastics Industry Association, "Bioplastics," https://www.plasticsindustry.org/supply-chain/recycling-sustainability/bioplastics; see also, https://www.grandviewresearch.com/industry-analysis/biodegradable-plastics-market.
64. Carly Fletcher, "What Happens to Waste PPE during the Coronavirus Pandemic?" *The Conversation*, May 12, 2020, https://theconversation.com/what-happens-to-waste-ppe-during-the-coronavirus-pandemic-137632.

16

Eco-Fascism and Alienation:
Plastics in a Post-COVID World

Sasha Adkins and Brittany Y. Davis

I thought plastic was the perfect enemy: toxic from the extraction of its fossil fuel feedstock until its eventual decomposition is complete.[1] Picture ominous plastic bags and bottles lingering in the environment for centuries, slowly and inexorably seeping carcinogenic, endocrine-disrupting, neurotoxic chemicals. This ubiquitous enemy has claimed as its territory everywhere from the bottom of the Mariana Trench to the top of Mount Everest. Nanoplastics lurk in the fresh mountain air over the Pyrenees. Even the child forming in the womb is not safe. After finding microplastics that had crossed from the maternal to the fetal side of the human placenta, researchers coined a new term: the plasticenta.[2] I suspect that next we will find that breast milk is spiked with microplastics. From conception until death, we all live in the shadow of plastics.

Plastics also function symbolically as a mirror reflecting our darkest impulses. Instant gratification and convenience for (certain) individuals take precedence over the long-term survival of the planet and its ecosystems. As materials, they condition us to think of the world as instrumentally valuable. Whatever no longer serves can be discarded. We have no lasting obligations. Elsewhere, I argue that this habit of the heart extends to our regard for each other.[3] Plastics are fungible, mass-produced, and devoid of any individuality. While some view this as democratization, it can also be seen as a descent into faceless conformity. We are dehumanized by our relationship to plastics.

That US society treats its workers, particularly BIPOC workers, as disposable is self-evident. As one employer put it, when defending himself in court against the claims of workers permanently debilitated by a toxic glue that he chose to continue to use rather than the safer, but somewhat more expensive alternative: "There are people lined up out there for jobs. . . . If they start dropping like flies, or something in that order, we can replace them today."[4] The employer views workers as disposable and not worthy of protection, even as he literally enjoys the profits of their labor. Is our reliance on plastics a cause or a symptom of this mentality? Or is it a self-reinforcing cycle?

Into such a world came COVID-19. Though public relations experts are using the pandemic to shift public perceptions of plastics, from villainizing it to relying on it for salvation, what we really have is more of the same. In spite of the fact that plastic surfaces harbor the virus longer than most other materials,[5] and that plastics are more difficult to sterilize, the industry has convinced millions that embracing single-use plastics is the only way to stay safe. Plexiglas sneeze guards turn cash registers, classroom desks, and even vice-presidential debate podiums into what Shannon Mattern terms "defensible spaces,"[6] without regard for the fact that the aerosolized virus can easily pass over and around these largely symbolic borders. And what of the disproportionately BIPOC and impoverished frontline workers stationed behind these "guards" and "shields"? Is the rift expanding between the workers and those they serve?

Mattern elaborates:

> The new architectures of (pandemic plexiglass) protection hastily installed across our built environments aim to keep us safe—pure, secure (but) not by addressing the virus's mode of transmission via vaccine or socially distanced quarantine, much less by reassessing relations between humans, animals, habitat loss, and environmental health, but … as part of … a means of maintaining social and biological order that, in turn, promises epidemiological and economic resilience. Yet the plexi shields and hoods are little more than the architectural equivalents of hydroxychloroquine, snake oil neatly packaged in capsules and vials—jury-rigged shells mocked up so that we can keep working and consuming and pretending that social space hasn't split open at its long-deepening fault lines; that the worker on the far side of the screen isn't standing there all day, at risk.[7]

Mattern highlights both the futility of using Plexiglas barriers against an airborne virus and the extent to which the public has collectively fallen into the trap of believing these barriers will be their salvation. This reliance on plastic for salvation requires the public to suspend their belief in what they've been taught about the harms of plastic or ignore those harms in a selfish, foolish bid for self-protection.

In the United States, the most effective barrier for the public, the face mask, remains stubbornly underutilized. My neighbor embodies the worst of this pernicious lie. For the past eight months, she has refused to emerge from her apartment without wearing plastic gloves. She proudly tells me that she uses plastic wipes to disinfect her groceries, the surfaces of her home, and even (during the first months when toilet paper could not be found) her own bottom. She is far less consistent, however, in her use of a face mask. Exasperated, I point out that aerosolized particles are much more concerning than fomites and implore her to wear a mask in the building's common spaces. To no avail.

Plastics have come to symbolize humanity's control of Nature. We can synthesize entirely anthropogenic molecules, in whatever pattern we fancy to serve our own ends, and these molecules are as close as we have come to approaching immortality. Perhaps subconsciously, this is a comforting symbol for these times. Even more relevant today, plastic allows us to control the permeability of our body's interface with the environment. Given that the dominant metaphor for contagion is a foreign invasion, it is worth looking more closely at how the narratives we are constructing around plastics and COVID intersect with longstanding ideas about Nature and the "Other."

A brief tour of racism and xenophobia in environmentalism

At the root of racism and xenophobia is the fear of being left out in the competition for scarce resources. Concern for the environment takes many forms and arises from widely divergent motivations. One entry-level variant is self-interest. At an activists' meeting years ago organized to get bisphenol-A (BPA) out of baby bottles and sippy cups, I tried and failed to include in our demands language advocating for occupational protections for cashiers who absorb BPA through their skin as they handle receipts. The proposal was dismissed as strategically misguided: "Occupational health just isn't sexy right now." One attendee was even more blunt: "I don't care about anyone except myself and my family." Given that we were discussing baby bottles, this apparent expression of self-interest might have been a new mother speaking out of sleep deprivation and overwhelm. This concern over what has been labeled toxic trespass, in the unfolding story of COVID, perhaps correlates to the belief that workers making PPE, along with other "essential" workers, must be sacrificed to preserve others' ability to police the molecules that will and will not be taken into their bodies. This positions the workers involved with (plastics) PPE production as instrumental, fungible, and disposable.

Existing simultaneously with self-interest and a desire for personal protection is the purported lack of self-interest of others. Environmentalism is often closely associated with whiteness. There has long been an idea among some (White) environmentalists that Black and Latinx communities "don't care about the environment," with a variety of reasons being given for this. These views exist despite ample evidence and statements to the contrary.[8,9] Moreover, the concerns of BIPOC communities—environmental racism, proximity to polluting facilities, access to affordable housing and healthy food—have been seen as outside of mainstream environmentalism, which focused on maximizing the protection and enjoyment of "pristine," "natural," or "wild" places and spaces.

Self-interest, however, can flow into mutual interest, as celebrated in the organizing principles of Standing Up for Racial Justice (SURJ).[10] It reminds us that we do not act on behalf of others whom we cast as victims, but instead that our liberation is bound up with that of others, to paraphrase an oft-cited quote from a 1970s-era Aboriginal activists' group in Queensland, Australia.[11] Air pollution increases susceptibility to COVID. The virus has the ability to "hitchhike" on microscopic airborne particles of metal or plastic and find its way deep into the lungs, and from there, into the bloodstream. Air pollution does not respect anthropogenic boundaries within bodies or between nation-states. Could this crisis lead to a convergence of the interests of frontline communities around plastics manufacture and mainstream environmentalists?

That is one possibility. A less desirable possibility is the resurgence of what Nils Gilman refers to as "avocado politics"—green on the outside, brown(shirts) on the inside.[12] In one variant of avocado politics, preservationists aim to preserve "untouched" wilderness areas for anthropocentric "beneficial uses" such as recreation. Those benefiting from these uses tend to be people with privilege, and the people who threaten the enjoyment of these lands to which they feel entitled, are not.

BIPOC folks can, quite conveniently, be thought of (by White environmentalists) as either closer to nature (than "civilized"—read White—people), or conversely as separate from and as a threat to nature, whichever proves most expeditious in the moment for advancing the settler-colonialist project. As will become clear, eugenics, xenophobia, and environmentalism have long been bedfellows.

To bolster the "closer to nature" argument, nonsensical theories have been invented that posit that, for example, Black Africans are evolutionarily closer to primates than are Whites.[13] In 1906, Ota Benga, a Mbuti man from the Congo, was kept in an exhibit of apes at the Bronx Zoo. Similar exhibits were popular among White visitors at fairs and exhibitions across the United States and Europe. As Phillip Atiba Goff and Jennifer L. Eberhardt demonstrate, the pernicious and often unconscious tendency to dehumanize Blacks as simian persists. Activation of this frame makes subjects in psychological studies more likely to perceive violence against African Americans as justified.[14]

Indigenous people are likewise disparaged as closer to nature (uncivilized) or as a threat to nature—in some instances, they manage to be both at the same time. John Muir described the Mono Indians as "most ugly, and some of them altogether hideous . . . they seemed to have no right place in the landscape." This view, of course, led to their violent dispossession and the creation of Yosemite National Park. To Muir, "most Indians I have known are not a whit more natural in their lives than we civilized whites," yet in the next breath: "As to Indians, most of them are dead or civilized into useless innocence."[15] One of Muir's primary complaints about both Indians and Blacks was that he appraised them as "dirty," ignoring the societal conditions which left their communities in "environmentally compromised spaces."[16] As has been well established, racism is closely linked to beliefs about purity and cleanliness.

Fortress conservation is a practice that derives from a belief that humans are not part of nature and that they represent a threat to the integrity or purity of the ecosystem. In order to "protect" nature, humans must be excluded from ecologically sensitive areas. In practice, this only applies to certain humans, and for the profit of others. As Kenyan carnivore ecologist Mordecai Ogada points out, environmentalists who position themselves on the political left in the United States find themselves aligned with the far right once they step off the plane in Kenya. Armed guards with little regard for human rights or for the law are assigned by the government to patrol "sensitive" lands to prevent the traditional inhabitants from gathering food, water, or firewood. Meanwhile, elites from abroad purchase permission to trophy hunt in these very same areas. In the name of conservation, of course.[17]

In the United States, it is those crossing from the southern border who are deemed a threat to Nature and the "American way of life." No Más Muertos / No More Deaths volunteers leave plastic jugs of drinking water in the desert so that people won't die of thirst. The Border Patrol reports that every day for the last twenty-two years, there has been at least one death a day. Over 8,000 deaths have been tallied in official counts so far, and there are likely many bodies that are never found.[18] A website has been set up to map where remains have been found, with as much identifying information as can be gathered, to help grieving families of the missing find closure.[19] It should be noted that the United States government, under the (Democratic) Clinton administration, engineered a plan to intentionally shift undocumented migrants away from urban crossing points into the desert, where the countless deaths were meant to serve as a deterrent to those contemplating undertaking the journey. What they did not understand, if I am being charitable, was that many migrants' chances of survival would not actually improve if they were to stay put. In the midst of this humanitarian crisis, volunteers who patrol the desert to offer water, food, and emergency medical care are prosecuted. The crime?

> The Defendants did not get an access permit, they did not remain on the designated roads, and they left water, food, and crates in the Refuge. All of this, in addition to violating the law, erodes the national decision to maintain the Refuge in its pristine nature.[20]

Madison Grant, an influential conservationist, was motivated by his love for "nature" (he and I would define nature differently) to limit the number of people in the US of non-Nordic ancestry. Toward this end, he cofounded the American Eugenics Society, served as president of the Eugenics Research Association and vice president of the Citizens' Committee on Immigration Legislation, and wrote a book Adolf Hitler referred to as "his Bible," *The Passing of the Great Race*.[21] Many leaders of the Sierra Club, including John Tanton, Joseph LeConte, and David Starr Jordan, openly tied their support for eugenics to both White nationalism and environmentalism.[22] The modern-day confluence of White nationalism and environmentalism shows up in places like Aspen, Colorado, where the City Council passed a resolution calling for federal action to restrict immigration to the United States in 1999.[23]

In "The Tragedy of the Commons," widely regarded as an indispensable part of the environmental canon, Garrett Hardin explains why society should privatize all of its natural resources: "Under a system of private property, the men who own property recognize their responsibility to care for it, for if they don't they will eventually suffer. A farmer, for instance, will allow no more cattle in a pasture than its carrying capacity justifies. If he overloads it, erosion sets in, weeds take over, and he loses the use of the pasture."[24] Hardin ignores that societies have thrived under collective property regimes, in part due to a sense of responsibility to future generations. Thus, it's no surprise that, according to the Southern Poverty Law Center, Garrett Hardin is "one of the intellectual pillars of modern scientific racism and white separatism."[25]

Hardin expands on this thinking with an analogy of society as an overcrowded lifeboat that can ill afford more passengers. In a 1974 essay entitled "Lifeboat Ethics: The Case Against Helping the Poor," he sets out his case that democracy cannot be trusted to ensure the survival of society. He calls for "a true world government to control reproduction and the use of available resources." The resources he worries about are not merely food, which he allows could, "perhaps, be significantly increased to meet a growing demand." He also worries about having access to pristine wilderness: "But what about clean beaches, unspoiled forests, and solitude?" In order to preserve his access to "unspoiled" nature, he contends that people living in poverty should be allowed to starve: "every Indian life saved through medical or nutritional assistance from abroad diminishes the quality of life for those who remain, and for subsequent generations."[26]

Pentti Linkola, a fisherman from Finland, moves Hardin's lifeboat analogy one step closer to genocide:

> What to do, when a ship carrying a hundred passengers suddenly capsizes and there is only one lifeboat? When the lifeboat is full, those who hate life will try to load it with more people and sink the lot. *Those who love and respect life will take the ship's axe and sever the extra hands that cling to the sides.*[27]

Today, we see his lifeboat analogy made real, as Western European nations turn away ships of migrants from Africa and the Middle East from their ports and refuse to participate in search-and-rescue missions, as a way of keeping those who are literally on lifeboats from reaching their shores. They express concern about the potential "drain" on resources if they were to allow these migrants in, using financial/resource concerns as a cover for fascist and nationalist ideologies.

Given the number of people expected to be displaced by climate change, Hardin's sentiments are shared by many who currently hold positions of power:

> By 2070, the kind of extremely hot zones, like in the Sahara, that now cover less than 1 percent of the earth's land surface could cover nearly a fifth of the land, potentially placing one of every three people alive outside the climate niche where humans have thrived for thousands of years. Many will dig in, suffering through heat, hunger and political chaos, but others will be forced to move on. A 2017 study in *Science Advances* found that by 2100, temperatures could rise to the point that just going outside for a few hours in some places, including parts of India and Eastern China, "will result in death even for the fittest of humans."[28]

Those in power are reluctant to accept climate refugees, even though they may themselves be in that position in the future. The United Nations Human Rights Committee ruled in January 2020 that governments cannot return people to their countries if their lives are threatened due to climate change.[29] This ruling is an effort to force governments away from their fascist, nationalist, and social Darwinist ideologies. They can no longer deport the impoverished to deal with the consequences of the climate change caused by fossil fuel emissions that did not originate in those countries.

Anti-toxic activism can veer to NIMBY ("not in my backyard") or NIABY ("not in anybody's backyard") values. However, (White) NIMBY activists have led to socially vulnerable communities facing a disproportionate burden of toxins due to their desire to keep such toxins and hazards away from themselves. The environmental justice movement draws our attention to this in the US and globally, and to the divergent political responses that complaints receive.

Since the wealthiest 1 percent of the world's population (those earning more than $100,000 USD annually) produce more than twice as much carbon dioxide as the poorer half of the world (according to an Oxfam & Stockholm Environment Institute report that looked at the period from 1990 to 2015),[30] would not the more sensible solution be to limit the number of super wealthy people?

The politics of plastic pollution

Just as they are scapegoated for environmental degradation overall, BIPOC people are singled out as scapegoats for plastic pollution. The American Council on Science and Health reassures US readers, "You aren't the problem. Asia is."

Consumers in the sachet economy (single-use packets typically made of layers of unrecyclable foil, paper, and plastic that contain a serving-size portion of essential goods like soap powder, marketed to customers for whom an entire bottle of the product would be unaffordable) are another frequent target. Some call for more sustainable packaging of the sachets, when the proper question is why so many people are living in poverty in the first place. Could their poverty be causally linked to the bloated profits of the fossil fuel companies that exploit their desperation?

Many of the chemicals in plastics disrupt reproduction. An "individual responsibility" approach to recognizing and avoiding these perils will result in impaired fertility among people with the least resources and the fewest choices. This is passive eugenics, and it is wrong. Using the climate emergency as justification for population control and using the insidious poisoning of some bodies via plastics is a twenty-first-century twist on social Darwinism.

The ecosystem may adapt and thrive without us, but the same cannot be said of us. Human health is absolutely dependent on a healthy ecosystem. No amount of money will replace potable water, breathable air, and nourishing food. These are our commons, and to privatize them so as to deny them to that majority of humanity without the ability to pay is unjustifiable.

Plastics and COVID-19

Despite all evidence to the contrary, plastics make people *feel* safer by conferring the ability to be impermeable. Ironically, having contact with diverse microorganisms actually strengthens the immune system by creating a more diverse microbiome, thus boosting our resilience to novel pathogens like COVID-19. Yet, like my neighbor, many are seeking out plastics that present the illusion that a product is *untouched* in order to preserve their separateness.

To make the connection more explicit, COVID has intensified, for many, a fear of contact with the stranger and/or the foreign, and the plastics industry is making the most of this marketing opportunity. Even dine-in restaurants switched to disposable plastic cutlery and glassware—not because sending stainless steel or glass through a dishwasher and sanitizer would pose any risk, but to soothe their misled customers. People felt an aversion to putting something in their mouth that had once been in the mouth of a stranger. No amount of scientific data can assuage this type of fear, because ultimately it is not rational. As Adler-Bell puts it, "The common theme [in eco-fascism] is this link between a yearning for purity in the environmental sphere and a desire for racialized purity in the social sphere."[31]

COVID-19 has also activated what George Lakoff would call frames (central metaphors that help us make sense of the world)[32] about the dirty and dangerous stranger. Certainly, disinfecting our hands and the surfaces we come into contact with is an effective tool that can move us safely through the pandemic. However, disinfection can quickly move beyond a proportional and rational response to the virus. Both historical and contemporary examples demonstrate the risks we are willing to impose on others who are deemed dirty.

In the early twentieth century, people crossing the border into the United States from Mexico were routinely disinfected by customs officials by being made to bathe in gasoline. Their clothing was fumigated with highly toxic pesticides including Zyklon B, which was later used in Nazi gas chambers.[33] In 1917, this treatment led Carmelita Torres, a young maid who endured it each morning on her way to work, to incite the Bath Riots.

More recently, Haitian refugees at the Krome Detention Center North allege that an insecticide called R&C Spray, which is only authorized for use on objects, was applied directly to their bodies. A subsequent outbreak of gynecomastia (growth of breast tissue in men) was attributed to a hormone imbalance resulting from exposure to the active ingredient in the spray, phenothrin.[34]

Today, workers in hazmat suits fumigate and disinfect public spaces with toxic chemicals, including quaternary ammonium cations (quats), which are likely to cause more harm than any fomites would. The fact that the virus is actually spread through airborne aerosols does not deter this latest iteration of hygiene theater. Acknowledging that plastic can only do so much (little) to protect us from the virus would remove this seeming line of defense, leaving people to deal with the harsh reality that, despite industry promises, plastics do not make everything possible.

Conclusion

COVID-19 has been used by the fossil fuel industry to oppose efforts to reduce or eliminate single-use plastics. The virus supersedes environmental concerns because, rather than making plastic use about protecting the planet, the discourse focuses on the importance of protecting oneself from harm first and foremost. Plastics' promotors take advantage of preexisting eco-fascist ideas and people's desire to keep their figurative (but not social) distance from others by proposing plastics as the solution to every conceivable problem presented by COVID-19. In the process, disadvantaged and BIPOC communities are placed at greater risk: of contracting the virus through exposure while doing their 'essential' work; of increased mortality if they do contract COVID-19; and of dealing with the accumulated plastic waste once it is disposed of, given the location of landfills/waste sites.

People's desire to distance themselves from others physically, socially, and emotionally means these risks go unseen, with the focus instead shifting to questions about how to return to "normal." This "normal" is one where the lives of BIPOC communities and ecosystems have long been taken for granted and not something to which any of us should seek to return. Rather, we must envision and achieve a post-COVID society which draws on our interconnectedness, abolishing the false divisions between self, nature, and other that plastics, eugenics, and racist ideologies have perpetuated for several decades.

The enemy is not plastics. It is a culture so impoverished in imagination that nature and humans appear to be separate entities in competition for survival. The way that we tell stories matters. When we craft a narrative in which economically marginalized people are the villains responsible for not "managing" their plastic waste as conscientiously as we in the United States claim to manage ours, we excuse the economically over-privileged people who profit from the fracking and cracking and production of these wasteful products. When we do this, we buttress disingenuous arguments that in order to protect

Nature, we must police and relocate Indigenous people from their ancestral homelands. We begin to think it plausible that there should be fewer of "them." Whether or not we intend it, we end up on the slippery slope to eco-fascism.

Notes

1. As co-authors, we have elected to use the first person singular to refer to ourselves and our experiences throughout this chapter.
2. Antonio Ragusa et al., "Plasticenta: First Evidence of Microplastics in Human Placenta," *Environment International* 146 (2021): 106274, https://doi.org/10.1016/j.envint.2020.106274.
3. Sasha Adkins, *From Disposable Culture to Disposable People: The Unintended Consequences of Plastics* (Resource Publications, 2018); Sasha Adkins, "Plastic and the State of Our Souls," *Sojourners* (2020).
4. Ian Urbana, "As OSHA Emphasizes Safety, Long-Term Health Risks Fester," *New York Times*, March 31, 2013, https://www.nytimes.com /2013/03/31/us/osha-emphasizes-safety-health-risks-fester.html.
5. Joana C. Prata et al., "COVID-19 Pandemic Repercussions on the Use and Management of Plastics," *Environmental Science & Technology* 54, no. 13 (2020): 7760–5. https://doi.org/10.1021/acs.est.0c02178.
6. Shannon Mattern, "Purity and Security: Towards a Cultural History of Plexiglass," *PLACES Journal* (2020), https://placesjournal.org/article/purity-and-security-a-cultural-history-of-plexiglass/.
7. Ibid.
8. Dorceta E. Taylor, *Race, Class, Gender, and American Environmentalism*, vol. 534 (US Department of Agriculture, Forest Service, Pacific Northwest Research Station, 2002).
9. Sylvia Hood Washington, "An Archaeology of the Modern Environmental Justice Movement," in *Packing Them In: An Archaeology of Environmental Racism in Chicago, 1865–1954* (Lanham, MD: Lexington Books, 2005).
10. "SURJ Values," *Showing Up for Racial Justice—SURJ*, https://www.showingupforracialjustice.org/surj-values.html.
11. Attributing Words, January 1, 1970, https://unnecessaryevils.blogspot .com/2008/11/attributing-words.html.
12. Nils Gilman, "Beware the Rise of Far-Right Environmentalism," *The World Post*, October 17, 2019, https://www.berggruen.org/the-world post/articles/beware-the-rise-of-far-right-environmentalism/.
13. "As OSHA Emphasizes Safety, Long-Term Health Risks Fester," *New York Times*, March 31, 2013, https://www.nytimes.com/2013/03/31/us/osha-emphasizes-safety-health-risks-fester.html.
14. Phillip Atiba Goff and Jennifer L. Eberhardt, "Race and the Ape Image," *Los Angeles Times*, February 28, 2009, https://www.latimes.com/la-oe -goff28-2009feb28-story.html.
15. Caitlin Schneider, "No Heroes: The Racist Legacy of John Muir and

American Conservation," *Discourse Blog*, July 10, 2020, https://discour seblog.substack.com/p/no-heroes-the-racist-legacy-of-john.

16. Washington, "An Archaeology of the Modern Environmental Justice Movement," 49.

17. John Mbaria and Mordecai Ogada, *The Big Conservation Lie: The Untold Story of Wildlife Conservation in Kenya* (Auburn, WA: Lens & Pens Publishing LLC, 2017).

18. James Verini, "How US Policy Turned the Sonoran Desert into a Graveyard for Migrants," *New York Times*, August 18, 2020, https://www.nytimes.com/2020/08/18/magazine/border-crossing.html.

19. "Migrant Death Mapping," *Humane Borders*, https://humaneborders.org/migrant-death-mapping/.

20. US Magistrate Judge Bernardo Velasco, qtd. in Rafael Carranza, "Aid Volunteers Found Guilty of Dropping off Water, Food for Migrants in Protected Part of Arizona Desert," *The Republic*, January 18, 2019, https://www.azcentral.com/story/news/2019/01/18/no-more-deaths-vo lunteers-found-guilty-dropping-water-food-migrants-cabeza-prieta-refu ge-arizona/2617961002/.

21. Jonathan Peter Spiro, *Defending the Master Race: Conservation, Eugenics, and the Legacy of Madison Grant* (Burlington: University of Vermont Press, 2009).

22. Susie Cagle, "'Bees, not refugees': The Environmentalist Roots of Anti-Immigrant Bigotry," *Guardian*, August 16, 2019, https://www.theguardi an.com/environment/2019/aug/15/anti.

23. Lisa Sun-Hee Park and David Naguib Pellow, *The Slums of Aspen: Immigrants vs. the Environment in America's Eden* (New York: New York University Press, 2011).

24. Garrett Hardin, "The Tragedy of the Commons," *Science* 162, no. 3859 (1968): 1243–8, https://doi.org/10.1126/science.162.3859.1243.

25. "Garrett Hardin," *Southern Poverty Law Center*, n.d., https://www.splc enter.org/fighting-hate/extremist-files/individual/garrett-hardin.

26. Garrett Hardin, "Lifeboat Ethics: The Case Against Helping the Poor," *Psychology Today*, September 1974, https://www.garretthardinsociety .org/articles/art_lifeboat_ethics_case_against_helping_poor.html.

27. Pentti Linkola, "The Doctrine of Survival and Doctor Ethics," *Could Life Win—and on What Conditions?*, http://www.penttilinkola.com/pe ntti_linkola/ecofascism_writings/translations/voisikoelamavoittaa_trans lation/VI%20-%20The%20World%20And%20We/; emphasis added.

28. Abrahm Lustgarten, "The Great Climate Migration," *New York Times Magazine*, July 23, 2020, https://www.nytimes.com/interactive/2020/07 /23/magazine/climate-migration.html.

29. Yvonne Su, "UN Ruling Could Be a Game-Changer for Climate Refugees and Climate Action," *The Conversation*, January 28, 2020, https://the conversation.com/un-ruling-could-be-a-game-changer-for-climate-refug ees-and-climate-action-130532.

30. Fiona Harvey, "World's Richest 1% Cause Double CO_2 Emissions of

Poorest 50%, Says Oxfam," *Guardian*, September 20, 2020, https://www.theguardian.com/environment/2020/sep/21/worlds-richest-1-cause-double-co2-emissions-of-poorest-50-says-oxfam.

31. Sam Adler-Bell, "Why White Supremacists Are Hooked on Green Living," *New Republic*, September 24, 2019, https://newrepublic.com/article/154971/rise-ecofascism-history-white-nationalism-environmental-preservation-immigration.

32. George Lakoff, "Why It Matters How We Frame the Environment," *Environmental Communication* 4, no. 1 (2010): 70–81, https://doi.org/10.1080/17524030903529749.

33. John Burnett, "The Bath Riots: Indignity along the Mexican Border" *National Public Radio*, January 28, 2006, https://www.npr.org/templates/story/story.php?storyId=5176177.

34. Steven A. Brody and D. Lynn Loriaux, "Epidemic of Gynecomastia among Haitian Refugees: Exposure to an Environmental Antiandrogen," *Endocrine Practice* 9, no. 5 (2003): 370–5, https://doi.org/10.4158/ep.9.5.370; "Haitians Suing US over Treatment in Detention," *New York Times*, October 8, 1987, https://www.nytimes.com/1987/10/08/us/haitians-suing-us-over-treatment-in-detention.html.

17

Plastic in the Time of Impasse

Mark Simpson

> Plastic cracks time.
>
> —Heather Davis

Had you happened to spend some time in early 2021 on the American Chemistry Council website, you might have come across "Lifecycle of a Plastic Product," a contemporary iteration (since removed) of that curious sort of biography known to literary historians of the eighteenth century as the *it narrative*.[1] Granted, the text presented under this heading did test the sense and strain the credibility of "lifecycle" as concept and promise: what you would have encountered was less an integral story of plastic's key moments, whether in life or in cycle, and more a congeries of disparate, discontinuous plastic factoids starting with "Overview," meandering through process and type, then landing, unceremoniously, at "End Life" and a time-saving flurry of bullet points. Midway through this haphazard congeries, however, came a section entitled "Plastic Uses" that, while offering no real purchase on the matter of "lifecycle," did manage in its opening paragraphs to throw into relief some telling symptoms of what one could call *the plastic condition*. "Whether you are aware of it or not," observed the American Chemistry Council's anonymous polymer biographer, "plastics play an important part in your life."

> Plastics' versatility allow [*sic*] them to be used in everything from car parts to doll parts, from soft drink bottles to the refrigerators they are stored in. From the car you drive to work in to the television you watch at home, plastics help make your life easier and better. So how is it that plastics have become so widely used? How did plastics become the material of choice for so many varied applications?
>
> The simple answer is that plastics can provide the things consumers want and need at economical costs. Plastics have the unique capability to be manufactured to meet very specific functional needs for consumers. So maybe there's another question that's relevant: What do I want? Regardless

of how you answer this question, plastics can probably satisfy your needs.

If a product is made of plastic, there's a reason. And chances are the reason has everything to do with helping you, the consumer, get what you want: Health. Safety. Performance. And Value. Plastics Make It Possible.®

"From car parts to doll parts": here, "the simple answer" regarding plastic as endlessly versatile helpmeet puts consumer desire at the center of things. And with good reason: consumption, as logic and lure, motivates the normative understanding—the common-sense—of plastic in modern life. Plastic exists, in this familiar story, because humans consume, the fungibility of its substance arriving as a welcome gift to meet and to solve the insatiable omnivoracity of human hungers.[2]

Animated by the imperative to consume, plasticity thus imagined unfolds through contradiction. Plastic alone can meet "very specific functional needs"—which is to say, the needs for ubiquity and capaciousness without regard. In this reckoning, the *whateverness* of plastic indexes its *uniqueness*. Plastic is unlike anything else because it alone can become anything at all. No wonder that by the terms of such a promise consumer "want" and "need" prove interchangeable. The plastic condition is effectively a plastic alchemy, with the power to transmute consumable objects and consuming subjects alike.

As distilled and promoted by the American Chemistry Council in this once-present, now-absent website tableau, the plastic condition issues from and depends upon the fossil-fueled energy regime at the heart of what Stephanie LeMenager calls *petromodernity*.[3] It constitutes a signal feature of petroculture in the contemporary moment—and, as such, holds decisive consequence for the ever-intensifying problem of energy impasse today. Inspired and provoked by recent critical theory on the plastic condition—especially work by Catherine Malabou and Alberto Toscano on plasticity as concept and by Amanda Boetzkes and Heather Davis on plastic as oil-born calamity—the present essay will elaborate this claim, arguing that plastic, as petromodernity's ubiquitous offspring, epitomizes while materializing energy impasse. In so doing, it will attend more particularly to *plastic time*: to the time signature and temporal imaginary instituted and normalized, in petroculture, through plastic. The time of plastic operates according to a contradictory rhythm that connects disposability with saturation and impermanence with perpetuity while seeding malleability or fungibility as the determining condition and abiding allure of objects and subjects alike. How might a reckoning with this plastic temporality illuminate the urgent problem of energy impasse today—and how might it open up some position or perspective against or beyond such impasse?

In the mire

To venture an analysis of the issues outlined above will require, by way of preface, some account of the meaning and significance of impasse as that concept bears on matters of energy. In the course of theorizing precarity in *Cruel Optimism*, Lauren Berlant offers a provocative meditation on impasse. Her observations are worth quoting at length:

> The impasse is a space of time lived without a narrative genre. Adaptation to it usually involves a gesture or undramatic action that points to and revises an unresolved situation. One takes a *pass* to avoid something or to get somewhere: it's a formal figure of transit. But the impasse is a *cul-de-sac*—indeed, the word *impasse* was invented to replace *cul-de-sac*, with its untoward implications in French. In a cul-de-sac one keeps moving, but one moves paradoxically, in the *same space*. An impasse is a holding station that doesn't hold securely but opens out into anxiety, that dog-paddling around a space whose contours remain obscure. An impasse is decompositional—in the unbound temporality of the stretch of time, it marks a delay that demands activity. The activity can produce impacts and events, but one does not know where they are leading. That delay enables us to develop gestures of composure, of mannerly transaction, of being-with in the world as well as of rejection, refusal, detachment, psychosis, and all kinds of radical negation. . . .
>
> Whatever else it is, and however one enters it, the historical present—as an impasse, a thick moment of ongoingness, a situation that can absorb many genres without having one itself—is a middle without boundaries, edges, a shape. It is experienced in transitions and transactions. It is the name for the space where the urgencies of livelihood are worked out all over again, without assurances of futurity, but nevertheless proceeding via durable norms of adaptation.[4]

Berlant's suggestive account of impasse contains numerous telling details relevant for the question of energy: transit; paradoxical movement; decomposition; ongoingness; the edgeless, shapeless middle. Most immediately pertinent in framing the argument here, however, is the prospect of transition raised by Berlant and the relation of that prospect to impasse. In Berlant's version, the "historical present" is "an impasse . . . experienced in transitions and transactions": an understanding that manages to fuse together impasse and transition (along with transaction) as conjoined symptoms of the now. And although Berlant is not contemplating petroculture when she advances this understanding, her argument distils the contradictory conjuncture of transition, transaction, and impasse decisive for yet occluded by capital's energy regime.

Let me unpack this claim. As I have argued elsewhere, a discourse of *energy transition* has become axiomatic in the present even as—or indeed to the extent that—*energy impasse* defines and determines the contemporary situation.[5] In normative discourse, energy is now transitional, which is to say that it names as it signifies a property and a prospect in motion or flux. For governments and extra-governmental organizations, banks and stock markets, media outlets and energy companies, the concept and prospect of transition sets the meaning—not to mention the transactional value—for energy today, serving to frame and delimit any envisioned shift from fossil fuels to other sources of energic power.[6] Transition's allure as an idea and idiom has everything to do with its capacity to intimate ordinariness alongside orderliness and clarity alongside certitude. Transitional change, as a kind of operational common sense, thus signals smooth change above all—precisely in order to obviate and occlude the turbulence of even partial departures from the fossil regime. As Christophe Bonneuil and Jean-Baptiste Fressoz argue in *The Shock of the Anthropocene*, "[t]o say 'transition' rather than 'crisis' [makes] the future less generative of anxiety, by attaching it to a planning and managerial rationality."[7] Thus configured, transition proves intimate with what elsewhere I have called "*lubricity*, the texture and mood requisite to neoliberal petroculture," which delivers "smoothness as cultural common sense."[8] At stake is nothing less than the dreamworld of capital, within which transitional (and transactional) shifts in energy source can protect the ruling order by ensuring and propelling its endless, relentless unfolding.

Bonneuil and Fressoz make vivid the devastating weakness that hobbles such transition-logic: it holds no purchase whatsoever on historical reality. "The bad news," they report,

> is that, if history teaches us one thing, it is that there never has been an energy transition. There was not a movement from wood to coal, then from coal to oil, then from oil to nuclear. The history of energy is not one of transitions, but rather of successive *additions* of new sources of primary energy.[9]

There has never been an energy transition—which means that any promise offered by the transitional idiom is fundamentally empty. Historically speaking, forms of energy never really disappear: they only and always *accrete*. Hence the actuality of energy impasse today: the accretive propensity ensures an interminable piling up of energy forms that, sedimenting dependence on fossil fuels, renders any desire for change functionally irrelevant.

So diagnosed, the problem of energy impasse is not simply a problem of blockage, no matter how tempting such an explanation might seem. To see impasse as a series of obstacles—malign state actors, greedy corporations, municipalities in denial, ignorant publics, and so on—standing in the way of a sustainable because renewable energy future is to misunderstand the dynamics of impasse. Which is not to say that a Jair Bolsonaro or a Royal Dutch Shell do not in some sense impede movement away from petroculture, since of course they do. But a blockage model would make impasse an impediment to overcome: troublesome, yet only temporarily so—and merely incidental to the current energic situation. In so doing, such a model would arguably compound impasse, precisely by misrecognizing its character and minimizing its consequence.

Against the blockage model, Imre Szeman and I have ventured to theorize impasse in terms of what we call *stuckness*:

> the texture or atmosphere setting the conditions of possibility for a given situation that, irrespective of any overcoming of actually-existing blockages, manages nevertheless to perpetuate the situation as it is. Impasse in this sense names a continuation of the same wherein the overcoming of blockages cannot solve—and may in fact compound—the abiding stuckness.[10]

To the extent, we propose, that the grammar of energy transition today works "less to provide viable means for a better future than to indicate our constitutional inability to imagine transformation itself," then transition as promise and logic effectively ensures "the conditions of our stuckness"—"[w]hich is to say that existing genres of energy transition are all too often forms of impasse."[11] Or, to recall Berlant's trenchant formulation: "the historical present" is "an impasse . . . experienced in transitions and transactions."[12]

Such is the problem of impasse to which *the plastic condition* corresponds and contributes. For plastic is manifestly not just an obstacle we can bypass on the way out of petroculture. Plastic mires us. We are stuck with and in the multifarious synthetic polymers that now set the very terms of our being and doing. And that impassable fix, as I hope to show, holds unnerving implications for the temporalities we have come to occupy—and need to endure.

(Im)Possibilities

In light of the foregoing account of impasse, I want to return to the American Chemistry Council's now-vanished "Lifecycle of a Plastic Product" so as to consider a narrative detail deliberately bracketed until now: the tagline phrase Plastics Make It Possible®. A grammarian might lament the pronoun ambiguity, demand the antecedent— "what I want"? "health"? "safety"? "value"?—and thereby miss the point entirely. Referential promiscuity is exactly the aim, since in the ACC's remaindered account the whateverness of plastic entails and so constitutes its promise and possibility. Grammatical rules notwithstanding, here the obscurity of pronoun reference accurately conveys plasticity's infinitude.

All the same: how, back in 2021, did the ACC expect its readers to define and delimit the possibilities of plastic? What exactly might this "it" have served to conjure?

Was the ACC imagining weather, for example, as an instance of plastic possibility? In 2019, the US Geological Survey published a study bluntly entitled "It is Raining Plastic."[13] Could one read a trace of exasperated insistence in this declarative, as though the study's authors were determined to persevere in making their case despite and against some prior incredulity, some absent rebuke from government ministry or funding council to the effect that *plastic rain is simply impossible*? While collecting "[a]tmospheric wet deposition samples"—that is, rain—at eight locales in Colorado's Front Range, the authors discovered plastic in more than 90 percent of the samples, which to them "suggests that wet deposition of plastic is ubiquitous and not just an urban condition."[14] Noting that since their "study was not designed for collecting and analyzing samples for plastic particles" so "[t]he results are unanticipated and opportune" (a phrasing that, for some, will test unto breaking the sense of that last term), the authors go on to deliver a conclusion as inauspicious as unnerving: "It is raining plastic. Better methods for sampling, identification, and quantification of plastic deposition along with assessment of potential ecological effects are needed."[15] Plastic rain is ubiquitous, we might infer, because plasticity is fully global—yet any "potential ecological effects" remain unclear and opaque, obscured by impoverished methodology. Meanwhile, the rhetorical circuit beginning titularly and ending conclusively with the same blunt declarative—*it is raining plastic*—will mimic the feedback loop that endlessly churns plastics from ocean into atmosphere into water table and back again.

No plastic, no plastic weather: plastics make it possible.

How about plastiglomerates? Did the ACC craft its tagline with them in mind? "Plastiglomerate" as term and concept was devised in 2012 by geologist Patricia Corcoran, artist Kelly Jazvac, and oceanographer Charles Moore to name a substance first encountered some six years prior by Moore on Kamilo Beach in Hawai'i. Corcoran, Jazvac, and Moore report that they

use the term plastiglomerate to describe an indurated, multi-composite material made hard by agglutination of rock and molten plastic. This material is subdivided into an in situ type, in which plastic is adhered to rock outcrops, and a clastic type, in which combinations of basalt, coral, shells, and local woody debris are cemented with grains of sand in a plastic matrix.[16]

In the course of investigating such material, Corcoran, Jazvac, and Moore determined that the "plastiglomerate fragments were formed anthropogenically"—as a result of humans "[b]urning plastic debris."[17] This determination leads them to conclude that "Kamilo Beach provides an example of an anthropogenic action (burning) reacting to an anthropogenic problem (plastics pollution), resulting in a distinct marker horizon of the informal Anthropocene epoch."[18] For Amanda Boetzkes, such "[p]lastiglomerates signal the extension of the oil industry into an ecosystemic predicament that demands a perspective of the future on a geological scale. . . . In their implicit geological scope, they disclose the human world as its own end."[19] Musing on the distinctive emergence of these novel formations, Kirsty Robertson hones in on the contradictions at issue in such endless ending:

> Plastiglomerate clearly demonstrates the permanence of the disposable. It is evidence of death that cannot decay, or that decays so slowly as to have removed itself from a natural lifecycle. It is akin to a remnant, a relic, though one imbued with very little affect. As a charismatic object, it is a useful metaphor, poetic and aesthetic—a way through which science and culture can be brought together to demonstrate human impact on the land. Thus, to understand plastiglomerate as a geological marker is to see it as unchanging. Plastiglomerate speaks to the obduracy of colonialism and capitalism. The melted veins of plastic that actually become the rock speak to how difficult it is to undo unequal relations of destruction.[20]

The permanence of the disposable: the persistent and sedimentary effect of destructive asymmetries as underscored by Robertson will suggest that the plastiglomerate-as-remnant is likewise a revenant.

No plastic, no plastiglomerate: and plastics make it possible.

To venture one more: what about infrastructure—did the ACC anticipate that plasticity might enable this ever-extending global carapace? In a 2019 article entitled "The Plastic Pipeline," Beth Gardiner documents a precipitous increase in plastic-manufacturing infrastructure: "while individuals fret over images of oceanic garbage gyres, the fossil fuel and petrochemical industries are pouring billions of dollars into new plants intended to make millions more tons of plastic than they now pump out."[21] The aim is to solve a curious petrocultural problem (one only exacerbated under COVID capitalism): the looming surplus of petrochemical feedstock.

Companies like ExxonMobil, Shell, and Saudi Aramco are ramping up output of plastic—which is made from oil and gas, and their byproducts—to hedge against the possibility that a serious global response to climate change might reduce demand for their fuels ... Petrochemicals, the category that includes plastic, now account for 14 percent of oil use, and are expected to drive half of oil demand growth between now and 2050. ... The World Economic Forum predicts plastic production will double in the next 20 years.[22]

At stake is a type of path dependency: a binding commitment to plastic manufacture and use, manifest in massive infrastructure that, in the words of former EPA official Judith Enck (as quoted by Gardiner), is effectively "'locking us into a plastic future'" all but impossible to reverse or undo.[23] That outcome will materialize what Hannah Appel, in her trenchant account of "Infrastructural Time," hauntingly calls "a particular kind of infrastructural futurity that is more akin to deferral."[24]

Without plastic, no plastic infrastructure: so plastics make ... Repeat.

From equivalence ...

The foregoing effort to identify potential referents for the *it* in the ACC's hauntingly spectral tagline will not only capture some less than welcome versions of plastic possibility but also begin to texture what Berlant calls the "thick moment of ongoingness" that, I am arguing, can characterize plastic impasse—the stuckness of plasticity.[25] Is it perverse, now, to veer away from the specific back toward the interchangeable? Perhaps so—but in any case, before leaving the ACC's tagline behind, I want to probe a little further the bothersome *it* in all its referential imprecision and promiscuity.

The purposeful withholding of reference here serves, as I have suggested, to signify the endless promise of plastic. Adaptable yet opaque, the *it* in ACC's 2021 tagline materializes grammatically the interchangeable, substitutable profligacy that plastic affords and encourages. This *it* is generic, blank—such that the tagline might as well read *Plastics make X possible. It* performs in language what plastic entails in life: a becoming-anything—which affords the determination of everything. As channeled by *it*, plastic is antithetical to foreclosure, given to an openness that indicates totality: the limitlessness of plastic possibility; its all-encompassing, all-subsuming power alongside its optative mood.

Monitoring the open *it* at the heart of ACC's redacted tagline is the ® at its end. The registered trademark symbol provides a graphic legal check on the reproducibility and circulability of this phrase as branded commercial property. Even as plastic possibility is limitless, we must infer, it can only become so through regulation. The *it* and the ® mark the contradictory, constitutive scripts of plastic license: a generic openness without reserve; a generic closure through tight restriction and restraint. The antinomy underwrites plasticity's generative power for petroculture: to materialize and modulate fixity alongside fungibility as coeval modes through which to set the given present as the perpetual future.

Such power, of course, comes at no small cost. Writing about Georges Bataille's theorization of excess, Amanda Boetzkes outlines the dangerous potential bound within restriction-as-license:

> While [bourgeois capitalism] still produces surplus energy, the rule of profit demands a continual rerouting of wealth into its system, while prohibiting any burn-off. But restricted economies inevitably defect to destructive forms of expenditure ... in a restricted economy, surplus energy is accumulated, one might even say, recycled into the system. But because of its irrepressible heterogeneous nature, it inevitably discharges in unexpected and highly destructive ways.[26]

Symptomatic of the dynamics of accumulation and recycling in a restricted economy as delineated by Boetzkes, the *it* and the ® prove reciprocal. Pronoun and trademark effectively constitute one another as necessary, necessarily countervailing principles of plastic possibility. Their feedback loop animates and amplifies the time of impasse, now and in future.

At stake in the foregoing analysis is an understanding of plastic as not just substance but switchpoint: both medium and mediator. Endlessly adaptable and given to shape-shifting, plastic constitutes a kind of general, generic equivalent—and thereby might cause one to think of that *other* universal equivalent, money. In a memorable parenthetical in the *Grundrisse*, Marx delineates money's fungible, mediating character:

> (The exchange value of a commodity, as a separate form of existence accompanying the commodity itself, is *money*; the form in which all commodities equate, compare, measure themselves; into which all commodities dissolve themselves; that which dissolves itself into all commodities; the universal equivalent.)[27]

Could Marx not as easily have written that *money makes it possible*? Or—in a kind of hallucinatory intuition circa 1858—that the plastic-yet-to-come approximates "a separate form of existence ... into which all commodities dissolve themselves" while likewise "dissolv[ing] itself into all commodities"? Conceivably, of course, such analogy risks what Alberto Toscano calls "the trap of analogical thinking"[28] precisely by overestimating or indeed overselling the similarities at issue: for while money is the abstraction that renders discrete materialities functionally the same or equitable, plastic is the substance that sets and materializes difference in semblance. Money expresses the relation in *value* between the beverage bottle and the carrier bag whereas plastic sets the relation in *matter* between that bottle and that bag. Still, the resonance is telling, not least because it intimates a likeness about likenesses connecting plastic to money: their shared capacity to transmute differences into sameness—and so into transactional equivalence.

The inexact analogy between plastic and money comes to life most obviously in the credit card, that modern instrument colloquially known as *plastic* that drives as it signifies the operations of consumer credit and debt. Joe Deville identifies "the reliable reproducibility of the consumer credit transaction" as "one of plastic's achievements,"[29] underscoring that "the success of plastic is its malleability: its willingness to hold its shape, to live more or less happily alongside a diverse range of other elements and materials, and to help to gather these together into a neat, robust, whole."[30] But in addition to the credit card one could also consider the advent in recent decades of polymer bank notes—an innovation epitomizing by materializing the redundancy of plastic money as both concept and form (see fig. 17.1).

Suggestive, here, is the coincident emergence of such new economic technologies—plastic credit cards from the mid-twentieth century,

Figure 17.1 Canadian polymer currency.

polymer currencies from the 1980s forward—alongside fungible and speculative modes of subjectivity under neoliberalism. As striking are the complementary time signatures on offer, whereby the credit card, with its capaciously abstract promise, seeds the inevitability of an indebted reckoning-to-come, while the polymer bill, with its resilient, forgery-resistant durability, promises a reliably durational certitude. Both temporal frames, it bears emphasizing, serve to counteract decisive risks—default, fraud, forgery—that attend the fungibility and plasticity of money and its currencies under capitalism.

In "Plasticity, Capital, and the Dialectic"—the same essay that reminds us about analogy's trap—Toscano supplies a provocative and compelling theorization of the constitutive malleability of the money-form. "Money," he observes,

> is abstraction made tangible and visible, the representative, equivalent, and medium of a fundamentally impersonal exchange, a relationship without qualities. . . . [M]oney is not just real community; it is also a *sensus communis*. Monetized exchange structures a whole socially transcendental aesthetic, which is not solely a matter of commensurability (and of its dialectical reliance on singularity or the appearance of uniqueness), but also that of a practical arrest of time and evacuation of space.[31]

Such tendencies, in Toscano's reading, mean that "the quantitative indifference of money can unleash the most varied and uncontainable of metamorphoses, making a mockery of stable identities or hallowed oppositions."[32] As Toscano's argument will signal, at stake is nothing less than money's plasticizing power: its contradictory capacity to form by deforming—to set and fix by perpetually liquifying the relations that compose social life as such.

. . . to explosiveness?

Toscano elaborates this reckoning of money as part of an engagement with Catherine Malabou's influential account of (neuro)plasticity with and against capital, one unfolded across a number of recent publications and laid out with elegant clarity in the succinct treatise *What Should We Do with Our Brain?*. There, after noting that "the word plasticity . . . means at once the capacity to receive form . . . and the capacity to give form," Malabou goes on to identify a crucial third sense shadowing these first two:

> But it must be remarked that plasticity is also the capacity to annihilate the very form it is able to receive or create. We should not forget that *plastique*, from which we get the words *plastiquage* and *plastiquer*, is an explosive substance made of nitroglycerine and nitrocellulose, capable of causing

violent explosions. We thus note that plasticity is situated between two extremes: on the one side the sensible image of taking form (sculpture or plastic objects), and on the other side that of the annihilation of all form (explosion).[33]

With respect to cerebral or neuroplasticity in particular, the implications are twofold: that "to talk about the plasticity of the brain means to see in it not only the creator and receiver of form but also an agency of disobedience to every constituted form, a refusal to submit to a model"; and accordingly that "the plasticity of the brain, understood in this sense, corresponds well to the possibility of fashioning by memory, to the capacity to shape a history."[34] Thus understood, plasticity names the determinant ontology of our making yet likewise, in an entirely generative contradiction, the agential possibility of our making-otherwise.

The prevailing ideological script of capitalism today—in its neoliberal, immaterial, or cognitive idiom—works to occlude such understanding, for Malabou, precisely by making plasticity and flexibility synonymous. "[F]lexibility," she contends,

is the ideological avatar of plasticity—at once its mask, its diversion, and its confiscation. We are entirely ignorant of plasticity but not at all of flexibility. In this sense, plasticity appears as the coming consciousness of flexibility. At first glance, the meanings of these two terms are the same. Under the heading "flexibility," the dictionary gives: "firstly, the character of that which is flexible, of that which is easily bent (elasticity, suppleness); secondly, the ability to change with ease in order to adapt oneself to the circumstances." The examples given to illustrate the second meaning are those that everybody knows: "flexibility on the job, of one's schedule (flex time, conversion), flexible factories." The problem is that these significations grasp only one of the semantic registers of plasticity: that of receiving form. To be flexible is to receive a form or impression, to be able to fold oneself, to take the fold, not to give it. To be docile, to not explode. Indeed, what flexibility lacks is the resource of giving form, the power to create, to invent or even to erase an impression, the power to style. Flexibility is plasticity minus its genius.[35]

Malabou's aim, in theorizing plasticity's contradictory power, is to disable this reductive equivalence: "to perturb flexibility" by unleashing the genius of plasticity—its potential not just to receive but also to break imposed form and thus to bring, through explosive fracture, new form, social and political as well as individual, into being.[36] "Between the upsurge and the explosion of form," she argues, "subjectivity issues the plastic challenge."[37] At stake, in such challenge, is for Malabou nothing less than the possibility of what she calls a "biological alter-globalism."[38]

Though Toscano concludes "that [Malabou's] conception of plasticity as a kind of form taking and form leaving that is not infinitely malleable, reversible, renewable can help us resist certain tendencies within contemporary philosophy whose celebration of novelty and change puts them into unwitting resonance with capitalist fantasies of novelty," he remains skeptical about the perfect congruence between neuroplasticity and contemporary capitalism—and therefore about plasticity's political potential.[39] "When Malabou assumes the ontological and social reality of [a] 'mirroring' between brain and capital," he contends,

> she risks, in order to bolster her argument about the political meaning of plasticity, to give excessive credence to one among the many apologetic discourses of capitalism. Managerial discourses on delocalization and (especially) the abolition of hierarchy are in the main thin veneers over practices of labor exploitation which have in many ways grown fiercer since the days of "Fordism." ... Though attention to different historical and political regimes of labor and subjectivity is of paramount significance, the premium on flexibility is not an invention of "post-Fordism," nor is it constitutively linked—despite the incessant temptation to naturalize capitalism that permeates management apologias—to our shifting understanding of the brain. It is a primordial, axiomatic imperative of capitalism, reflected in the metamorphic capacities of money, in the "bad infinite" of capital accumulation and in the transformation of all workers, as Marx notes in the *Grundrisse*, into "virtual paupers" whose "organic presence" is a matter of "indifference" to the system of production and exchange.[40]

The contradiction is striking: capitalism's driving imperative, across its history, constitutes a fixity that requires the ceaseless fungibility and flexibility of modes of value as of exploitation. Hence the importance, for Toscano, of reckoning what he terms "discontinuity" when confronting capital's social and political challenge.[41]

One of the signal passages within capital's durative, continuously discontinuous pursuit of the imperative of flexibility involves the advent of what Andreas Malm memorably terms *fossil capital*: the turn to coal from water as a means of powering millworks in early nineteenth-century Britain. In Malm's account, this shift had nothing, in its moment, to do with energic cost or efficiency. The appeal of coal derived instead from its consequences for *social* power—for the political economy of labor. By replacing flow (water) with stock (coal), factory owners could resituate manufacture in burgeoning urban spaces abundantly populated with surplus labor, thereby intensifying the exploitation of workers in space and time so as to maximize profit. The dynamic at issue, serving flexibility's imperative so well, characterizes the subsequent history of fossil capital, with its accumulating array of fossil fuels—coal joined by kerosene, then petroleum, then diesel, then shale gas—and its relentless pattern of energy deepening. As Malm observes, "[c]apitalist growth ... did not become welded to fossil fuels because it is a linear, neutral, incremental addition of wealth, output or productive forces. ... That growth is a set of relations just as much as a process, whose limitless expansion *advances by ordering humans and the rest of nature in abstract space and time* because that is where most surplus-value can be produced."[42]

When, in *What Should We Do with Our Brain?*, Malabou claims that "[e]nergetic explosion is the idea of nature," we might infer that she does not have this explosive material and social history of fossil-fueled energy in mind.[43] To the extent, though, that energy has exploded over the past two centuries, it has done so as a material, social, and political condition of petroculture—thereby serving *to form*, rather than in any way *deforming*, the structure and system of capital. Relevant here is the proliferating ubiquity of polymer plastics as a signal feature of such fossil-fueled explosion over the last half-century. In spite—or is it precisely because?—of their debased mundanity, these synthetics do haunt, as a kind of irresistible supplement, Malabou's "genius" of plasticity, all the more so as we begin to understand the extent to which they have come to infiltrate our very bodies. As reported in August, 2020, by the *Guardian*, "[m]icroplastic and nanoplastic particles are now discoverable in human organs," effectively rendering *the human* a plastiglomerate species.[44] If, as a recent commentator has argued, "Malabou's materialism describes plasticity 'made flesh,'" then in a disturbing turn of the screw petroculture has managed quite literally and materially to make flesh plastic.[45] What happens to neuroplasticity when polymer plastics invade the brain? While scientists do not yet know the consequences for human health of this plasticization of tissue, it remains difficult to envision the potential explosiveness of such intrusions as anything other than gravely damaging to human alongside social form. And the specific example in view, especially pertinent because vivid in the ironies it will hold for Malabou's theory, is but one among countless instances of damage done by plastics to the *bios*. Whatever new history or "alter-globalism" might arise through the genius of plasticity will not manage to escape the toxic legacy—the slow-roiling explosion—of petrocultural plastics: an unwelcome fact that raises once more the specter of impasse.

Plastic time

As the onset of microplastic weather and the discovery of nanoplastic organs make dramatically clear, today's global conjuncture is mired in petrocultural plasticity.[46] A key feature of this stuckness involves what, near the start of this essay, I have called *plastic time*: the discontinuous, contradictory, vexing temporal imaginary that polymer plastics enable and enforce. Can the work of confronting plastic's time signature serve to clarify or reframe the challenge of energy impasse? If so, might the resulting perspective help us to find some way through the mire?

From the theoretical perspective advanced by Malabou, the very prospect of plastic time as I am invoking it here might constitute a redundancy or, worse, a nonsense. In *The Future of Hegel*, the work in which she begins to develop her theory of plasticity, Malabou insists on the pivotal importance of that concept for any philosophically rigorous and convincing account of time and the future. "To understand the future otherwise than in the ordinary immediate sense of 'a moment of time,'" she argues, "requires by the same token an opening-out of the meaning of time: an extension made possible by the very plasticity of temporality itself."[47] Such temporal plasticity indicates difference—a dialectic of *substance* and *accidents*: "[t]ime is not always (simultaneously, successively, and permanently) the same as itself. The concept of time has its own moments: it differentiates itself and thus temporalizes itself."[48] Malabou's theoretical framing issues from a provocative premise: that "[b]y 'plasticity' we mean first of all the excess of the future over the future."[49]

Malabou's target here is the commonplace understanding of time as a perpetual sequence of *nows*—a progressive procession confounded or indeed exploded, in its tenability, by "the very plasticity of temporality itself." Is polymer plastic immaterial to or redundant within this plasticity of temporality? Or might petroculture's gifted curse supplement or contravene temporality's differentiation to vexatious effect? What does the ever-growing surfeit of plastics that now saturate the globe mean for "the excess of the future over the future" signaled, in Malabou's theory, by plasticity?

In raising such questions, I take my cue from the remarkable accounts of petrocultural plastic recently devised by Amanda Boetzkes and Heather Davis. Read together, the sharp insights they advance manage to limn the vexing time signatures materialized by what I have been calling the plastic condition. Boetzkes characterizes this condition in terms of its pervasiveness, both spatial and temporal:

> plastic . . . is a pervasive condition that produces conflicted relations, behaviors, and affective modes. . . . Its existence as an incorporated waste—a waste that is never eliminated but which continually returns to disrupt ecosystems—is the expression of its fundamental attribute of convenience, anticipated and tailored by its chemical makeup, economic deployment, and the cultural meanings it procures in and through its aesthetic form.[50]

So understood, the *nowness* of plastic—the immediacy and instanta-neity of the convenience it supplies and signifies—works to guarantee, paradoxically, its *foreverness*, its uneliminability and perpetual return. Thus, to the extent that "[t]here is no outside to which plastic can be relegated, only a 'recycling' within a closed system," there is likewise no beyond, no after, for its consignment.[51]

To Boetzkes, this curious, convoluted time signature means that plastic, far from indicating the sort of futural excess that Malabou attributes to plasticity, is rather "a condition that forecloses the pos-sibilities of the future."[52] The implications of such foreclosure, in Boetzkes's understanding, go to the heart of plastic's constitutive tem-poral contradiction. "[P]lastic," she observes, "is a distinctly present future form."[53] Recalling Timothy Morton's "explanation of the mas-sively distributed temporality of hyperobjects," she continues:

> we would have to acknowledge that because of plastic's robust and expan-sive topological reach, it appears in the present *from* the planet's future. In Morton's terms, plastic is an *attractor*, which is not to say that plastic exists in the future and pulls the present toward it, but rather that its tem-poral span is so great that it always already radiates from the future to the present. As an attractor, plastic refutes being understood as a *telos* that proceeds through linear development. Instead, it is a "strange stranger," a flitting appearance from the future inserted into the present.[54]

Thus reckoned, plastic contravenes the temporal differentiation Malabou associates with plasticity by instead generating *in*difference—the inextricability and fusion of future to present—within time. Here "the revenge of time" Malm attributes to carbon dioxide comes at us from the other direction:[55] just as we cannot escape the arrival of long-burned CO_2 into the current moment, so too we must bear the obdurate presence and persistence of an inevitably, durably futural plastic today—and forever. The diptych is grim.

Boetzkes underscores the disturbing implications of this calamitous temporality. "Crucially," she observes,

> for Morton, the strangeness of the hyperobject is due to the fact that it gives a glimpse of a future that is potentially without us; it carries a terrify-ing thought of radical otherness. It is for this reason that plastic perturbs the familiar thinking of causality and teleology. It brings about the full dilemma of the Anthropocene: the supreme power of humans to master the planet has generated a reality in which humans have annihilated them-selves and planetary life. The present of complete dominance is fundamen-tally bound to a future of total impotence—nonbeing altogether.[56]

The perturbance by plastic of commonplace precepts about causality and teleology—a perturbance materialized in the temporal convolution whereby plastic enjoins the planetary present from its future—would seem to confound any hope one might hold about the capacity to effect, through the plasticity of our brains, some kind of historical change beyond impasse.

Davis's analysis complements, powerfully and provocatively, the perspective offered by Boetzkes. Malabou calls time plastic. In tacit reply, Davis says that "[p]lastic cracks time"—a fracturing capacity keyed to material recalcitrance and environmental refusal:[57]

> Plastic has an unfortunate metaphorical connotation. For although plastic is often thought of as a malleable material, as in the common use of the term "plasticity," or in the case of Catherine Malabou's conceptualization of the functioning of the brain, it is perhaps the hardest material there is. It is hard, because it refuses its environment, creating a sealant or barrier that remains impermeable to what surrounds it. It influences its environment while remaining mute to that environment's influence.[58]

This infiltrating, saturating impermeability constitutes a double bind that registers, for Davis, in temporal terms: as nothing less than a sort of plastic undeath or life-beyond-life. "Not only are the lifespans of plastic products often extremely short," she notes,

> synthetic polymers, derived from oil, are a kind of living dead among us. After digging up the remains of ancient plants and animals, we are now stuck with the consequences of these undead molecules, the ones that refuse to interact with other carbondependent life forms. For although plastics photodegrade and break apart, they do not biodegrade. That is, the pieces may get smaller and smaller, but they do not turn into something else. They do not go away. The molecules themselves remain intact, holding onto their identity. ... [I]n its proliferation and accumulation, [plastic] does indeed extend death outwards, transforming the ecologies that it now composes. ... Plastic survives, lives on, and accumulates for a projected 100,000 years.[59]

The lure of disposability that renders plastic so appealingly useful and ensures its ceaseless, proliferating abundance only consolidates and deepens the underlying impasse: for however much we might believe we could leave plastic behind, plastic—undead—will always remain, never done with us.

The contradiction at issue in what Davis calls "[t]his recalcitrance of matter, plastic's non-plasticity" is monstrous: a "materialization of the horror of identity, of the stability of form, of a futurity without change."[60] And the implications of such changeless futurity are calamitous ones, as Davis makes clear:

> Plastic, in this sense, represents the fundamental logic of finitude, carrying the horrifying implications of the inability to decompose, to enter back

into systems of decay and regrowth. In our quest to escape death, we have created systems of real finitude that mean the extinguishment of many forms of life.[61]

So reckoned and rendered, the terrifying genius of petrocultural plastic inheres in the obdurate malignancy of its impasse: a perpetual endurance that, paradoxically enough, proves temporally as well as ontologically terminal.[62]

Coda: Beyond the mire?

I began the previous section by wondering whether the work of confronting plastic's time signature could serve to clarify or reframe the challenge of energy impasse and, if so, whether the resulting perspective could illuminate some way through that fossil-fueled mire. The bracing accounts of plastic time given by Boetzkes and Davis certainly provide a clarifying, terrifying new frame for understanding and analysis, but in so doing they also bring us to conclude that the prospect today of any movement beyond plastic's impasse is, at the very least, quite balefully daunting. Remnant and revenant, interminable yet therefore terminal, plastic would seem to leave—because it serves to ensure—no way out of its temporal imaginary and time signature. Plastic—exactly against the ACC's enthusiastic refrain—makes it impossible.

Given these grave challenges attending the plastic condition, what might its endurance involve and entail? Boetzkes and Davis have some provocative thoughts. Framing the issue as a matter of dwelling, Boetzkes advocates a posture and practice of indifference: "To rethink the body's plasticity through its indifference to plastic leverages a perspective that perturbs the ideology of its own anthropogenic procedures."[63] Davis, for her part, proposes instead a narrative of "extinguishment" that "embraces both the fecundity of life as well as the complete randomness of its systems, while proposing a model within which humans can begin to take responsibility for what we have done—but without tying this to the destiny of humanity."[64] Both versions of endurance—indifference against anthropogenesis; responsibility without destiny—intimate a mode of compromised, resilient agency athwart plastic and its temporality.

Could endurance instead or additionally activate some form of rupture? For though plastic might now constitute an obdurate, even perpetual presence in the natural landscape, the plastic condition is not a force of nature. It is instead a function and symptom of fossil capital—which is to say an aspect and facet of energy as social relation. This insight is key to critical work on petroculture, which tends to insist that a meaningful transformation in energy regime must involve not just a shift in the dominant sources of energic power but also and crucially a rending and remaking of the social fabric itself. Brent Bellamy and Jeff Diamanti summarize the case incisively:

> The critique of energy is the critique of our structural dependence on an environmental relation inherited from the industrial revolution; it is a critique of the facile faith in a technological fix to climate change; it is a critique of the many barbarisms that flow from the contradictions of late fossil capital; and it is a critique of a fossil-fueled hostility to the very notion of social revolution—and hence of the very notion of structural dependence too.[65]

Although plastic's convoluted temporality will make—because it has already made—life without plastic materially impossible, that fact does not necessitate or prescribe the unimaginability of life beyond capital. Quite the opposite: the former's impossibility must drive home the latter's urgency. The very necessity of enduring the plastic condition as an inheritance of petromodernity only serves to underscore the unendurability of fossil capital as an entirely contingent (if massively sedimented and naturalized) social form. How might such recognition and reckoning serve to energize a transformative refusal of the social here-and-now, despite and beyond the endurance of impasse?

Notes

1. See Mark Blackwell, ed., *The Secret Life of Things: Animals, Objects, and It-Narratives in Eighteenth-Century England* (Lewisburg, PA: Bucknell University Press, 2007) and Christina Lupton, *Knowing Books: The Consciousness of Mediation in Eighteenth-Century Britain* (Philadelphia: University of Pennsylvania Press, 2012).
2. The introductory claim about plastic's importance appeared under "Lifecycle of a Plastic Product," *American Chemistry Council*, https://plastics.americanchemistry.com/Lifecycle-of-a-Plastic-Product/;   the ensuing passage about plastic's versatility appeared under "Lifecycle of a Plastic Product: Plastic Uses," *American Chemistry Council*, https://plastics.americanchemistry.com/Life-Cycle/#uses. As the past tense employed to frame this opening reading will intimate, the text in question is no longer on the ACC website. Taking its place are accounts of plastic that, revealingly enough, foreground *sustainability* as an aspirational ideal (see https://www.americanchemistry.com/chemistry-in-america/chemistry-in-everyday-products/plastics and https://www.americanchemistry.com/better-policy-regulation/plastics), although the "Plastics Make It Possible®" mantra does persist through its own dedicated website (see https://www.plasticsmakeitpossible.com/). It seems alto-

gether appropriate that normative corporate discourse *about* plastic is every bit as disposable and fungible as the substance on which that discourse focuses—a resemblance holding resonance for the temporal dynamics of impermanence and perpetuity raised toward the end of this introductory section and theorized more fully in the ensuing analysis of *plastic time*.

3. Stephanie LeMenager, *Living Oil: Petroleum Culture in the American Century* (Oxford: Oxford University Press, 2014), 67: "modern life based in the cheap energy systems made possible by oil."

4. Lauren Berlant, *Cruel Optimism* (Durham, NC: Duke University Press, 2011), 199–200.

5. Mark Simpson and Imre Szeman, "Impasse Time," *The South Atlantic Quarterly*, special issue on "Solarity," ed. Darin Barney and Imre Szeman, 120, no. 1 (January 2021): 77–89, https://doi-org.login.ezp roxy.library.ualberta.ca/10.1215/00382876-8795730; Mark Simpson, "Energy Impasse," in *Routledge Handbook of Energy Humanities*, ed. Graeme Macdonald and Janet Stewart (forthcoming).

6. See for instance: BP, *Advancing the Energy Transition*, n.p., https:// www.bp.com/content/dam/bp/country-sites/nl-nl/netherlands/home/ documents/energy-economy-other/Advancing-the-energy-transition.pdf; "What is Energy Transition?" *S&P Global* (February 24, 2020), n.p., https://www.spglobal.com/en/research-insights/articles/what-is-energy-transition; "Canada's Energy Transition: Historical and Future Changes to Energy Systems—Update—An Energy Market Assessment," *Canada Energy Regulator*, n.p., https://www.cer-rec.gc.ca/nrg/ntgrtd/mrkt/cnd snrgtrnstn/ntrdctn-eng.html; *1.5° C: Aligning New York City with the Paris Climate Agreement*, City of New York (2017), n.p., https:// www1.nyc.gov/assets/sustainability/downloads/pdf/publications/1point 5-AligningNYCwithParisAgrmt-02282018_web.pdf.

7. Christophe Bonneuil and Jean-Baptiste Fressoz, *The Shock of the Anthropocene: The Earth, History and Us*, trans. David Fernbach (London: Verso, 2017), 102.

8. Mark Simpson, "Lubricity: Smooth Oil's Political Frictions," in *Petrocultures: Oil, Politics, Culture*, ed. Sheena Wilson, Adam Carlson, and Imre Szeman (Montreal and Kingston: McGill-Queen's University Press, 2017), 289.

9. Bonneuil and Fressoz, *The Shock of the Anthropocene*, 101.

10. Simpson and Szeman, "Impasse Time," 80.

11. Ibid. 80.

12. Berlant, *Cruel Optimism*, 200.

13. Gregory Wetherbee, Austin Baldwin, and James Ranville, "It is Raining Plastic," *US Geological Survey*, n.p., https://pubs.usgs.gov/of/2019/1048 /ofr20191048.pdf.

14. Ibid.

15. Ibid.

16. Patricia L. Corcoran, Charles J. Moore, and Kelly Jazvac, "An

Anthropogenic Marker Horizon in the Future Rock Record," *GSA Today: A Publication of the Geological Society of America* 24, no. 6 (June 2014): 4–8, https://www.geosociety.org/gsatoday/archive/24/6/article/i1052-5173-24-6-4.htm.

17. Ibid.

18. Ibid.

19. Amanda Boetzkes, *Plastic Capitalism: Contemporary Art and the Drive to Waste* (Cambridge, MA: MIT Press, 2019), 196, 198.

20. Kirsty Robertson, "Plastiglomerate," *e-flux* 78 (December 2016): n.p., https://www.e-flux.com/journal/78/82878/plastiglomerate/.

21. Beth Gardiner, "The Plastic Pipeline: A Surge of New Production Is on the Way," *Yale Environment 360* (December 19, 2019): n.p., https://e360.yale.edu/features/the-plastics-pipeline-a-surge-of-new-production-is-on-the-way.

22. Ibid.

23. Ibid.

24. Hannah Appel, "Infrastructural Time," in *The Promise of Infrastructure*, ed. Nikhil Anand, Akhil Gupta, and Hannah Appel (Durham, NC: Duke University Press, 2018), 45, https://doi-org.login.ezproxy.library.ualberta.ca/10.1215/9781478002031-002.

25. Berlant, *Cruel Optimism*, 200.

26. Boetzkes, *Plastic Capitalism*, 12–13.

27. Karl Marx, *Grundrisse*, trans. Martin Nicolaus (Harmondsworth: Penguin, 1993), 142.

28. Alberto Toscano, "Plasticity, Capital, and the Dialectic," in *Plastic Materialities: Politics, Legality, and Metamorphosis in the Work of Catherine Malabou*, ed. Brenna Bhandar and Jonathan Goldberg-Hiller (Durham, NC: Duke University Press, 2015), 105, https://doi-org.login.ezproxy.library.ualberta.ca/10.1215/9780822375739-006. In an earlier take on analogy, Toscano amplifies the terms of the risk at issue. The problem is not so much that analogy is categorically flawed as that, when poorly deployed, it can deform the reckoning of history: "Analogy . . . involves a calibration of the degree of identity and difference between now and then; a bad analogy can obscure the singularity of the present by subsuming it under some paradigmatic past, or distort the contours of history through rear-projection of the present." "The Spectre of Analogy," *New Left Review* 66 (November/December 2010): 153.

29. Joe Deville, "Paying with Plastic: The Enduring Presence of the Credit Card," in *Accumulation: The Material Politics of Plastic*, ed. Jennifer Gabrys, Gay Hawkins, and Mike Michael (London: Routledge, 2013), 100.

30. Joe Deville, "The Matter of the Credit Card," *Transactions: A Payments Archive* (2013), n.p., https://transactions.socialcomputing.uci.edu/post/58430974481/american-express-one-of-the-first-plastic-credit.

31. Toscano, "Plasticity," 94–5, 96.

32. Ibid., 101.

33. Catherine Malabou, *What Should We Do with Our Brain?*, trans. Sebastian Rand (New York: Fordham University Press, 2008), 5, https://hdl-handle-net.login.ezproxy.library.ualberta.ca/2027/heb.08589.
34. Ibid. 6.
35. Ibid. 12.
36. Ibid. 56.
37. Ibid. 82.
38. Ibid. 78.
39. Toscano, "Plasticity," 107.
40. Ibid. 106.
41. Ibid. 105.
42. Andreas Malm, *Fossil Capital: The Rise of Steam Power and the Roots of Global Warming* (London: Verso, 2016), 308; italics in original.
43. Malabou, *What Should We Do*, 73.
44. Damian Carrington, "Microplastics Now Discoverable in Human Organs," *Guardian*, August 17, 2020, n.p., https://www.theguardian.com/environment/2020/aug/17/microplastic-particles-discovered-in-human-organs.
45. Jennifer A. Wagner-Lawlor, "The Persistence of Utopia: Plasticity and Difference from Roland Barthes to Catherine Malabou," *Journal of French and Francophone Philosophy* 25, no. 2 (2017): 80, http://jffp.pitt.edu/ojs/index.php/jffp/article/view/804/776.
46. This global mire is arguably a prime symptom of what Ulrich Brand and Markus Wissen memorably term "the imperial mode of living": one "based on exclusivity . . . [that] can sustain itself only as long as an 'outside' on which to impose its costs is available." *The Imperial Mode of Living: Everyday Life and the Ecological Crisis of Capitalism*, trans. Zachary Murphy King, ed. Barbara Jungwirth (London: Verso, 2021), 5. For recent accounts of plastic resonant for this trenchant concept, see for instance "'Waste colonialism': World Grapples with West's Unwanted Plastics," *Guardian*, December 31, 2021, https://www.theguardian.com/environment/2021/dec/31/waste-colonialism-countries-grapple-with-wests-unwanted-plastic, and Karen McVeigh, "'Oil spills of our time': Experts Sound Alarm about Plastic Lost in Cargo Ship Disasters," *Guardian*, February 9, 2022, https://www.theguardian.com/environment/2022/feb/09/cargo-ship-disasters-are-oil-spills-of-our-time-because-of-health-risk-from-plastic. Life under pandemic conditions has, moreover, only compounded the mire. See for instance Sandra Laville, "About 26,000 Tonnes of Plastic COVID Waste Pollutes World's Oceans—Study," *Guardian*, November 8, 2021, https://www.theguardian.com/environment/2021/nov/08/about-26000-tonnes-of-plastic-covid-waste-pollutes-worlds-oceans-study.
47. Catherine Malabou, *The Future of Hegel: Plasticity, Temporality and Dialectic*, trans. Lisabeth During (London: Routledge, 2004), 13.
48. Ibid. 14.
49. Ibid. 6.

50. Boetzkes, *Plastic Capitalism*, 182.
51. Ibid. 182.
52. Ibid. 184.
53. Ibid. 201.
54. Ibid. 201–2.
55. Malm, *Fossil Capital*, 6.
56. Boetzkes, *Plastic Capitalism*, 202.
57. Heather Davis, "Plastic: Accumulation without Metabolism," in *Placing the Golden Spike: Landscapes of the Anthropocene*, exhibition catalogue, ed. Dehlia Hannah and Sara Krajewski (Milwaukee: INOVA, 2015), 71, http://heathermdavis.com/wp-content/uploads/2014/08/Libra ryScanFile_2.pdf.
58. Heather Davis, "Life and Death in the Anthropocene: A Short History of Plastic," in *Art in the Anthropocene: Encounters among Aesthetics, Politics, Environments and Epistemologies*, ed. Heather Davis and Etienne Turpin (London: Open Humanities Press, 2015), 352, http:// openhumanitiespress.org/books/download/Davis-Turpin_2015_Art-in-the-Anthropocene.pdf.
59. Ibid. 352.
60. Ibid. 352, 353.
61. Ibid. 353.
62. Davis has extended and elaborated her bracing theoretical analysis of plastic in her superb recent book *Plastic Matter* (Durham, NC: Duke University Press, 2022), a work appearing too late in the development of this essay to receive more than passing mention.
63. Boetzkes, *Plastic Capitalism*, 222.
64. Davis, "Life and Death," 356.
65. Brent Bellamy and Jeff Diamanti, "Materialism and the Critique of Energy," in *Materialism and the Critique of Energy*, ed. Bellamy and Diamanti (Chicago and Alberta: MCM Publishing, 2018), xxvi, http:// www.mcmprime.com/files/Materialism_Energy.pdf.search?q=rubber+d uckie+song&oq=rubber+duckie+son&aqs=chrome.0.0j46j69i57j0j46j0 j46j0.5945j0j7&sourceid=chrome&ie=UTF-8.

Index